Provenance of Arenites

NATO ASI Series

Advanced Science Institutes Series

A series presenting the results of activities sponsored by the NATO Science Committee, which aims at the dissemination of advanced scientific and technological knowledge, with a view to strengthening links between scientific communities.

The series is published by an international board of publishers in conjunction with the NATO Scientific Affairs Division

A	Life Sciences	Plenum Publishing Corporation
B	Physics	London and New York
C	Mathematical and Physical Sciences	D. Reidel Publishing Company Dordrecht, Boston and Lancaster
D	Behavioural and Social Sciences	Martinus Nijhoff Publishers
E	Engineering and Materials Sciences	The Hague, Boston and Lancaster
F	Computer and Systems Sciences	Springer-Verlag
G	Ecological Sciences	Berlin, Heidelberg, New York and Tokyo

Series C: Mathematical and Physical Sciences Vol. 148

Provenance of Arenites

edited by

G. G. Zuffa

Department of Earth Sciences,
University of Calabria,
Castiglione Cosentino Stazione,
Cosenza, Italy

D. Reidel Publishing Company

Dordrecht / Boston / Lancaster

Published in cooperation with NATO Scientific Affairs Division

Proceedings of the NATO Advanced Study Institute on
Reading Provenance from Arenites
Cetraro, Cosenza, Italy
June 3-11, 1984

Library of Congress Cataloging in Publication Data

NATO Advanced Study Institute on Reading Provenance from Arenites (1984: Cosenza, Italy)
   Provenance of arenites.

   (NATO ASI series. Series C–Mathematical and physical sciences; vol. 148)
   "Proceeding of the NATO Advanced Study Institute on Reading Provenance from Arenites, Cetraro, Cosenza, Italy, June 3—11, 1984"-T.p. verso.
   "Published in cooperation with NATO Scientific Affairs Division."
   Includes indexes.
      1.   Arenites-Congresses.   2.   Sedimentation and deposition-Congresses.
I.   Zuffa, G. G. (Gian Gaspare), 1943—   .   II.   North Atlantic Treaty Organization.
Scientific Affairs Division.   III.   Title.   IV.   Series.
QE471.15.A68N37   1984   552'.5   84-29797
ISBN 90-277-1944-6

Published by D. Reidel Publishing Company
P.O. Box 17, 3300 AA Dordrecht, Holland

Sold and distributed in the U.S.A. and Canada
by Kluwer Academic Publishers,
190 Old Derby Street, Hingham, MA 02043, U.S.A.

In all other countries, sold and distributed
by Kluwer Academic Publishers Group,
P.O. Box 322, 3300 AH Dordrecht, Holland

D. Reidel Publishing Company is a member of the Kluwer Academic Publishers Group

CONTENTS

INTRODUCTION

MAIN FACTORS CONTROLLING THE COMPOSITION OF ARENITES

PREFACE

Gian Gaspare Zuffa

Dipartimento di Scienze della Terra
Università della Calabria
87030 Castiglione Cosentino Stazione, Cosenza, ITALY

The tradition has been to consider Sedimentology and Sedimentary Petrology as quite separate areas of research. It is however impossible to arrive at an optimal description of sedimentary rocks without integrating the two fields since sedimentary processes and compositional aspects are strongly intertwined.

The study of arenites is of particular importance in obtaining paleogeographic and paleotectonic reconstructions aimed at determining the geodynamics of the earth's crust. It also has important implications in exploration for and exploitation of hydrocarbons.

At the NATO ASI Meeting on Reading Provenance from Arenites held in Calabria (Italy), June 3-11, 1984, field sedimentologists and sedimentary petrologists were given opportunity to pool their resources in order to obtain better analyses of both source areas and depositional basins.

The papers collected in this volume represent an edited version of the lectures given and provide a comprehensive picture of the present state of the art since they include such important topics as: 1) the climate and relief of the source areas, 2) mechanical transport of sediments and depositional processes, 3) postdepositional processes, and 4) the methodology adopted for petrographic optical analyses. Particular attention has been paid to the limitations and errors introduced into paleogeographic and paleotectonic reconstructions by incomplete and incorrect data.

This volume also includes some special topics on analytical techniques and on how to deal with quartz, feldspar, green particles and heavy minerals in provenance studies. Finally an overall view is presented of the use of arenites in studying ancient and modern tectonic settings.

The uniqueness of this book lies in the fact that it combines

sedimentology and petrology of arenites, two fields which have hith-
erto been considered quite separate research areas.

    The editor is very grateful to NATO for enabling him to or-
ganize such an interesting and profitable meeting. Thanks for their
financial support are also due to UNESCO, Italian National Research
Council (CNR), University of Calabria, and to the following Oil
Companies, AGIP, SNIA BPD, TOTAL Mineraria. Sponsorship was grate-
fully received from Italian Geological Society, Italian Society of
Mineralogy and Petrology and from International Association of Sed-
imentologists.
Ms. Shirley Hill deserve particular mention because of her constant
support both in the preparatory stage and during the meeting itself.
Last but not least I should like to thank Tarquin Teale and Albina
Colella for their collaboration in preparation the field excursions,
Ms. Denise Scott, Claudia Spalletta, Gian Gabriele Ori and Alessan-
dra Failla for help with the editing and the members of the Earth
Science Department of the University of Calabria, Maurizio Sonnino,
Gianluca Ferrini, Franco Gagliardi, Eugenio Gori and Emilio Tocci
whose invaluable help I will never be able to repay.

Castiglione Cosentino Stazione (Cosenza), October 1st, 1984.

INFLUENCE OF CLIMATE AND RELIEF ON COMPOSITIONS OF SANDS RELEASED AT SOURCE AREAS

Abhijit Basu

Department of Geology, Indiana University
Bloomington, IN   47405, U.S.A.

ABSTRACT

Mass balance requires that any study of the relation between provenance and detrital sediments must take into account all the processes, such as climatic and biochemical, that contribute to any modification of the parent material at the beginning of a sedimentary cycle. In addition, exceptions to the relationship between tectonic setting and sand-stone composition may be traced to climatic proces-ses. Despite an enormous amount of research on atomic level dissolution phenomenon, little work has been done to characterize the sandy residue of weathering. Available modal data, albeit meagre, show that the mineralogic composition of the sand size fraction of soils is similar to that of first order stream sands. This indicates that pedogenic processes largely control the composition of first cycle sands derived from similar bedrocks. Further, the data also suggest that modal compositions of first cycle sands are broadly indicative of both parent rock type and climate. Recalculation of data from only one available study indicates that steep hill slopes exceeding the angle of repose can obscure climatic effects on first cycle sand composition. One may infer that slope angle, which controls the duration of pedogenic processes, not relief, has more significance in overcoming climatic effects. Evalua-tion of the relative importance of dissolution and disintegration of minerals, especially polycrystal-
1

G. G. Zuffa (ed.), Provenance of Arenites, 1–18.

line quartz, is difficult because lattice dislocation increases solubility as well as brittle strength. Given the extreme paucity of data from controlled studies on the effects of climate and relief and the seemingly significant compositional diversity brought about by pedogenic processes, we must conclude that this is a potential area of much fruitful research.

INTRODUCTION

It is desirable to define provenance at the onset of this "Institute" so that all of us can use the same vocabulary, and more importantly, focus on the concept. The word provenance has been derived from the French word provenir and the Latin word proveniens meaning origin or the place where produced. However, in sedimentary petrology, provenance refers specifically to the nature, composition, identity and dimensions of source rocks, relief and climate in the source area, and to some extent includes a transportation factor which is mostly understood to convey a sense of the distance and rigour of transport of sediments until deposition (Suttner, 1974). Thus one can write:

provenance = f (source rock, relief, climate,
                          transportation)

The objective of all provenance studies should be to work out the quantitative paleogeology of a part of the earth's crust. Sedimentary facies and basin analysis, aided by seismic stratigraphy, provide paleogeographic reconstructions; and, we may infer the relative positions of land and water, the shorelines, perhaps the width of the shelf, and also the relative locations of some land features such as mountains, river valleys, swamps, etc. with respect to the location of the main basin of deposition. Detrital remanent magnetic studies of sediments can provide paleolatitude. Structural analysis of folded units may, but not necessarily, provide some information of the tectonic setting of the depositional basin. However, only provenance studies can bring all results together to provide a broad picture of paleogeology. We strive to infer not only the location of a mountain shedding sediments or to infer the rocks that made up the mountain, but also to find the relative proportions of these rocks as the highlands were eroded to the base level. The task is not easy

and it may take years before tangible results are
obtained from research with such an aim.   After all,
"the question of provenance is one of the most difficult the sedimentary petrographer is called on to
solve" (Pettijohn et al., 1972).

PROVENANCE

It is now common knowledge that the plate
tectonic regime of a highland, i.e. of an area which
can and does shed sediments, controls the petrologic
assemblage of the highlands (Dickinson, 1972, 1980).
The rate of erosion of the highland is controlled by
relief and climate in the area; uplift and erosion
gradually unroof the rocks and eventually expose
those in the core of the highland.   Thus exposure and
the duration of exposure of a rock type in the source
area is also a function of relief and climate which
provide a certain degree of topographic maturity to a
highland.   The effect can be understood best from an
example.   Both the Alps in Europe and the
Appalachians in North America are products of
continent-continent collisions.   However, because of
the difference in degrees of unroofing, and because
of the difference in their topographic maturities,
these two mountain chains now shed sands of slightly
different compositions. Exposure age and the plate
tectonic regime of highlands also control to a large
extent the relief and the rigour of transport.
Usually, rigour of transport is often associated with
distance of transport which may be dependent on
tectonic setting.   For example, slow moving big
rivers usually drain large continental blocks and
flow on passive trailing edges of continents. In
contrast, rapidly flowing short streams drain mobile
belts in active continental margins (Potter, 1978).
However, there are exceptions.   For example, relatively short rapidly flowing streams drain the trailing edge of the continental block of southern India.
Overall, it appears that the variables source rock,
relief and transport are functions of plate tectonic
regimes and exposure ages of the highlands.

Climate of an area is, however, independent of
any direct tectonic control.   The more important
parameters in rock weathering, viz. rainfall and
temperature are dependent on latitude, elevation and
orographic effects, and on land-water distribution.
Rainfall and temperature have a profound effect on

the rate of rock-weathering and, as pointed out
earlier, determine the rate of erosion to a large
extent. Rainfall and temperature also control, to a
large extent, the growth of vegetation and the
attendant biomass in a highland area, and thus
significantly contribute to biochemical alteration of
original rock material. Many exceptions to the now
well established relationship between tectonic settings and sandstone composition (e.g. Dickinson et
al., 1983) may be traced to the vagaries of climate
in the source areas (Mack, 1984).

Genesis of sediments begins with regolith and/or
soil formation on a bedrock. The composition of the
bedrock is controlled primarily by plate tectonics;
however, the processes of alteration of the bedrock
material are driven primarily by climate. In this
paper we shall (i) try to examine the control of
climate (rainfall + temperature) on the composition
of sand released by common source rocks, (ii) briefly
comment on our ignorance about the effect of relief,
and (iii) try to identify areas of potential research
in provenance interpretation.

COMPOSITIONS OF MODERN SANDS

The only way to evaluate the control of climate
on the composition of sands released at source areas
is to adopt a sampling technique which keeps all
other variables constant. Therefore, in a desirable
sampling technique one would try to keep the source
rock, relief, and transport as invariable as possible
and sample modern sands from different climatic
zones. Although many writers have characterized
modern sands derived from different tectonic regimes
and rock associations in different climatic zones,
except for the Indiana group no one has adopted the
sampling plan described above. Therefore, we shall
initially consider only the Indiana data (Suttner et
al., 1981; Basu, 1976; Young, 1975; Darnell, 1974).

One important result of the Indiana group indicates that the modal compositions of the sand size
fraction of hill slope soils and regolith are very
similar to that of the first order streams draining
the same hills where slope angle < angle of repose
(fig. 1). Data were collected separately for
plutonic igneous, i.e. granitic bedrocks, and high
rank metamorphic bedrocks, i.e. high grade schists

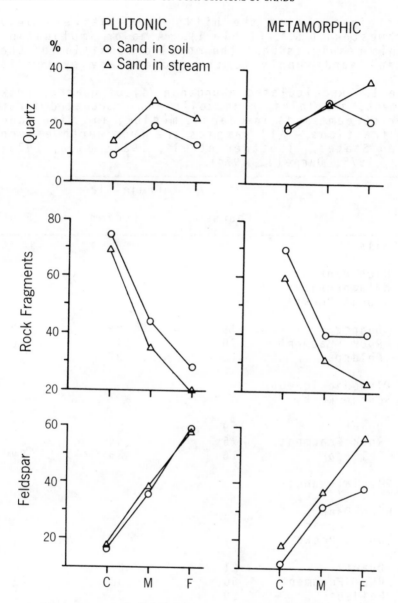

Fig. 1. Modal compositions of coarse (c), medium
(m) and fine (f) sand from hillslope soils and
adjacent streams draining either plutonic
igneous or metamorphic bedrocks (after Suttner
et al., 1981). Note the remarkable similarity
between the compositions of sands from two subenvironments.

and gneisses, from the hills in relatively arid
northwestern U.S.A. (Table 1). A major implication
of this result is that the modal composition of the
initial sand supply to the sedimentary system is

Table 1. Recalculated abundance (%) of quartz, rock
fragment, and feldspar in soils and associated first
order stream sands in coarse, medium, and fine sand
size fractions. All samples are from northwestern
United States. (Suttner et al., 1981; Basu, 1976;
Young, 1975; Darnell, 1974).

	Grain Size		
	Coarse	Medium	Fine

I. Soils

(a) High Rank
     Metamorphic
     Source Rocks

Quartz	19	28	22
Rock Fragment	70	40	40
Feldspar	11	32	38

(b) Plutonic Igneous
     Source Rocks

Quartz	9	20	13
Rock Fragment	75	44	28
Feldspar	16	36	59

II. Fluvial sands

(a) High Rank
     Metamorphic
     Source Rocks

Quartz	21	27	36
Rock Fragment	60	32	23
Feldspar	19	31	41

(b) Plutonic Igneous
     Source Rocks

Quartz	14	28	23
Rock Fragment	69	35	20
Feldspar	17	37	57

determined by the degree of weathering in the soil zone for any given bedrock (Suttner et al., 1981). Comparable data are not yet available for other climatic zones. It is quite likely that the proportion of clay relative to the sand fraction in the regolith or soil developed in hot humid climate would be much higher. However, there is no evidence yet to show that the relationship between modal compositions of sand sized fractions of soils and adjacent stream sediments would be any different from that found in northwestern U.S.A.

The second result of the Indiana group shows that modal compositions of stream sands derived from similar parent rocks are controlled by climate, i.e. precipitation in the source area. The results, strictly limited to medium sand size fraction (0.25-0.50 mm) to avoid size-composition effects, are good for common parent rocks, viz. plutonic igneous granites and metamorphic rocks i.e. schists, gneisses, and granulites (fig. 2; table 2).

Table 2. Average abundance of quartz, feldspar, and rock fragments in medium size sand (0.25-0.50 mm)from first order streams draining exclusively plutonic or metamorphic bedrock in humid and arid climates. The numbers in parenthesis denote one standard deviation.

Source		Daughter Sands(0.25-0.5 mm)		
Rock	Climate	Quartz %	Feldspar %	Rock Fragment %
Plutonic	Humid	60(5)	27(4)	13(3)
Plutonic	Arid	27(4)	39(4)	34(5)
Metamorphic	Humid	74(12)	6(6)	20(9)
Metamorphic	Arid	29(8)	3(3)	68(10)

Note that the Indiana group adopts the traditional point-counting technique and not the Gazzi-Dickinson method (Ingersoll et al., 1984; Zuffa, this volume). The population of rock fragments in the Indiana data effectively separates plutonic and metamorphic source rocks (fig. 2). If counted otherwise, we suspect, not only the parent rock distinction but also the

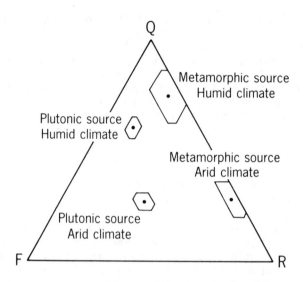

Fig. 2. Average compositions of medium sand-sized (0.25-0.50 mm) fraction of first cycle stream sand derived from plutonic igneous and metamorphic sources but produced under different climatic conditions are plotted in a standard QFR diagram (after Suttner et al., 1981). One standard errors on Q, F, and R are used to construct the error polygons around each average point (after Ingersoll and Suczek, 1979); actual scatter of the raw data plot in overlapping fields (Suttner et al., 1981; their fig. 1). Modal analyses were performed in the traditional fashion. Note that climate produces distinctive compositional effects on first cycle sediments. If counted according to the Gazzi-Dickinson method (Ingersoll et al., 1984) most of the distinctions seen on this plot would have been lost.

climatic distinction might be obliterated to a large extent.

Many scientists have investigated the chemistry of rock weathering induced by the biosphere, especially by plants. However, only one paper explicitly considers the effect of land plants on detrital mineralogy. Todd (1968) argued that depending upon the type of vegetation, the Na^+/H^+ and K^+/H^+ ratios in soil-waters could vary as much as to affect the relative stability of andesine and orthoclase. Todd (1968) showed that the nature of plants, rainfall and temperature determined if andesine would be more "weathered" than orthoclase, or vice versa in a soil horizon. Relative stability of K- and Na-feldspars were also invoked by Basu (1981) who argued that in well vegetated areas alteration of K-bearing minerals is faster than in areas without much vegetation because "potassium appears to move more quickly from the soil into the plant than do other common cations" (Paton, 1978). Therefore, before the advent of land plants, K-feldspars in pre-Silurian sediments should have had suffered less alteration than K-feldspars in post-Silurian sediments because of the added uptake of K^+ from soil-waters by the biomass of vascular plants in post-Silurian times. Although compilation of chemical data by Maynard et al. (1982) support the hypothesis, Graustein and Velbel (1981) disagree. Clearly, this is an open field and many new interesting research could be performed to evaluate the control of biochemical processes in determining the modal composition of sands.

On the other hand, a vast amount of geochemical research has been and is being done on the nature and the kinetics of reactions that lead to the dissolution of rock forming minerals and precipitation of new minerals in soil horizons (Nesbitt and Young, 1984). One area of very interesting research evaluates the role of newly formed minerals (e.g. kaolinite) in coating and mechanically protecting the original mineral (e.g. feldspar). Wollast (1967) in a now classic paper propounded this shielding idea. Since then many have tested it with dissolution kinetics experiments (e.g. Siever and Woodford, 1979). In a series of experiments Berner and his associates showed how dissolution actually takes place initially at points of crystal defects and then progresses inward apparently uninhibited by any protective coating (Berner and Holdren, 1979; Holdren

and Berner 1979; Schott and Berner, 1983; etc.). All
of these works are extremely important to understand
the aqueous geochemical processes that control the
fundamental weathering reaction whereby fresh rock is
altered and new minerals and solutions are produced
(cf. Tardy, 1971). However, a bridge remains to be
built between the aqueous geochemists and the sedimentary petrologists. We need to know the
mineralogic compositions of the weathering residues
in sand sizes to better estimate the control of
climate (rainfall + temperature) on the initial
material of a sedimentary cycle.

Unfortunately and surprisingly, no comprehensive
work has been done to evaluate quantitatively the
effect of relief on sand composition. Textbooks are
replete with ad hoc statements which, in general,
imply that the rate of mechanical disintegration
would be enhanced with increased relief. However, no
single study exists which demonstrates that this
enhanced mechanical weathering would lead to any
difference in the modal compositions of sands in the
source region of detrital sediments, if variables
other than relief remain constant. One marginally
relevent study (Garner, 1959) shows that climate is
far more important than relief (up to ~ 6000 m) in
controlling clast size in the Andes Mountains. The
data of Franzinelli and Potter (1983) from the Amazon
basin show that a combination of very low relief, hot
humid climate, and ample vegetation can produce sands
of quartz arenite composition primarily from granitic
bedrocks.

Ruxton (1970) has described some quartz-poor
sands from Papua; we have recalculated some of his
data to understand the extent of control of relief on
sand composition. The bedrocks in Ruxton's area of
study comprise pristine and metamorphosed basic and
ultrabasic igneous rocks and metagreywackes. The
relief is moderate (365m) but the slopes are steep
(up to 35^0-40^0) and frequently exceed the angle of
repose. All sand sized material coarser than 0.21 mm
consists of rock fragments. Discrete mineral grains
show up only in the fine grained fraction (0.075-0.21
mm). The petrographic data from this size fraction
show that the material on strongly weathered crests
of hills is very different from that on the slopes
(Table 3; fig. 3). However, the compositions of the
sands that are found on floodplains or on the beaches
of Papua are very similar to each other. The beach

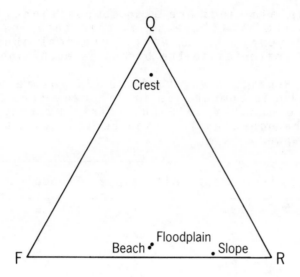

Fig. 3. Recalculated modal data of Holocene sand samples from relatively flat hill crests, steep (up to 35°-40°) hill slopes, and the adjacent floodplains and beaches in Papua (after Ruxton, 1970) are plotted in a standard QFR diagram. Note that the tropical climate has produced a relatively quartz-rich composition in the lateritic soil at hill crests; however, most of the sands, rich in accessory minerals, are compositionally very immature. Hill slope soils must also overwhelmingly contribute to the floodplain and beach sands as seen from the similarity in their compositions. Extreme immaturity of the sands is attributed to the steepness of the hill slopes which must cause a very short residence time of soils, if any, on the slopes.

and floodplain sands are also compositionally much
closer to the hillslope sands than they are to the
hillcrest regolith.   Clearly, the contribution of
hillslope material to the basins is much higher than

Table 3.   Recalculated abundance (in volume percent)
of mineralogic components in Holocene sands in the
island arc complex in Papua (after Ruxton, 1970).
Data on the non-clay fraction pertain to 0.075-0.21
mm size range.

	Hill crest	Hill slope	Floodplain	Beach
Quartz	19	1.5	5	3
Feldspar	2	14	37	31
Rock Fragments	2	46.5	38	32
Accessary Minerals	38	27.5	18	32
Clay	38	10.5	2	2

the hilltop material.   Note that in Ruxton's study
area the slopes are steeper than the angle of repose
and the residence time of any soil on these slopes
would be extremely low.   If there were other studies
with similar source rocks on relatively flat lying
areas in both arid and humid climates, comparison
with Ruxton's data would have been extremely useful.
Unfortunately, such is not the case.   However, the
data in hand do show that slope angle, as a control
on the residence time of soil horizons, not relief,
indirectly controls the initial composition of the
sand sized material released from bedrocks.   It
appears that low relief might only aid in increasing
the residence time of soil thereby increasing the
degree of chemical weathering in a source region.
Thus, low relief in a humid highland may actually
contribute more towards affecting sand composition
than high relief in any climate.   Although we might
suspect that high relief (not to be confused with
elevation) is relatively insignificant in controlling
detrital mineralogy of sands, we just do not have any
quantitative data even to form an opinion.   It
appears, however, that volumetrically significant

contribution of sands from hillslopes, if slopes are greater than the angle of repose, could virtually bypass all possible climatic effects documented by the Indiana group. I am convinced that the relative importance of climate, relief, slope angle of hills, and elevation in determining the composition of first cycle sands will form the basis of some extremely significant research in very near future.

SURVIVABILITY OF QUARTZOSE PARTICLES

This introduction to provenance interpretation would remain incomplete if no comment is made on the survivability of detrital particles. I shall not repeat what may be found in standard textbooks on the chemical stabilities of various rock forming minerals under atmospheric conditions (e.g. Garrels and Mackenzie, 1971; Pettijohn et al., 1972; Blatt et al., 1980; Leeder, 1982; etc.). I shall, however, comment on the theoretical aspect of survivability of strained i.e. undulose quartz grains and polycrystalline quartz grains with sutured contacts between subgrains.

Under normal atmospheric conditions and in common soils, quartz does not react with water to form new minerals but does go into solution although the solubility of crystalline silica is very low. Solubility of quartz increases with increasing density of lattice dislocation (Wintsch and Dunning, 1985). Therefore, strained quartz (undulose monocrystalline quartz and polycrystalline quartz grains with sutured contacts between subgrains) is apt to suffer greater degress of dissolution than nonundulose monocrystalline quartz. According to Blatt (1967) this increased solubility makes strained quartz thermodynamically less stable than unstrained quartz grains. Note that any increased dissolution along fractures and cleavage helps the mechanical disintegration of any particle. Blatt's (1967) data do show that the durability of strained and polycrystalline quartz in Holocene sand seem to agree with the above model. However, strain and dislocation of lattices of crystals increase their internal energy and provide greater brittle strength. The phenomenon is that of "work hardening" as referred to in metallurgy. Indeed, Pettijohn et al. (1972) use the term "cold rolling" in describing polycrystalline quartz with sutured contacts of subgrains. At this time, we do not know if increased solubility of strained quartz is more important than the additional brittle strength, in determining the survivability of detrital quartz. This is a problem that needs more careful attention, perhaps with well designed laboratory experiments. It is possible that solubility of quartz grains is important only in soil horizons, whereas the brittle strength of quartz particles continue to be the prime factor in determining the survivability of quartz in fluvial transport (cf. Mack, 1978). Data of Basu (1976) seem to bear out the latter.

CONCLUSIONARY COMMENTS

It should be apparent to the readers that we really have very scanty usable information on the effects of the interactions between climate and relief on compositions of sand released at source areas. Data at hand do show that, if the traditional point counting technique is applied, it is possible to recognize the effects of arid and humid climate in first cycle sediments. However, if hill slopes exceed the angle of repose and the residence time of all slope soil is low, even hot and humid climate would not be as effective to achieve advanced decomposition of bedrocks. Compositions of ancient sandstones believed to have been derived from magmatic arcs (Dickinson and Suczek, 1979; Ingersoll and Suczek, 1979; Yerino and Maynard, 1984; Valloni and Mezzadri, 1984; etc.) are very similar to the modern beach sands of Papua (Ruxton, 1970). Is it possible that the residence times of the regolith/soils in active magmatic arcs have always been very low? Climate has a tremendous influence on the ecology of land organisms. Vascular land plants are notorious for secreting acid at their roots. Some data suggest that the extent and nature of biochemical weathering can be very significant in determining the initial sand composition. Unfortunately, very little follow up work has been done. In short, we sedimentary petrologists have a vast area of potential research open to us.

ACKNOWLEDGEMENTS

I am grateful to G. G. Zuffa and L. J. Suttner

for comments on a previous version of the paper. This paper was prepared with the support of Indiana University Foundation and through the dedication of the staff of the Department of Geology.

REFERENCES

Basu, A., 1976, Petrology of Holocene fluvial sand derived from plutonic source rocks: implications to paleoclimatic interpretation: Jour. Sed. Petrology, v. 46, p. 694-709.

Basu, A., 1981, Weathering before the advent of land plants: evidence from unaltered detrital K-feldspars in Cambrian-Ordovician arenites: Geology, v. 9, p. 132-133.

Berner, R. A., and Holdren, G. R., Jr., 1979, Mechanism of feldspar weathering--II. observations of feldspars from soils: Geochim. Cosmochim. Acta, v. 43, p. 1173-1186.

Blatt, H., 1967, Original characteristics of clastic quartz grains: Jour. Sed. Petrology, v. 37, p. 401-424.

Darnell, N., 1974, A comparison of surficial, in situ sediments overlying plutonic rocks of Boulder Batholith and gneissic rocks of the southern Tobacco Root Mountains in Montana: Unpub. AM Thesis, Dept. Geol, Indiana University, 126 p.

Dickinson, W. R., 1972, Evidence for plate-tectonic regimes in the rock record: Am. Jour. Science, v. 272, p. 551-576.

Dickinson, W. R., 1980, Plate tectonics and key petrologic associations, in Strangway, D. W., ed., The Continental Crust and Its Mineral Deposits, Geol. Soc. Canada, Special Paper No. 20, p. 341-360.

Dickinson, W. R., and Suczek, C. A., 1979, Plate tectonics and sandstone compositions: Bull. Am. Assoc. Petroleum Geologists, v. 63, p. 2164-2182.

Dickinson, W. R., Beard, L. S., Brakenridge, G. R., Evjavec, J. L., Ferguson, R. C., Inman, K. F.,

Knepp, R. A., Lindberg, F. A., and Ryberg, P. T., 1983, Provenance of North American Phanerozoic sandstones in relation to tectonic setting: Geol. Soc. Am. Bull., v. 94, p. 222-235.

Franzinelli, E., and Potter, P. E., 1983, Petrology, chemistry, and texture of modern river sands, Amazon river system: Jour. Geology, v. 91, p. 23-39.

Garner, H. F., 1959, Stratigraphic-sedimentary significance of contemporary climate and relief in four regions of the Andes Mountains: Geol. Soc. Am. Bull., v. 70, p. 1327-1368.

Garrels, R. M., and McKenzie, F. T., 1971, Evolution of Sedimentary Rocks: Norton, N.Y., 397 p.

Graustein, W. C., and Velbel, M. A., 1981, Comment on Weathering before the advent of land plants: evidence from detrital K-feldspars in Cambrian-Ordovician arenites: Geology, v. 9, p. 505.

Holdren, G. R., Jr., and Berner, R. A., 1979, Mechanism of feldspar weathering--I. Experimental studies: Geochim. Cosmochim. Acta, v. 43, p. 1161-1171.

Ingersoll, R. V., and Suczek, C. A., 1979, Petrology and provenance of Neogene sand from Nicobar and Bengal fans, DSDP sites 211 and 218: Jour. Sed. Petrology, v. 49, p. 1217-1228.

Ingersoll, R. V., Bullard, T. F., Ford, R. L., Grimm, J. P., Pickle, J. D., and Sares, S. W., 1984, The effect of grain size on detrital modes: a test of the Gazzi-Dickinson point-counting method: Jour. Sed. Petrology, v. 54, p. 103-116.

Leeder, M. R., 1982, Sedimentology: Allen and Unwin, London, 344 p.

Mack, G. H., 1978, The survivability of labile light mineral grains in fluvial, aeolian, and littoral marine environments: the Permian Cutler and Cedar Mesa Formation, Moab, Utah: Sedimentology, v. 25, p. 587-604.

Mack, G. H., 1984, Exceptions to the relationship

between plate tectonics and sandstone composition: Jour. Sed. Petrology, v. 54, p. 212-220.

Maynard, J. B., Valloni, R., and Yu, H.-S., 1982, Composition of modern deep-sea sands from arc-related basins, in Legget, J. K., ed., Trench and Fore-arc Sedimentation, Geol. Soc. London, p. 551-561.

Nesbitt, H. W., and Young, G. M., 1984, Prediction of some weathering trends of plutonic and volcanic rocks based on thermodynamic and kinetic considerations: Geochim. Cosmochim. Acta, v. 48, p. 1523-1534.

Paton, T. R., 1978, The Formation of Soil Material: Allen and Unwin, London, 143 p.

Pettijohn, F. P., Potter, P. E., and Siever, R., 1972, Sand and Sandstones: Springer-Verlag, N.Y., 618 p.

Potter, P. E., 1978, Petrology and chemistry of modern big river sands: Jour. Geology, v. 86, p. 423-449.

Ruxton, B. P., 1970, Labile quartz-poor sediments from young mountain ranges in northeast Papua: Jour. Sed. Petrology, v. 40, p. 1262-1270.

Schott, J., and Berner, R. A., 1983, X-ray photoelectron studies of the mechanism of iron silicate dissolution during weathering: Geochim. Cosmochim. Acta, v. 47, p. 2233-2240.

Siever, R., and Woodford, N., 1979, Dissolution kinetics and the weathering of mafic minerals: Geochim. Cosmochim. Acta, v. 43, p. 717-724.

Suttner, L. J., 1974, Sedimentary petrographic provinces: an evaluation: Soc. Econ. Paleontologists Mineralogists, Spec. Pub. No. 21, p. 75-84.

Suttner, L. J., Basu, A., and Mack, G. H., 1981, Climate and the origin of quartz arenties: Jour. Sed. Petrology, v. 51, p. 1235-1246.

Tardy, Y., 1971, Characterization of the principal weathering types by the geochemistry of waters from some European and African crystalline massifs: Chem. Geol., v. 7, p. 253-271.

Todd, T. W., 1968, Paleoclimatology and the relative stability of feldspar minerals under atmospheric conditions: Jour. Sed. Petrolgoy, v. 38, p. 832-844.

Valloni, R., and Mezzadri, G., 1984, Compositional suites of terrigenous deep-sea sands of the present continental margins: Sedimentology, v. 31, p. 353-364.

Wintsch, R. P., and Dunning, J., 1985, The effect of dislocation density on the aqueous solubility of quartz and some geologic implications: a theoretical model: JGR, in press.

Wollast, R., 1967, Kinetics of the alteration of K-spar in buffered solution at low-temperature: Geochim. Cosmochim. Acta, v. 31, p. 635-648.

Yerino, L. N., and Maynard, J. B., 1984, Petrography of modern marine sands from the Peru-Chile Trench and adjacent areas: Sedimentology, v. 31, p. 83-89.

Young, S. W., 1975, Petrography of Holocene fluvial sand derived from regionally metamorphosed source rocks: Unpub. Ph.D. dissertation, Dept. Geol., Indiana University, 144 p.

INFLUENCE OF TRANSPORT PROCESSES AND BASIN GEOMETRY ON SAND COMPOSITION

Franco Ricci Lucchi

Istituto di Geologia, Università di Bologna
Via Zamboni, 67, 40127 Bologna, Italy

ABSTRACT

This contribution is an attempt to bridge some gaps between source-minded and emplacement-minded students of clastic sediments. The first part outlines a simple conceptual frame of basin topography, hydrodynamic base levels, transport-deposition processes and the resulting modification of sand before and after final deposition (pre-emplacement vs pre-burial). The second part emphasizes geometry, fill and dispersal pattern of elongated basins in orogenic foreland settings. Factual information from the mediterranean area (particularly the Apennines and Po Plain subsurface) is summarized.

INTRODUCTION

By tradition, petrographers focus their attention on provenance of clastic sediments, sedimentologists on emplacement. Both families of students acknowledge that transport is the crucial link between source and depositional areas. Sediment transport is studied for different goals by geomorphologists, hydraulic engineers and sedimentologists. However, a complete treatment of this matter is not available in textbooks. There is not systematic body of knowledge shared by all students of sediments: missing links are still numerous between source area and burial site.

By mechanics of sediment transport, we usually mean processes operating at or near the depositional site. This is only the final part of the problem: in the case of a turbidite sand, for example, the time elapsed between particle production

19

G. G. Zuffa (ed.), Provenance of Arenites, 19–45.

and final emplacement can be short or long, the pathway straight
or tortuous, with or without parking areas, and so on.
The availability of a certain volume of sand at the depositional
site results from both production of clastics in the source area
and suitable geomorphic settings and vectors for transport. Pro=
duction and delivery can be studied separately for convenience,but
are in fact parts of a system that processes material and modifies
it to various degrees. It will be referred to as the transport sys
tem. Although the output of this machine can be different from the
input,the representativity of the former is sometimes overemphas=
ized for source detection. This introduces a bias in basin analysis.
Modification of sand produced by transport must be distinguished
from that produced by diagenesis, of course. Modification is
qualified as pre- and post-depositional in the two cases. We should
not neglect, however, that significant modification of sand can
occur after deposition but before burial. Mechanical processes
are thus responsible for two phases of modification: pre-emplacement
(main transport phase) and pre-burial (mechanical processes opera=
ting in the depositional environment).
The transport system includes geomorphic elements of various scales,
shapes and orientations, material and energy flows, stops-and-go,
temporary or permanent sinks. I do not attempt to chart a flow
diagram or a budget of this system, but only to outline a general
framework and make some points in order to put transport and depos=
ition of sand in the perspective of basin analysis.
Particularly significant is the presence or absence, in the system,
of sediment filters (as the littoral belt where the fine material
is washed out and fragile particles are ground) and areas of "par=
king" or temporary storage, such as deltas, estuaries, minor basins.
Storage areas can be qualified as secondary sources, i.e. sources
for resedimentation processes, land areas representing the primary
sources. A substantial modification of original detritus can occur
between the two, whereas little is produced by resedimentation, or
mass remobilization of sediment. Resedimentation can be regarded
as a cannibalistic process affecting basin margins and their preser=
vation in the stratigraphic record.
Secondary sources and cannibalistic supply pose problems for usage
or provenance terms: for example, are they intra- or extra-basinal?
(location is within the basin but sand was produced outside).
This and other questions are addressed in next sections as a basis
for discussion; no answers, no exhaustive discussions, no thorough
review of existing opinions are provided, however, by the author
himself.

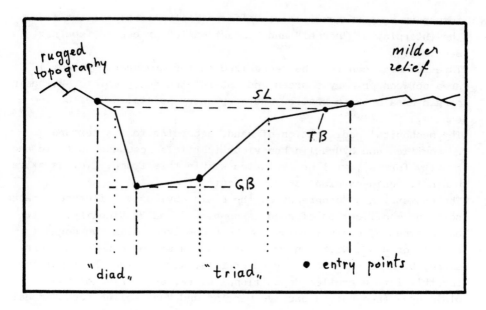

Fig.1 – Schematic profile of basin margins. SL, sea level; TB, traction base; GB, gravity base.

GEOMETRIC FRAMEWORK (Fig.1)

Entry points (loci where sediment get access to the depositional site) are here regarded as the main milestones of the transport system. They are at the intersection of the topographic profile with some significant base levels of mechanical processes: 1. the sea level SL (or continental water level WL), 2. the base of traction TB (maximum depth at which sand can be transported by "normal" proc= esses), closely associated with 1, and 3. the gravity base level GB where deposition by gravity processes starts (slumps, slides, tur= bidity currents and other sediment gravity flows sensu Middleton and Hampton,1973,1976).

The two first baselines are related to the water body and its dynamics (tides, waves): in this respect, they are hydrological datums. GB depends mostly on topographic conditions: it is effective only where sufficient depth and slope gradients exist.

Correspondingly, two types of entry points can be found, in the nearshore belt and at the base of slopes, respectively: river mouths and delta fronts are examples of shallow entry points, while mouths of submarine canyons represent deep e.p.

Entry points may or may not be sites of secondary sources; sand can accumulate there or bypass them. When storage occurs, it can be

temporary or permanent. These different situations come out from
the interplay of "normal" and "catastrophic" processes (see next
section).
Three domains can thus be recognized in the transport system: 1.
area between primary sources and SL-related entry points, 2. basin
margin area between shallow and deep entry points, and 3. basin
center.
The mechanical modification of sand, according to its texture,
composition and hydraulic behaviour (particle response to fluid and
gravity forces), will be evaluated within this frame, i.e. as extra=
basinal, basin-marginal or basin-central.
The geometry and topography of the basin obviously influences trans=
port and modification of sand in domains 2 and 3. Simplifying, two
basic types of basin margins can be recognized: a) a two-segment
profile or diad, with a narrow shelf and a steep, non-depositional
slope; b) a tripartite or triad profile, with a wide shelf passing
smoothly into a gentler depositional slope, and a rise or basin
plain area (see mature passive margins and "unda-clino-fondo" model
of Rich,1950). This scheme is here introduced with particular re=
ference to orogenic basins where deformation occurs during sedim=
entation and produces cross-sectional asymmetry.
Sea level changes do not radically alter the basic shape of the
profile, but rather the function of its segments in the transport
system. During low stands, for example, the shelf edge is sharpened
by erosion and mass wasting, but this effect is less important than
the dramatic change from submerged to subaerial conditions in the
shelf area. Moreover, storage areas are abandoned and/or eroded,
and resedimentation at GB enhanced.
Comparatively, the diad margin is less sensitive to relative SL
fluctuations because of the smaller area exposed. Lowering of SL
determines erosion of the shelf and entrenchment of streams in land
areas; consequently, the total sediment supply is increased. Coastal
entry points are bypassed and sand is directly introduced at deep
e.p. by gravity flows. Sea level rise restores sediment traps in
land and coastal areas and reduces supply to basin center. These
general trends are attenuated on diad margins, which remain closer
to sources and liable to resedimentation even during high stands.
Triad-type margins are less prone to gravity failure but more sen=
sitive to SL changes, whose effects are here amplified: more sand
is retained in the shelf during high stands and eroded during low
stands, as storage areas tend to be larger. Basin center can be
starved during extreme high stands.

TRANSPORT PROCESSES

A first fundamental distinction can be made between selective and
massive processes. Sand particles have an individual behaviour in
the first case, a collective one in the second; also, differences
in particle and fluid behaviour (relative velocities) are maximized
and minimized, respectively.
In selective transport, sand moves in close contact with the
bottom, whereas mass transport creates more or less expanded
sediment-fluid admixtures (dispersions) with well defined bounda=
ries (density barriers) with respect to ambient or still fluid.
Most dispersions, however, are also bottom-hugging.
Particle sorting (by size, shape and density) is a typical expres=
sion of selective processes of transport. It can occur also in
mass   flows, but only where turbulence prevails, i.e. in the
more dilute dispersions (turbidity currents). Turbulence acts as
a supporting force that stimulates individual behaviour of grains.
This results in differential settling, with coarser and heavier
particles moving closer to the bottom. In concentrated, non-tur=
bulent dispersions, sorting is absent or poor.
Particle abrasion is the other main effect of selective transport.
Disaggregation of less resistant material and differential rounding
of harder grains results from strong grain-to-grain and grain-to-
boundary intercations (collisional solid friction). In mass flows,
collisions are much less numerous and often cushioned by high-den=
sity fluids. Consequently, solid friction is less effective by
several orders of magnitude.
The two groups of processes can also be regarded from the viewpoint
of energy dissipation, and be called normal and catastrophic, res=
pectively. Somebody discards the term catastrophic for phylosophical
reasons and prefers episodic, or spasmodic (see Dott, 1983).
This is not the same distinction as that between "low-energy" and
"high-energy", because high-energy (though too vague a concept)
characterizes also normal processes. Rather, it expresses different
time-energy distributions (Fig.2). Normal processes are almost
continuous and vary gradually in intensity, catastrophic processes
are discontinuous and instantaneous ("flashy"). Mechanical energy
is both temporally and areally less concentrated in normal proc=
esses. The transport energy of normal currents and waves is best
expressed by velocity of flow and maximum size of carried particles
(competence), whereas mass, density, momentum and inertia are more
indicative of catastrophic transport.
Another distinction, independent from the previous one, is made
between fluid-driven and gravity-driven processes. Although all

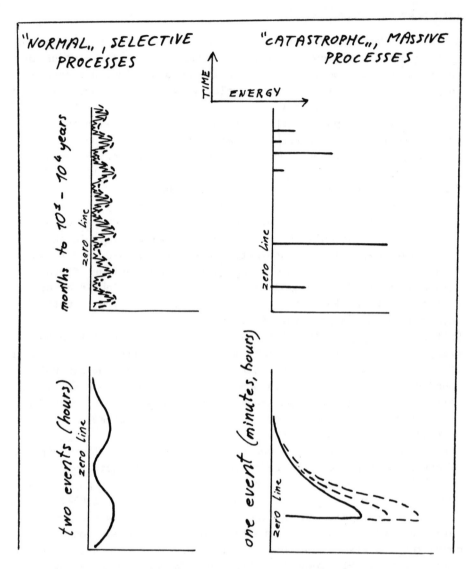

Fig.2 – Qualitative graphs of energy dissipation in transport processes. The lower diagrams are matched by grain-size logs.

transport is ultimately driven by gravity, sedimentologists conven= tionally regard as gravitative the processes in which the fluid is a passive phase and the gravity pull is exerted directly on solid particles or the body of the dispersion.

Most fluid flows are normal, but some are not: storm currents and flash floods are examples in point. Neither are all mass flows gravity-driven (mudflows and debris flows can move under fluid

drag, too) or all gravity processes massive (see individual fall
and sliding downslope).

Selective processes

A fluid able to move sand particles in a differential mode is a
tractive current in a broad sense (including oscillatory movements
of waves). Moving grains form the bedload or traction carpet, i.e.
a mm or cm-thin layer. The gravity-driven equivalent of the traction
carpet is the avalanche carpet flowing down local slopes at the
angle of repose.
The efficiency of selection varies considerably according to the
distance travelled, the flow velocity, the nature of the medium
(air or water) and particle properties (hydraulic behaviour).
The distance that counts is not the geographic but the cumulated
one, thus waves are more efficient than unidirectional currents
because of their longer and more continuous action. In a thin
medium, like air or hot volcanic gases, both abrasion and sorting
are more effective as fluid cushioning is reduced by the low density
and viscosity.
As for flow velocity, there is no simple relationship with selective
efficiency. For a given velocity, there is a critical size (or
weight) of particles: particles larger than the threshold value
cannot be removed. This is a first cause of sorting; then, when
flow speed is sufficient to carry all particles, they travel at
different velocities. At peak flow velocities, however, a transition
can occur between tractive and mass transport, with suppression of
differential grain movements.
In downflow direction, sorting first separates the bedload from
the suspension load and, within the bedload, gravel from sand. The
segregation of different sand fractions is quite more laborious.
A good visual indication of the flow sorting ability is the pres=
ence of lamination in the sand, i.e. its organization in depositional
laminae and laminasets (Campbell, 1967). Several authors demonstra=
ted by experimental work in flumes that tractive lamination (not
to be confused with settling lamination occurring in finer mate=
rials) does not need pulsations in the mean flow velocity or in the
supply of sand; it rather reflects a differential response of the
bedload to continuous traction in uniform and steady flow conditions.
For sampling purposes in provenance studies, it is not necessary
to pick up grains from distinct laminae: a laminated layer is one
event, a single lamina is not. In the field, it is thus important
to discriminate between bedding surfaces and lamination surfaces,
as only the former ones separate distinct depositional events.

Lamination represents a form of geometrical order, and we say that a laminated sand is an <u>organized</u> or structured sand, which is one more indication of selection and hydraulic maturity.

The efficiency of sorting is decreased in boundary conditions, as defined both spatially and temporally, i.e. where and when fluid energy rapidly decays. Examples are the mixing of mud and sand during falling stages of fluvial or tidal floods, the bulldozing of unsorted debris by wave surges on beaches and lagoons, and the churning of sand-mud couplets by burrowers.

The finding of poorly sorted sand interbedded with well sorted and structured levels may suggest the spasmodic occurrence of mass flows. Sorting alone, however, is not enough for recognizing mass deposits; the most characteristic signature consists of <u>thick</u>, <u>un=structured</u> (or massive) beds, indicating hydraulic immaturity.

Selective processes can rework mass deposits, and mass transport can remobilize tractive deposits; the two sets of processes can as well act alternatively on the same materials. Clues to these phenomena are the examples of "maturity inversions" that can be found in sands (texturally mature particles within hydraulically imature beds, and viceversa).

A base level for traction was previously introduced: the picture must now be refined. First of all, traction can also occur in deep water, but not everywhere. Thermohaline oceanic circulation and suitable topographic conditions produce <u>bottom currents</u> capable of entraining sand and accumulating it in sediment drifts. Set=tings favourable to such currents are continental rises and topo=graphic constrictions of the sea floor in modern oceans; in an=cient basins, they are not easily recognizable or predictable.

The signature left by bottom or contour currents consists of thin beds of well laminated and sorted sand, either terrigenous or biogenous. These so-called contourites represent mature sedim=ents in an environment usually dominated by pelagic settling or less mature turbidite sands.

Secondly, the traction base TB is an average limit; in shelves dominated by wave energy, it is comprised between <u>fair-weather</u> and <u>maximum storm</u> wave bases, and more or less corresponds to the limit of <u>average storms</u>. Sand can be transported to greater depths, as far as the shelf edge, by exceptional storm events, but in this case it is mobilized en masse.

Littoral and shelf sands are separated by a mud belt from deep-water sands; moreover, the deep mode is not supplied by the trac=tive agent. Bottom currents only rework sediment previously em=placed by gravity flows.

Mass flows

Relative amounts of fluid and solid particles and grain sizes are
quite variable in massive processes. At the highest solid concen=
trations (more than 30% by volume), the dispersion forms a fluid
by itself with density and viscosity higher than those of ambient
fluid. The flow of this rheological fluid is characterized by two
main facts: 1. particles and surrounding, or interstitial fluid
are still distinct components but move with the same velocity; 2.
there is only limited separation of them during deposition, which
consists of "freezing" of the whole mass or quick settling with
water expulsion. This prevents or strongly limits sorting.
At the lowest concentrations and densities ($\Delta \varrho$ as low as 10^{-3} g/cm^3),
the flow is newtonian, i.e. a turbulent suspension. Sand particles
eventually settle individually from them producing vertical and
lateral grading in the resulting bed. Massive dispersions can
segregate from the base of turbulent suspensions, where the high
grain concentration can suppress turbulence.
High-density flows can be schematically subdivided into inertial
(or granular) and viscous (or muddy) types. The former is made of
sand for more than 90% of the solid fraction, the latter contains
at least 10% mud (Hampton, 1972). The clast-mud mixture forms a
debris flow, in which clasts may consist mostly of sand or larger
grades. In the first case (sandy debris flow) the difference with
a cleaner sand flow (less than 10% mud) can seem trivial but this
is not so: a "dirty" sand can travel more as the mud is an efficient
lubricant.
Inertial flows (actually, they are given different names and inter=
pretations: grain flows, fluidized flows, liquefied flows, fluxo=
turbidite flows, etc.) do not go much beyond the base of the feeding
slope because of the strong intergranular friction, and even to get
there they need help by interstitial overpressure or liquefaction.
Without overpressure, only thin carpets can form, and critical
slopes are very steep (at least 18°; see Lowe, 1976).
In debris flows, the clasts are supported by mud strength and
buoyancy (Hampton, 1972). A poor selection can occur with larger
particles pushed upwards by vertical components of solid or solid-
mud friction: this results in inverse grading. Also, a faint or
diffuse lamination can be produced, not by bottom traction but by
internal shearing. In comparison, tractive laminae are thinner
and more evident (better textural contrast).
In general, high-density mass flows leave massive beds with sharp
base and top; sorting is good to moderate in granular flows, poor

to very poor in debris flows. It is important to remark that the
good sorting of massive sands does not derive from the transport
mechanism but is inherited from the immediate source (storage
area), and thus reflects the previous history of the sand.
The prototype of a suspension flow is the turbidity current.
This density flow can originate nearshore as in the head of a
submarine canyon, but develops fully on deeper slopes.
Finer particles are uniformly distributed by eddies within the
flow, whereas sand grains tend to concentrate near the bottom.
Where deposition starts, this sand forms a massive or graded bed
by more or less fast settling; later settling of finer particles
completes the layer with structured and massive portions. A clas
sical turbidite, therefore, is not simply a graded sand but a
graded sequence going from basal sand to topmost mud (Bouma seq=
uence).
A graded sand, or a graded layer, has long be regarded as the
"trade mark" of the turbidity current. This rigid interpretation,
however, is not correct. Grading is the mark of turbulence and,
consequently, of suspension flows; but not all suspension flows
are turbidity currents, i.e. gravity flows active in deep water.
Examples of non-turbidite, fluid-driven flows are known in fluvial
and deltaic environments, where they are caused by sudden invasion
of subaerial or shallow submerged areas by mechanisms operating
during floods (unconfined sheetflows, overspilling, crevassing);
similar effects are produced by storms in protected coastal areas
(bays, lagoons) and offshore. All these "shallow" mass flows are
originated by meteorological catastrophic events.
Changes of composition can be associated with grading: they point
out hydraulic sorting within the flow. Typical is the case of
carbonate or mixed, brecciola-type turbidites, where benthic
remains are concentrated in the basal portion and plankton prevails
in the upper one.
As already pointed out, the turbulent suspension is the most se=
lective among mass flows. It can also produce tractive lamination
on top of, or within the graded sand. Traction in this case is
accompanied by rapid bed aggradation, and superposed laminae can
grow very numerous. If some laminae are more marked than the others,
this can give the impression of several beds (distinct events).

Other important characteristics of turbulent mass flows are:
1. very low internal and boundary friction, which enables them
to spread their load of sand over wide surfaces: sheetlike deposits
thus result, in contrast with more localized, lens- or tongue-
shaped deposits of high-density flows;

2. a high momentum can result from both large mass and high velo=
city; a considerable inertial energy is still stored in the flow
when it reaches the base of the feeding slope. It can thus travel
on horizontal bottom (for tens of kilometres in the case of so=
called megaturbidites, whose deposited volume is more than 1 km^3: see
Mutti et al. 1984). By contrast, granular and debris flows are
mostly restricted to the slope toe, which makes massive beds
good indicators of proximality, i.e. proximity to secondary sand
sources and entry points;
3. great erosional power in slope areas, where the flow accelerates.
Excavation of canyons and gullies ensues, and previous slope depos=
its plus deeper rocks (substratum of basin margin) can provide
clastics in addition to the primary or external sources. High-den=
sity flows, on the other hand, have a limited erosional power;
sliding and mass wasting feed them with intrabasinal and substra=
tal material but mostly in the form of mud or fragments larger than
sand.
One more implication of point 2 is that coarse material (up to
gravel size) can be transported well into basin central areas
and included in organized deposits such as turbidites. This empha=
sizes the fact that grain size per se, in mass flow deposits, is
not the best indicator of proximality or marginality. Relative
proportions of disorganized to organized deposits in the section
are more significant.
As for point 3, the problem of "around-vs-below" in source identif=
ication (primary-subaerial vs substratal-submarine) is common to
selective mechanism in certain settings, e.g. river entrenchment
in regressive phases or wave reworking in transgressive phases.
Substratal sources should be distinguished from more distant extra=
basinal sources; areally, they are intrabasinal. As they are acti=
vated by basin dynamics (sedimentary, tectonic, eustatic), sub=
stratal sources could be qualified as (extrabasinal) cannibalistic,
to stress analogies with the temporary storage of loose sediment.

After discussing differences among mass flows, some common features
are worth stressing:
- a mass flow picks up everything available at its source,
whether it be primary or secondary; whatever the result of pre=
vious transport and modification, it is preserved in the final
deposit as resedimentation introduces very little modification;
- a mass flow abandons its load very quickly, which enables preser=
vation of delicate materials even in environments where normal
circumstances are hostile to preservation. Pondering this point

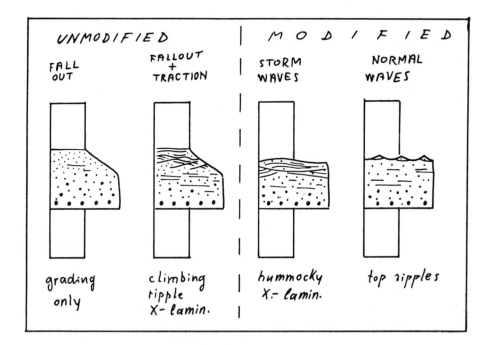

Fig.3 - Graded layers emplaced by suspension mass flows: emplacement
vs modification mechanisms.

can avoid misinterpretations such as to infer anoxic conditions
from the presence of vegetal matter or calm water from fragile
and delicate remains.

DISTRIBUTION OF PROCESSES IN THE SYSTEM

Selective processes are important in terrestrial and nearshore
areas while mass flows take over in submerged slopes and as far
as central basin areas. This general division of work is obviously
simplified: on one hand, a certain degree of selection is operated
by turbidity currents in deep-sea domains (it is more textural
than compositional, anyway). On the other hand, climatic-meteoro=
logical reasons make also subaerial and shallow environments prone
to catastrophic phenomena: see flash floods in semi-arid or steep-
gradient regions, crevassing in alluvial and deltaic plains, storm
washover in beach-lagoon complexes, storm waves and currents in
offshore areas of subtropical and cold belts.
Thus far we have examined synthetic criteria for identifying
different types of mass flow deposits on the basis of mechanisms,
textures, structures and geometry. Now we are faced with the task
of identifying the environment where a given mass flow took place.

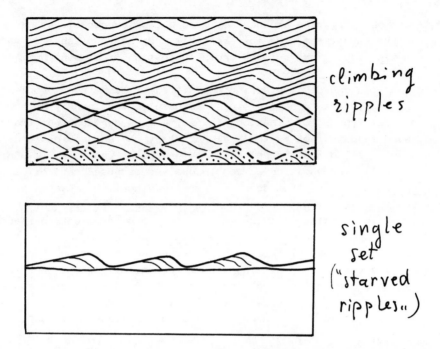

climbing ripples

single set ("starved ripples")

Fig.4 - Different ripple cross-lamination geometry related to emplacement (above) and modification (reworking) of sand. Modified from Jopling and Walker 1968.

Although typical of emplacement studies, environment recognition is not unrelated to the problem of source positioning (subaerial or subaqueous location, distance).

For this purpose, we need evidence that is often contained not in the mass deposits themeselves but in associated sediments.

For example, selective levels alternating with massive and graded deposits can be the key to discriminate between subaerial and subaqueous emplacement, or between a storm-generated and a deep-water, gravity-driven turbid flow. In this respect, selective sands can be called modification deposits, to point out the fact that they originated from reworking of mass flow deposits when normal conditions were restored at the depositional interface (Fig.3). Modification deposits commonly consist of thin sand units with tractive structures indicating shallow or nil water depth. They are members of a broader family of indicators, both mechani= cal, chemical and biological such as plant roots, specific trace fossils, pedogenic cements, nodules and crusts, oxidation, mud cracks, and so on. If these features are found on top of massive beds, a deep-water environment is excluded (with the possible

exception of ripples produced by deep bottom currents). Among
tractive structures, the most reliable indicators of shallow water
are wave ripples and related cross-lamination, including hummocky
c.l.
As traction can develop during deposition of sand from turbulent
mass flows, it is crucial to discriminate between purely selective
and mixed types of lamination. The latter one derives from traction
of sand supplied not by the bedload but by the suspension load: it
has been called traction plus fallout (Jopling and Walker, 1968;
see Fig.4), and shows up as continuously superposed sets of climbing
ripples. A current that does not carry sediment but reworks pre=
viously deposited sand organizes it in a single set of cross
laminae (more, discontinuous sets separated by erosional surfaces
indicate bedload supply). Differences in the texture of sand (more
matrix in the case of fallout) match the two lamination types.
Selective removal and elimination of mechanically weaker particles
can also characterize "modified" sands. As in all selected deposits,
the potential of sedimentary dynamics in changing composition must
not be neglected in favour of changes in input and provenance.

OROGENIC BASINS

Unlike passive margins or rift basins involved in orogeny after
sedimentation, the mobile basins that set up in active orogens are
characterized by syn-sedimentary tectonism. These basins develop
in the compressional setting of a fold-thrust belt, and also in
the extensional stages of the uplifted chain. Only the first group
is considered here.
The fills of compression-related basins have a wedge-shaped cross-
section and are here called orogenic clastic wedges to distinguish
them from other similar bodies accumulated in anorogenic settings.
In the Tethyan region, the original geometry and spatial relations
of pre- and syn-collisional wedges (Mesozoic to early Tertiary age)
are poorly known and little comparable with present active margins.
What is the modern equivalent, for instance, of Helminthoid-type
flysch units with their thick-bedded, fine-grained turbidites made
of intrabasinal carbonates? Their tectonic fragmentation prevents
a reliable reconstruction.
Better preserved is the geometry of foreland wedges deposited on
a shortening continental margin such as was the Apenninic-Adriatic
orogen from Oligocene to Present. This is a post-collisional belt
characterized by ensialic, or A-type subduction (Bally and Snelson
1980; Boccaletti et al. 1980). The subsidence, as effect of thrust
movement and load, migrated stepwise toward the foreland up to the

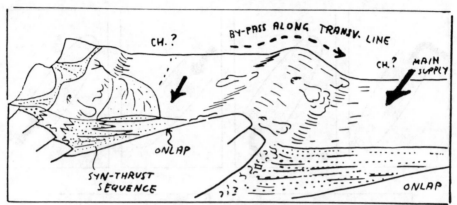

Fig.5 - Foreland basin with a "passive" side (right) and associated
"piggy-back" basin in the thrust belt. In both, gravity
base is effective and the main fill is represented by tur=
bidites of longitudinal dispersal and lateral slides (oli=
stromes). CH.? shows the possibility that sand be trapped
in a sub-trough at the base of the inner slope. These "chan=
nels" could form during stronger compression. The two
basins receive independent supplies from the same or dif=
ferent sources.

Pleistocene. The resulting geometric association of thrusts and
elongated basins has been altered in the uplifted chain but fossil=
ized in the subsurface of the Po Plain, where generalized subsidence
buried it under a thick sediment cover (Pieri and Groppi 1981; Dondi
et al. 1982). The Apenninic front extends under the Adriatic and
Ionian seas (Casnedi 1983; Rossi and Sartori 1981; Mascle et al.
1982).
In this pattern, main depocentres are located within a major fore=
land basin or foredeep, more or less complicated by transversal
structures along its axis. Other accumulations are found in inter-
thrust or "piggy-back" basins (Fig.5; see Ori and Friend 1984,
Ricci Lucchi 1984b), usually of smaller size. Subsurface evidence
corroborated by outcrop analysis shows that all these thrust-related
basins are asymmetrical in cross-section, with a steeper and unsta=
ble side on the thrust front and the opposite gentler flank on the
back of a more advanced thrust or on dipping foreland.
The typical orogenic basin is thus bounded by an inner margin of
the diad type and an outer margin of the triad type.
Foreland wedges represent two main stages of orogenic sedimentation:
the flysch stage, which is pre-paroxysmal, i.e. precedes the defi=
nitive uplift and emergence of the thrust belt, and the molasse
stage that follows it. Molasse is fed by the adjacent, emerged
chain, flysch is not, because at the time of its deposition the
orogen is mostly submerged.

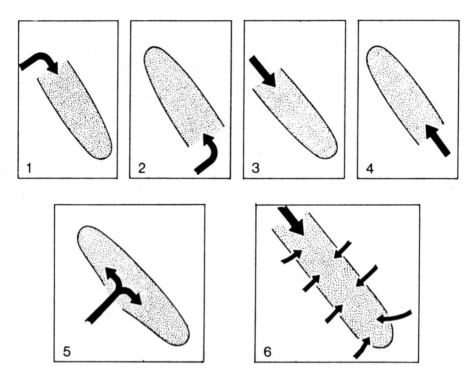

Fig.6 – Various types of dispersal in elongated foreland basins
of northern Apennines. Irrespective of scale and depos=
itional stage (flysch vs molasse), lateral input is
deviated axially in basin center. From Ricci Lucchi,1984b.

In the Apennines, flysch and molasse are different from their
typical expressions, as defined in the Alps. The true Alpine Flysch
is an oceanic, abyssal turbidite facies; the Apenninic Flysch is
not so deep, more terrigenous, and accumulated much faster (10–80
cm/ky as compared with 0.1–10 cm/ky). Furthermore, it is a post-
collisional product; in other words, it is fed by the same sutural
chain that originated the coeval Alpine Molasse. As for the latter,
the well known Alpine type is a shallow-marine to continental dep=
osit, whereas the Apenninic type is mainly a turbiditic-bathyal
facies with a relatively thin paralic to alluvial cover. Other
things being equal, just a different balance between sedimentation
and subsidence can explain the two different molasses. In the case
of flysch, the matter is more complicated as different geodynamic
settings are involved ("oceanic" vs "continental").
In any case, flysch and molasse cannot be defined unambiguously
on grounds other than geodynamic. The literature shows that purely

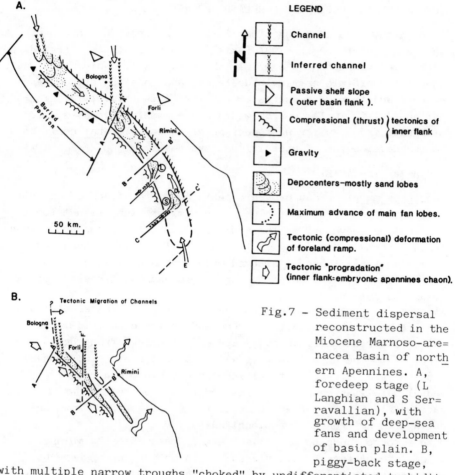

Fig.7 - Sediment dispersal reconstructed in the Miocene Marnoso-are=nacea Basin of north ern Apennines. A, foredeep stage (L Langhian and S Ser=ravallian), with growth of deep-sea fans and development of basin plain. B, piggy-back stage, with multiple narrow troughs "choked" by undifferentiated turbidite sands.From Ricci lucchi 1984a, courtesy of Springer-Verlag.

lithological and sedimentological criteria fail to provide opera= tional definitions; consequently, these terms are not useful in the field or at the microscope for description purposes. They convey more information as tectofacies, but definition relies heavily on interpretation in this case. This means that a critical amount of field and petrographical data is needed to characterize adequately flysch and molasse deposits.Moreover, the concepts are relative to the regional geologic context and geographic location.
Regardless of nomenclature, facies analysis of clastic orogenic wedges should answer two basic "where-and-when" questions: were they deposited on oceanic or continental crust? and before or after the emergence of the thrust belt to which they are physically con= nected?

THE DISPERSAL PROBLEM IN FORELAND BASINS

In both the Apennines and other Tethyan orogens, foreland basins
show a dominant longitudinal (axial) dispersal pattern, regardless
of tecto-depositional stage (flysch or molasse), sedimentary envir=
onment (turbiditic or fluvio-deltaic), and location of sediment
sources and transport routes around the basin (Fig.6, 7). Elon=
gation (2:1 to 10:1), narrow width and asymmetric cross-section
of basins favour a strong topographic control (lateral confinement)
on the development of depositional systems.

Basin margins (Fig.8)

The inner margin of an orogenic basin can be the flank of an enti=
rely submerged structural high,   a similar high (thrust top)
culminating above SL, or the newly emerged chain (composed of many
thrust sheets). Accordingly, the amount of available sand varies
from almost nil (case 1, starved margin with only slides) to
scarce (case 2, small drainage areas) to abundant (case 3, larger
areas and rate of erosion).
In cases 1 and 2, the high acts as a dam for lateral supply; sand
can   reach the basin only from one end or the other side. Damming
can be important also in early stage 3, causing stream deviation
parallel to the shoreline (Fig.9). As uplift continues, subsequent
streams take over and lateral input becomes dominant; fan delta
and alluvial fan coarse clastics are then deposited on top of deeper
facies (Fig.5). Beyond the base of the slope gravity flows are
deflected and transport is longitudinal.
Case 1 corresponds to the flysch stage, case 3 to the molasse
stage; the different role played by a diad margin can here be
appreciated: a barrier to transport in the former, a fast route
in the latter. This refers in particular to the foredeep; in an
adjacent satellite basin of piggy-back type the inner margin is
viable in the flysch stage too; sources are available in colli=
sional reliefs or craton area. Sporadic connections with the fore=
deep can be provided where transversal tectonic lines are activated
(Fig.5,9). In the Apennines (Fig.7), these structures provided
both pathways for lateral transport and trigger mechanisms (earth=
quakes) for huge gravity failures and flows. The occurrence of
laterally fed marker beds and sand bodies interbedded with the
"axial facies" has been extensively documented (Ricci Lucchi 1975,
1978; Ricci Lucchi and Valmori 1980). In the overall basin budget,
the lateral supply remains of minor importance, but individual
events (layers) can reach impressive size (up to more than 100 cubic

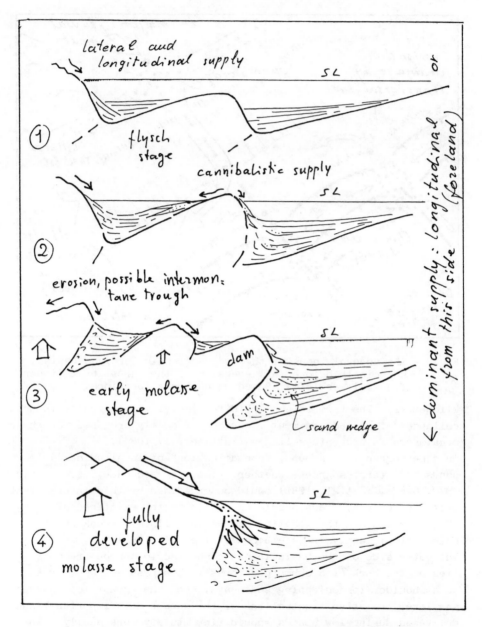

Fig.8 – Evolving topography and supply patterns in foreland
    orogenic basins. Conceptual scheme for collisional
    orogen (suture belt and craton to the left, outside
    the figure).

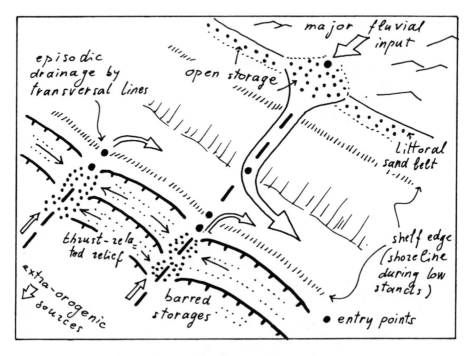

Fig.9 – The unequal input of the two margins in foreland basins,
        especially before full emersion of the thrust belt (inner
        flank, lower left). Based on northern Apennines.

kilometres). The term seismoturbidite, implying seismically induced
collapses of stored sediment , has been recently proposed for these
megalayers in orogenic settings (Mutti et al. 1984).
On a petrographical ground, "contamination" of longitudinally fed
sands by lateral sources, whether primary or secondary (see Sestini
1970; Gandolfi et al. 1981) could be explained by this intermittent
tectonic activity, thus providing a clue to the existence of mega=
shear systems and the location of the main lines of weakness in
paleotectonic analysis (see Boccaletti et al. 1982).
The outer side of the basin is not connected to source areas if it
lies on the back of a submerged thrust. This passive ramp serves
as a shoulder for onlapping sediments. When the thrust top reaches
sea level, relative fluctuations of the same become important: sand
delivered during low stands should find its way more easily down
the ramp than the other, steeper flank of the high. The ramp
provides, in fact, a larger area for drainage. Such intermittent
activation of transport systems on thrust tops during regressive
stages is testified in seismic sections (e.g., Po Plain) by uncon
formities between converging sedimentary sequences.

The outer flank of the foredeep can be a well developed triad
connected to emerged foreland areas. In the case of collisional
orogens, this is not simply a flat and geomorphically mature
area. Instead, rugged reliefs, formed by previous collisional events,
can exist, and actually existed in front of the Apennines foredeep,
particularly in the northern sector. They were represented by the
Alps, whose uplift preceded and accompanied the apenninic orogeny.
The arcuate alpine chain surrounded the apenninic-adriatic "prom=
ontory", whereby sutural reliefs were present also along the inner
side of the apenninic basins. It was the foreland side, however,
that gave the major clastic contribution (Fig.7,9). The detritus
of inner provenance, in fact, was trapped by thrusts in piggy-back
basins that formed closed storage areas. On the opposite margin
of the foredeep, more open repositories such as deltas were exposed
on shelf-slope areas to gravity failure and resedimentation.
The apenninic setting exemplifies one of the introductory state=
ments of this paper and the risks of overlooking it: the rate
of clastic supply in a basin depends on both its production and
delivery in source areas and its accessibility to basin entry
points.

Basin center

Foredeeps were mainly sites of turbidite accumulation during the
whole history of the apenninic orogen. Their filling occurred by
axial transport and mantling of the deepest, central area. Owing
to the asymmetry of the basin, the depocenter was near the inner,
steeper side (Fig.5, 10). The thickness of the sedimentary section
was increased here by slides and olistostromes.
Lithological and seismic evidence in sections show that the depos=
itional area contracted and expanded according to shifts in the
balance between sedimentation, subsidence and tectonic movements.
It can be assumed, for example, that when thrusting was less inten=
se or dormant, sedimentation overpaced subsidence and a wider,
flat basin plain developed (Fig.10, upper part). Turbidity currents
of opposite direction, coming from different entry points, ran
along this plain confirming its horizontality (Ricci Lucchi 1975,
1978; Ricci Lucchi and Valmori 1980). Sheetlike sands were spread
over miles and miles.
When thrusting was active, the stronger subsidence and its markedly
differential character, probably aided by tilting of the outer ramp,
deepened GB and changed the profile to a more channel-like shape
(Fig.10, lower part). The flow section was thus reduced and flow

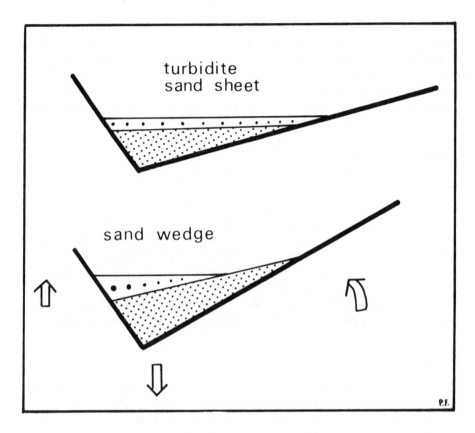

Fig.10 – Inferred changes of turbidite sand bodies and basin center
        geometry related to pulses of thrust activity. Current
        direction normal to section shown.

velocity increased. Coarser sand could be carried to basin center
and "stuffed" in narrow wedge-shaped bodies.

This case shows that lateral wedging and shaling of turbiditic
sandstones can be an indication of topographic control instead of
provenance and downcurrent energy decay. The geometry looks the
same as that produced in the current direction, and can suggest
proximal to distal changes where the section is instead across-
current. Consequently, the geometry of deposits must be checked
against paleocurrents and sand composition in order to establish
whether the dispersal is longitudinal or lateral. Also, sections
of differnt orientation have to be cross-checked for a reliable
geometrical reconstruction. In longitudinal sections, for example
(Fig.11), entry points can be visualized by thickening and abrupt
lateral truncation of sand beds. Slides and debris flows are also
more frequent near entry points.

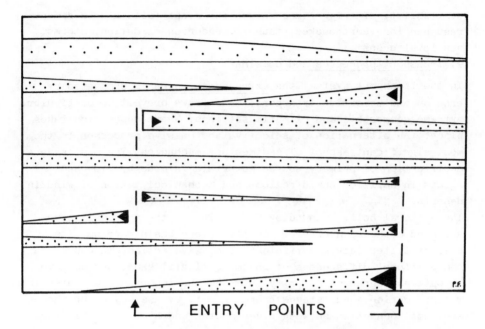

ENTRY   POINTS

Fig.11 - Schematic longitudinal section of a foredeep fill.
        Correlatable turbidites end abruptly at entry points
        (which actually mark the starting points of basinal
        deposition).

SUMMARY AND CONCLUSIONS

The main points presented for discussion in this paper are summa=
rized under three headings:
1. Accessibility of sand to entry points

Shallow and deep entry points must be separated as they are related
to different hydrodynamic base levels: sea level-wave base and
gravity base, respectively.
Availability of sand to the former, i.e. to the shoreline, depends
uppon three main factors: rate of erosion of land areas, extent
and slope of drainage basins, presence and persistence of physical
barriers (dam effect). The effect of climate is included in the
first factor.
Accessibility of sand to the base of slope (GB) is controlled by
previous factors, plus: topography of the basin margin (average
gradient, presence of submarine canyons, sills, etc.), effectiveness
of gravity action (sufficient depth and steepness), distance of
wave base from shelf edge and heads of canyons, presence and type
of storage areas or marginal repositories (barred vs open), and

availability of trigger mechanisms for slope failure or sediment
resuspension (earthquakes, tsunami, underconsolidation, storms,
sea level drops).

2. Transport processes and domains

In the transport system, the extrabasinal area is dominated by
erosion and selective transport; mass flows prevail in particular,
extreme, geomorphic and climatic settings. Where deposition takes
place, the alternation of selective and massive processes is the
most significant aspect of sedimentary mechanics. The most severe
modification of primary source materials is made by bio-chemical
agents in soils of humid regions and mechanical action of wind in
deserts.

The littoral belt, bounded by SL and TB, is the most selective
area and occupies a crucial position along basin margins. Its
role of filter (given sufficient time, wave action can cancel many
compositional differences of incoming fluvial sands and achieve
an optimal degree of maturity) stresses the importance of determining
whether marine sands stayed there or not. Bypass may occur through
a direct connection between river mouths and heads of submarine
canyons.

The deep water is the domain of mass flows, which are mostly gravi=
ty-driven. Their main sources are secondary, i.e. sand stored in
deltas, beaches a.s.o. Peculiar characters of mass flows are the
undiscriminating removal of sediment and the lack of modification
during transport (except for some hydraulic sorting by turbidity
currents).

Deltaic, littoral and shelf sands are delivered to the deep sea
especially in erosional regressive phases (SL low stands), when
funneling by canyons and slope failures are greatly incremented.
Regardless of SL position, earthquake of great magnitude can trig=
ger slides and sediment flows of exceptional size in orogenic
settings. Marginal sand repositories are here destroyed and their
evidence left in basin-wide megalayers (turbidites, homogenites: see
Cita and Ricci Lucchi 1984). These outstanding markers of both out=
crop and seismic sections are very significant in deep-water domains.

3. Cannibalistic supply

In the same mobile setting, where growing structures produce an
alternation of reliefs and sags of various position with respect
to SL, the term terrigenous is, perhaps, more convenient than
extrabasinal to connotate sand arriving at the burial site after
subaerial transportation. This sand, in fact, can be introduced
from several places and/or geologic bodies: 1. land areas outside

the orogen, 2. the newly emerged fold-thrust belt, 3. restricted
land areas on top of structural highs, 4. marginal areas of the
basin where the substrate is exposed to erosion, either subaerial
(lowering of SL) or submarine (canyon cutting), 5. marginal repos=
itories of sediment (parking or storage areas).
While sources of type 1 and 2 are straightfordwardly extrabasinal
(or primary), complications begin with the other types, all of
which could be qualified as cannibalistic. The concept of canni=
balism is not univocally defined in the literature; it overlaps
with those of resedimentation and recycling but coincides with
none of them, as exemplified by the above distinction in three
varieties.
Intrabasinal highs and exposed margins can deliver both previously
deposited superficial sediment and underlying rocks; these sources
can be regarded either as intra- or extra- according to the weight
given to areal or vertical reasoning. They are here called
substratal.
In the case of storage areas, cannibalism consists in the resedim=
entation of intrabasinal bodies of sand whose origin can be indif=
ferently extra- or intrabasinal.
We are thus faced with two problems: univocal definition of pro=
venance terms, and recognition of cannibalism and the form in
which it occurred. These problems are obviously complicated by
post-depositional changes, including tectonics. Thrust and nappe
movements, in particular, can profoundly alter the original paleo=
geography (inversions of relief, superposition of sediments and
source rocks, etc.).

Needlessto say, the solution to these and other problems of basin
analysis needs more integrated efforts of "emplacement" and "prov=
enance" people. The increasing use of the geometrical tool in
sediment analysis underscores that also sedimentology and strati=
graphy are being more and more integrated in a dynamic approach.

AKNOWLEDGEMENTS

This paper is based on research carried out in the last years
with the financial support of Italian Government (Ministero Pub=
blica Istruzione, Consiglio Nazionale Ricerche). The author is
grateful to G.G.Zuffa and NATO for the invitation to the Cosenza
Meeting, and to colleagues who participated in it and provided
useful criticism and suggestions for this draft.

REFERENCES

BALLY, A.W. and SNELSON, S., 1980, Facts and principles of world
    petroleum occurrence: realms of subsidence: Can. Soc. Petr.
    Geol. Mem. 6, p.9-94.
BOCCALETTI, M., CONEDERA, C., DECANDIA, F.A., GIANNINI, E. and
LAZZAROTTO, A., 1980, Evoluzione dell'Appennino settentrionale
    secondo un nuovo modello strutturale: Mem. Soc. Geol. It.,
    v.21, p.358-373.
BOCCALETTI, M., CONEDERA, C., DAINELLI, P. and GOCEV, P., 1982,
    The recent (Miocene-Quaternary) regmatic system of the western
    Mediterranean Region: Jour. Pet. Geol., v.5 (1), p.31-49.
CITA, M.B. and RICCI LUCCHI, F., 1984, eds., Seismicity and sedim=
    entation. Marine Geology, spec. issue, v.55 (1/2), 161 p.
CASNEDI, R., 1983, Hydrocarbon-bearing submarine fan system of
    Cellino Formation, Central Italy: Amer. Ass. Petr. Geol.Bull.,
    v.67, p.359-370.
CAMPBELL, C.V., 1967, Lamina, laminaset, bed and bedset: Sedimen=
    tology, v.8, p.7-26.
DONDI, L., MOSTARDINI, F. and RIZZINI, A., 1982, Evoluzione sedi=
    mentaria nella Pianura Padana. In: CREMONINI,G. and RICCI
    LUCCHI,F., eds., Guida alla geologia del margine appenninico-
    padano: Guida geol. reg. Soc. Geol. It., p. 47-58.
DOTT Jr., R.H., 1983, Episodic sedimentation. How normal is normal?
    Does it matter?: Jour. Sed. Petrol., v.53, p.5-23.
GANDOLFI, G., PAGANELLI, L. and ZUFFA, G.G., 1981, provenance and
    detrital mode dispersal in Marnoso-arenacea Basin
    (Miocene, northern Apennines): I.A.S. 2nd Reg.Mtg.,
    Abstr. Vol., p.65-68.
HAMPTON, M.A., 1972, The role of subaqueous debris flow in generat=
    ing turbidity currents: Jour. Sed. Petrol., v.42,
    p.77(-793.
JOPLING, A.V. and WALKER, R.G., 1968, Morphology and origin of
    ripple-drift cross-lamination, with examples from the Pleis=
    tocene of Massachusetts: Jour. Sed. Petrol., v.38, p.971-984.
KASTENS, K.A. and CITA, M.B., 1981, Tsunami-induced sediment trans=
    port in the abyssal Mediterranean Sea: Geol. Soc. Amer.Bull.,
    v.92 (Part I), p.845-857.
LOWE, D.R., 1976, Subaqueous liquefied and fluidized sediment
    flows and their deposits: Sedimentology, v.23, p.285-308.
MASCLE, J., AUROUX, C. and ROSSI, S., 1982, Neogene evolution of
    the Apulian Swell: Proc. C.I.E.S.M. Congr., Cannes.
MIDDLETON, G.V., 1977, Mechanics of sediment movement: S.E.P.M.
    Short Course No.3.

MIDDLETON, G.V. and HAMPTON, M.A., 1973, Sediment gravity flows:
    mechanics of flow and deposition:In: S.E.P.M. Short Course
    No.1, p.1-38.
MIDDLETON, G.V. and HAMPTON, M.A., 1976, Subaqueous sediment trans=
    port and deposition by sediment gravity flows: In: STANLEY,
    D.J. and SWIFT, D.J.P., eds., Marine sediment transport and
    envitonmental management, John Wiley & Sons, p.197-218.
MUTTI, E., RICCI LUCCHI, F., SEGURET, M. and ZANZUCCHI, G., 1984,
    Seismoturbidites: a new group of resedimented deposits:In:
    CITA, M.B. and RICCI LUCCHI, F., eds., Seismicity and sedim=
    entation, Marine Geol. spec. issue, v.55, p.103-116.
ORI, G.G. and FRIEND, P.F., 1984, Sedimentary basins formed and
    carried piggy-back on active thrust sheets. Geology, v. 12,
    p. 475-478.
PIERI, M. and GROPPI, G., 1981, Subsurface geological structure of
    the Po Plain, Italy: Progetto Final. Geod. C.N.R., Publ.
    No.414, 24 p.
RICCI LUCCHI, F., 1975, Miocene plaeogeography and basin analysis
    in the Periadriatic Apennines: In: SQUYRES, C., ed., Geology
    of Italy, Tripoli (1977), v.2, p.129-236.
RICCI LUCCHI, F., 1978, Turbidite dispersal in a Miocene deep-sea
    plain: Geol. en Mijnbouw, v.57, p.559-576.
RICCI LUCCHI, F., 1984a, The deep-sea fan deposits of the Miocene
    Marnoso-arenacea Formation, Northern Apennines: Geo-marine
    Letters, v.3, p.203-210.
RICCI LUCCHI, F., 1984b, Flysch, molassa, cunei clastici: tradizio=
    ne e nuovi approcci nell'analisi dei bacini orogenici dell'
    Appennino settentrionale: Mem. Soc. Geol. It., in press.
RICCI LUCCHI, F. and VALMORI, E., 1980, Basin-wide turbidites in a
    Miocene, "over-supplied" deep-sea plain: a geometrical ana=
    lysis: Sedimentology, v.27, p.241-270.
RICH, R.L., 1950, Flow markings, groovings and intra-stratal crum=
    plings as criteria for recognition of slope deposits, with
    illustrations from Silurian rocks of Wales: A.A.P.G. Bull.,
    v.34, p.717-741.
ROSSI, S. and SARTORI, R., 1981, A seismic reflection study of the
    External Calabrian Arc in the northern Ionian Sea (Eastern
    Mediteranean): Marine Geoph. Res., v.4, p.403-426.
SESTINI, G., 1970, Flysch facies and turbidite sedimentology. In:
    SESTINI, G., ed., Development of northern Apennines geosyn=
    cline, Sedim. Geol. spec. issue, v.4, p.559-597.
YALIN, M.S.,1972, Mechanics of sediment transport: Pergamon Press,
    N.Y., 290 p.

LOCAL MORPHOLOGIC CONTROLS AND EFFECTS OF BASIN GEOMETRY ON FLOW
PROCESSES IN DEEP MARINE BASINS

William R. Normark

U.S. Geological Survey, Menlo Park, CA 94025

ABSTRACT

Unravelling provenance for deep-water turbidite systems is
complicated by a relatively poor understanding of the effects of
basin morphology and large-scale bedforms and channel
configuration on the distribution of sediment within these
turbidite deposits. Outcrop studies of ancient deposits are
generally limited to scales of observations that are neither large
enough nor sufficiently closely-spaced to recognize the
complexities seen on modern turbidite systems. Examples from
several relatively well-studied modern submarine fans illustrate
examples of (1) headless channels, (2) hanging tributary and
distributary channels, (3) channel abandonment leading to lobe or
fan bypass, (4) meandering (or sinuous) channels, (5) sediment
waves, and (6) erosional scours of channel dimensions. These and
related morphologic features help to identify the complexities of
depositional processes on modern submarine fans with implications
for provenance interpretation in ancient turbidite systems.

INTRODUCTION

As provenance studies and sedimentologic interpretations
progress to finer detail and greater complexity of sedimentary
(sub)environments, a better knowledge of modern depositional
processes and their deposits becomes essential (Ricci Lucchi, this
volume). Work during the last decade or two has provided
significant insights to our understanding of deltaic and other
nearshore environments. Advances in our understanding of deeper
water deposits, including submarine fans, have been slower because

G. G. Zuffa (ed.), Provenance of Arenites, 47–63.
© *1985 by D. Reidel Publishing Company.*

of the difficulties in mapping deposits obscured by as much as
five kilometers of water. This paper reviews several relatively
well-studied modern submarine fans located offshore of the state
of California, U.S.A., to (1) illustrate some of the effects of
basin morphology on sediment distribution, (2) suggest that some
large-scale morphologic features that develop in turbidite systems
might be missed or misinterpreted if observed in ancient deposits,
and (3) show that external factors, such as oscillations of sea
level, can further complicate interpretation of provenance by
providing alternating sources of sediment.

The first-order effect of basin morphology on deep-sea
sediment deposits appears trivial: the shape of the deposit will
mimic the shape of the basin. When modern turbidite systems are
examined in detail, however, unexpected complexities in local
morphologic features and in sedimentary facies distributions can
be related to the effects of basin shape on controlling
depositional patterns. These effects include headless channels
that are not connected to the main feeder channel system and the
resulting development of depositional sequences that are "out of
place" with respect to that predicted by most sedimentation models
(e.g., leveed channels on "lower" fan areas).

The fine-scale morphology of most modern deep-water turbidite
depositional systems is poorly known. Features less than several
hundred meters in lateral extent will not be detected and/or
adequately resolved with conventional marine survey techniques
(Normark et al., 1979). This fine-scale morphology includes
depositional bedforms and erosional features that result from the
flow processes themselves but that, in turn, can have pronounced
effects on later flows of scales different from those creating the
morphologic relief. With continued advance in the development of
deep-water multibeam echo-sounding and side-scanning sonar
systems, it has become clear that we must pay more attention to
the finer-scale morphologic features to interpret the depositional
environments. A striking example is the apparently common
occurrences of highly sinuous (meandering?) channels on submarine
fans shown by new side-scanning sonar systems (Garrison et al.,
1982; Damuth et al., 1983; Colella and Normark, 1984; Kastens and
Shor, in press).

The purpose in reviewing several examples of the effects of
basin configuration and the nature of local (=fine scale)
morphologic features on modern turbidite depositional systems is
to learn more about the nature of turbidity-current flow
processes. This will, in turn, provide important implications for
mapping in ancient turbidite sequences where many of these effects
either have not been recognized or have been misidentified.

APPROACH

The effects of basin configuration in controlling

depositional facies and morphologic characters of turbidite deposits tend to be unique to each system. The distribution of facies reflects a complex relation involving sediment type, rate of sediment supply, grain size of sediment supplied, variability of depositional processes, and basin configuration as well as other variables. For most modern systems, there is insufficient information to begin to sort out the specific effects of each variable. Therefore, this paper draws heavily on well-studied fans --- primarily Navy and Monterey, both off California --- where some of the effects of basin configuration on the history of fan growth can be recognized. Thus, no new field investigations are involved but rather a discussion of numerous curious observations from the last decade of deep-water research on these two fans and other modern turbidite systems is included in this report.

Throughout the discussions to follow, a major problem will be to understand the implication of the scale of features or events. Local relief as defined by conventional shipboard marine-geologic tools may be of similar size to large channels seen in outcrop by a land geologist working on ancient sequences. Thus, most of the local morphologic features described in this paper are much larger than features described from outcrops. The modern examples have a great range in scale. The range in size of morphologic features on modern submarine fans makes direct comparison between different basins (modern and/or ancient) a dangerous exercise --- for example, a 100-m-thick turbidity current in Monterey fan valley would only occupy the lower fourth of the channel depth on the upper fan while the same turbidity current on Navy Fan would exceed channel depth by several times. Yet, it is now clear that any given turbidite basin probably receives turbidity currents that range from thin, completely channelized flows to those that might overtop the levees with a thickness two or three times channel depth.

EXAMPLES

Monterey Fan is reviewed first because it has a complicated history of channel development that can be related to the effects of basin morphology. The timing of events is not known, and relatively little is known about the sediments and depositional processes. Navy Fan, which is reviewed next, is sufficiently small that existing core samples can be used to discuss turbidity current processes for this system, and the effects of overall basin morphology are subordinate to those of local morphologic relief.

Monterey Fan

Morphology. Monterey Fan is a deep-water deposit on oceanic

crust off the central California coast (Fig. 1).   Two main
submarine canyon systems have provided much of the sediment for
the fan, which probably started growing in late Oligocene or early
Miocene time (Normark et al., 1984).   The fan is nearly 400 km in
length, almost an order of magnitude greater than Navy Fan.
Relatively little of Monterey Fan has been surveyed with deep-tow
or swath sonar echo-sounding systems until this year, when the
GLORIA system was used to map the U.S. west coast Exclusive
Economic Zone (Gardner, J. V., et al., in press).   The GLORIA data
will not be fully processed until 1985, however, so the bulk of
the presently available data base consists of seismic reflection
profiles and sediment cores (Normark, 1970a; Hess and Normark,
1976; Normark et al., 1984).

        The present morphology of the upper Monterey Fan is dominated
by the major valley that extends from Monterey Canyon.   This
valley is joined from the north by a prominent leveed valley
extending from Ascension Canyon, which is a set of prominent
tributary slope valleys (Fig. 1).   The Ascension Canyon system,
which heads on the outer shelf, is relatively inactive during high
sea level periods.   In contrast, Monterey Canyon and its
tributaries to the south cut across the shelf to near the

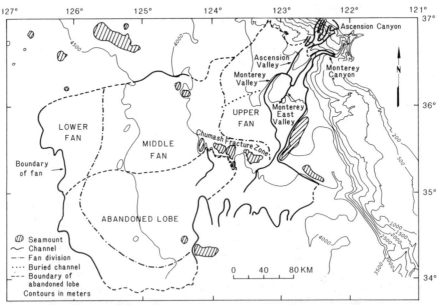

Fig. 1. Simplified bathymetry and morphologic interpretation of
Monterey Fan. Sediment waves are common throughout the upper and
middle fan segments north of the abandoned lobe; the distinction
between the two segments is that levee relief dominates the upper
fan.

shoreline and remain active during high sea level. Overbank
deposition on the upper fan is probably much slower now than for
much of the fan history because of a relatively recent incision of
the Monterey Fan Valley (see below).

Although Monterey Fan is growing into an unconfined basin
(the eastern Pacific) in contrast to those developed on active
margin basins (Barnes and Normark, 1984), the effects of
topographic barriers have resulted in an irregular fan shape and
periodic shifts in the locus of deposition. The history of fan
development depicted in Figure 2 illustrates the major features.
Prominent seamounts that rise above the fan surface today serve as
benchmarks to follow fan growth during each stage (A through E,
Fig. 2).

(A) Ascension and perhaps the Monterey systems were active during
the initial period of fan deposition. Seismic reflection profiles
show that the prominent levee relief of the Ascension system
apparently persisted through deposition of most of the sedimentary
section on the western side of the Ascension valley both upfan and
downfan of the confluence with the present Monterey valley. The
early history of the shorter Monterey system is less clear. Both
Ascension and Monterey(?) valleys fed sediment toward the
southwest and south into a one-kilometer-deep trough associated
with the Chumash Fracture Zone. The sediment moved southward
through valleys between the abyssal hills of the oceanic crust on
the way to the Chumash trough.

(B) Aggradation and extension of the Ascension valley toward the
southwest allowed turbidity currents to flow around the western
end of the Chumash Fracture Zone ridge. Menard (1955) first
suggested the diversionary effect of this ridge.

(C) When the fracture zone trough was filled, sediment from the
Monterey Valley system was still dammed north of the ridge and may
have contributed some sediment to the westward growth of the
fan. The topographic lows between individual seamount peaks on
the fracture zone ridge were breached as the fan continued to
aggrade toward the southwest.

(D) At this stage, a prominent leveed valley was formed that
headed in the pass between the two western seamounts of the
fracture zone ridge. This leveed valley complex is about 15 km
wide just south of the ridge but it is not connected to any
channel (buried or at the surface) north of the ridge. The leveed
valley apparently was formed by the reconfinement of overbank
flows from the Ascension and Monterey valleys as these flows were
dammed against the ridge itself and flowed through the pass.

Two major events occurred during this stage that markedly
affected fan development. First, the valley extending from
Monterey Canyon pirated the lower part of the Ascension Valley. A
large meander-shaped loop (Shepard, 1966) developed in the
Monterey Valley system as a result of this piracy. Second, the
Chumash Fracture Zone was breached near its eastern end. The sea
floor was several hundred meters lower on the south side of the

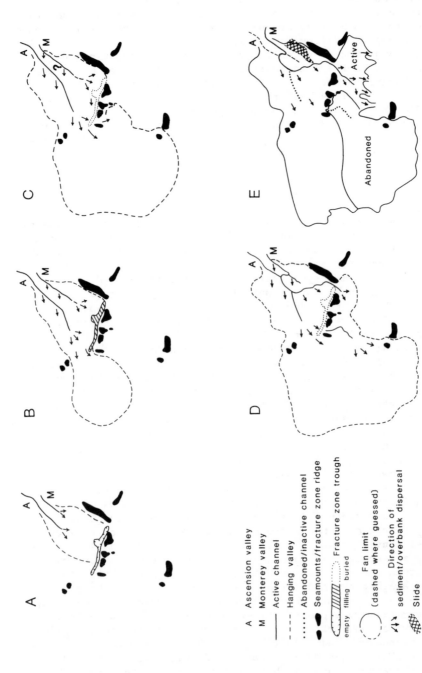

Fig. 2. Schematic sequence for Monterey Fan growth. See text for details. Basement highs standing above fan are in black and correspond to those shown in Figure 1 to provide fixed reference points. From Normark et al., 1984.

ridge, and northward erosion cut a channel headward into the pirated Ascension valley. This cutoff led to the eventual abandonment of the southwest-trending segment of the Ascension valley.

(E) Continued erosion within the Monterey Valley system, now in its present configuration, deepened the original valley by 200 m or more. This erosion left both a hanging tributary, the head of the original Ascension valley, and a hanging distributary, the lower reach of the original Monterey valley. The southwestward-trending leveed valley segment of the Ascension system now is nearly completely buried by overbank sediments from the incised Monterey valley.

The deeply incised Monterey valley confines more of the turbidity currents entering the two canyon systems than did the pre-erosion valleys. Flow stripping (the loss of the upper part of turbidity currents at channel bends; see next section and Piper and Normark, 1983) is not as effective, resulting in diminished overbank flow on the upper and middle fan, and the southwestern part of the fan now receives little turbidite sediment. This sector of the fan is basically abandoned. The sediment moving southward through the Monterey valley has formed an active channelized lobe of sand deposition. This active lobe area is losing sediment through valleys between abyssal hills to the south. It does not appear, from existing data, that Monterey Fan has re-established a local base level.

The reconstructed history for Monterey Fan illustrates the influence of basement topography on controlling turbidite systems even where deposition is not confined within a topographic basin. As a combination of the results of channel piracy and erosional deepening of the major valley, the upper fan now grows more slowly. The upper (hanging tributary) part of the Ascension valley has appreciable overbank deposition only during lowered sea level when the canyon is active. The eastern part of the upper fan receives sediment by flow stripping through the hanging distributary south of the meander loop or by mass movement from the adjacent slope. This change in channel pattern has resulted in the effective abandonment of more than a third of the fan area.

Relatively little is known about the local (fine-scale) morphology on Monterey Fan with the exception of large sediment waves commonly found on the righthand levees of large turbidite valleys. On Monterey Fan, the wavelengths of these features average about 1.7 km and amplitudes about 25 m. These large sediment waves could be formed by thick (100-800 m), low velocity, low concentration turbidity currents overflowing the righthand levee.

Sediment waves are recognized on other parts of the Monterey Fan (Fig. 1) where they appear to preferentially occur on locally steeper slopes that may cause acceleration of the turbidity currents.

Navy Fan

   Morphology. Navy Fan occupies part of the South San Clemente
Basin in the Southern California Borderland (Fig. 3). The fan is
relatively small (an order of magnitude less than Monterey Fan)
and through a series of cruises has probably become the best-
studied modern turbidite system (Normark and Piper, 1972; Normark
et al., 1979; Piper and Normark, 1983). Much of the fan upslope
from the smooth basin plain areas has been surveyed with a deep-
tow vehicle incorporating side-scan sonar, photography, and
precision echo-sounding systems. In addition, over 100 sediment
cores from the basin area have been used to correlate and date
depositional events in the basin (Piper and Normark, 1983).
   The morphologic representation of the fan shows the
pronounced effect of the basin shape (Fig. 3). A low fault-
bounded ridge forms a topographic barrier across nearly half of
the upper fan. Initial turbidite deposition in the present basin
(Late Pleistocene?) was confined behind this ridge. Subsequently,
turbidity currents flowed around the north end of the ridge and
into the northwest-trending area of South San Clemente Basin.
Only after this part of this basin was partially filled by several
hundred meters of sediment could turbidity currents move across
the sill into the southern part of the basin (Smith and Normark,
1976).
   The local morphologic relief on Navy Fan, which has been
discussed in detail elsewhere (Normark et al., 1979), has shown
several "unexpected" facets of turbidity current deposition.
These characters are briefly reviewed below.
(1) At the eastern end of the basin, the channel relief is
greatest (50 m) but it is cut into muddy parallel-bedded sediments
with no expression of levee morphology (Normark and Piper,
1972). High resolution deep-tow reflection profiles show that
sand is deposited on the channel floor but not overbank. Levees
are developed a short distance down the channel indicating that at
least some of the turbidity currents entering the basin must pass
through an hydraulic jump.
(2) There is only one active distributary channel and associated
depositional lobe (system B, Fig. 3). Two older distributary
systems are now separated from the active channel by low (2 to 3
meters relief) levees. There is no clear evidence for an older
bifurcating distributary network in the past, and it has been
suggested that channel migration results from aggradation of an
active lobe/channel system that is eventually cut off by a levee
breach immediately upstream. The channel system is not sinuous,
but exhibits sharp bends at the upslope end of each aggraded
channel/lobe system (Normark et al., 1979).
(3) The surfaces of the depositional lobes are smooth and convex
upward; the feeding channels terminate at the upslope part of the
lobe. The lower (down fan) margins of the lobes, however, are
bounded by channel-like features that begin and end along the

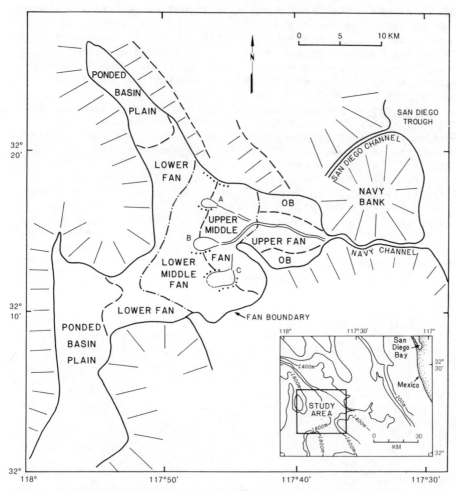

Fig. 3. Morphologic representation of Navy Fan. A, B, C refer to channel/lobe systems (see text) and OB indicates upper fan area of overbank sediments but without levee relief. Dotted lines show headless channels. Inset location-map contours are in meters.

lobes. These are one of two types of headless channels that are found on Navy Fan. The other type lies along the bend in the northern flank of the basin (Fig. 3). Both of these headless channel types probably result from re-channelization of turbidity currents that spread across the fan surface (Normark et al., 1979).

(4) In the area between channel remnants and over the areas of levee/channel termination, Navy Fan has numerous, large (up to 400 m wide) scour-shaped depressions. Most occur on the fan in the

gently sloping area between the main leveed valley and the upslope
ends of the lobes (denoted as "upper middle fan" in Fig. 3).
Because of their pronounced flute shape, their occurrence in areas
of maximum overbank channel spill, and a lack of evidence for
slump origin, the features are thought to result from turbidity
current erosion in an area of relatively sandy fan sediments
(Normark et al., 1979).

Sediments. An examination of the core samples from the fan
and basin shows direct evidence for the effect of basin
configuration on flow processes. Several turbidity current events
can be traced across much of the fan and can be related to basin
shape and the local morphology.

The most recent turbidite on the upper and middle parts of
Navy Fan (turbidite I, from Piper and Normark, 1983) is a mud unit
3-10 cm thick, locally with a thin sand division at the base.
Downslope from the channel/lobe system A (Fig. 3), the unit
thickens to 25 cm and it reaches at least 50 cm thickness in the
northwest trending basin. Near the bend in the northern slope of
the basin, turbidite I is found on the slope 30 m above the edge
of the fan. It was not observed in a core only 5 m above the fan

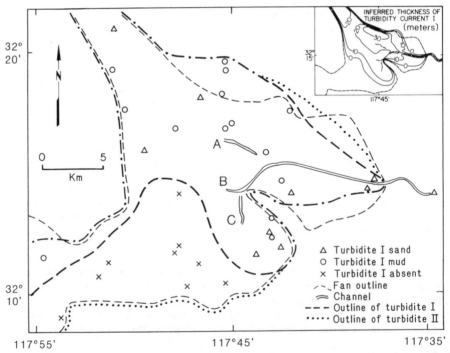

Fig. 4. Outline of Navy Fan showing extent of turbidites I and II
(Piper and Normark, 1983). Inset shows thickness (in meters) of
turbidity current I inferred from core data.

just 4 km northwest along the same basin margin. On the upper
fan, turbidite I is a gravelly sand bed in the main valley. More
significant is the observation that turbidite I is absent in all
cores south and west of lobe B (Fig. 4) and it occurs only on the
proximal part of lobe C (see Piper and Normark, 1983, for
details).

This distribution of turbidite I is used to estimate the
thickness of the turbidity current as it flowed across the fan
(inset, Fig. 4). The turbidity current (I) overflowed the levees
on the upper fan several kilometers downslope from the fan apex.
The current was dominantly muddy. Its thickness was about twice
the channel depth upstream of the first sharp bend in the
channel. At the bend itself, most of the turbidity current
continued flowing westward as its thickness greatly exceeded the
3-m height of the levee between the active channel and
channel/lobe system A; this loss of the upper part of a turbidity
current is an example of flow stripping (Fig. 5; see Piper and
Normark, 1983).

NAVY FAN

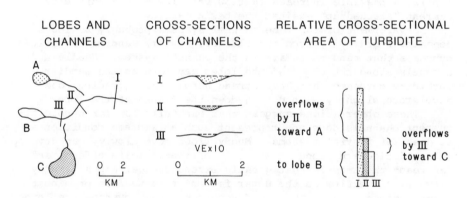

Fig. 5. Schematic illustration of flow stripping for a turbidity
current that just fills the main channel at cross-section I.

The flow-stripped part of turbidity current I continued
downslope to the west turning into the northwest basin. The flow
sloped upward to the north and never spread to the west or south
(except for the bottom-most part that wasn't flow-stripped at the
first bend and that stayed in the active channel; see distribution
of turbidite I in Fig. 4). The flow attained maximum thickness
near the bend in the northern basin margin where it reached 30 m
up the slope (Fig. 4). The flow was deflected to the north by
Coriolis force. Once the turbidity current entered the long,
narrow northwest arm of the basin, the entire basin became the

effective channel. The flow slowed, ponded, and deposited much of
the sediment of turbidite I in the basin.

Although the core documentation is less complete, sandier
turbidite units below turbidite I are found over more of the fan
surface. Flow stripping apparently was less effective in
diverting these sandier turbidity currents from the channel system
(Piper and Normark, 1983). This implies that sandy turbidity
currents are more concentrated (Bowen et al., 1984). Sand-
carrying turbidity currents that are thinner and faster, and of
higher density, than muddy flows like turbidity current I, can
efficiently transport sand to the depositional lobes and lower fan
areas because more of the flow remains in the channel.

For Navy Fan, the basin configuration has more influence on
the sedimentation pattern of thicker, muddier turbidity
currents. The small size and irregular shape of the basin results
in much of the overbank flow being deflected by the Coriolis force
toward the northern basin margin. As a result, the turbidity
currents thicken appreciably along the northern basin margin
before the margin bends northwestward. The headless channel
formed along the base of the basin slope under the general area
where turbidite I reached maximum thickness (Fig. 4, inset). This
implies a possible increase in flow velocity as the turbidity
current "stacked up" against the basin margin.

Development of the lobe/channel systems probably reflects
deposition primarily from the thinner, faster, sandier turbidity
currents that tend to remain in the channel system. Headless
channels along the edges of the lobes may form as the sandier
turbidity currents that have spread across the aggrading lobe
accelerate slightly when moving down the flanks of the lobe.

These observations suggest that bankfull flow for turbidity-
current channels does not represent such an extreme condition as
it does for fluvial systems. Muddy flows move slowly, can develop
a marked sideward slope as a result of the Coriolis effect, and
can reach thicknesses that greatly exceed channel depth. Muddy
flows will overflow on the upper fan and tend to form prominent
levee systems. Thus, these muddy currents that overtop the levees
seem to be the common events for aggradation on the upper fan.
Thinner, faster, sandier flows can effectively transport sand to
the mid-fan areas through the channel systems. Very large
turbidity currents that are thick and that carry appreciable
amounts of sand would be capable of eroding muddier levee sediment
and producing the large scour features such as those seen on Navy
Fan. These exceptional turbidity currents would be most effective
in transporting sand to the lower fan/basin plain environment.
These concepts are summarized in Figure 6 (from Piper and Normark,
1983).

Other Fans

Sinuous or meandering channel patterns on modern fans have

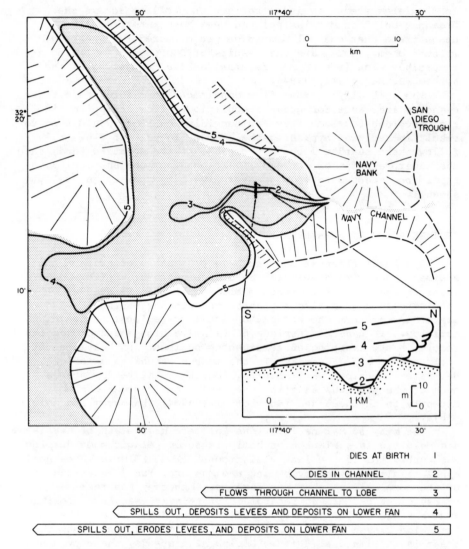

Fig. 6. Conceptualization of turbidity currents' size in relation to channel cross-section and area of fan reached (from Piper and Normark, 1983). Surface of fan is shown in stipple pattern.

been commonly noted for several decades (Shepard, 1966; Shepard and Buffington, 1968; Normark, 1970b). Long-range side-scanning sonar records, however, have shown that highly sinuous channel systems appear to dominate some upper fan regions (Garrison et al., 1982; Damuth et al., 1983; Kastens and Shor, in press). These systems remarkably resemble the meandering pattern developed

in fluvial systems. Seismic reflection profiles across the
"meander belt" on Mississippi Fan have been interpreted to
demonstrate that channel migration has occurred (C. Stelting,
written comm., 1984) and the results of DSDP Leg-96 drilling are
compatible with (but do not require) active meandering of the
system (Bouma et al., 1984).

The turbidity current flow conditions that form these tight
meanders are cause for great speculation. Flow stripping like
that recorded for some Navy Fan turbidity currents would tend to
result in sediment bypassing the meander loops. If flows that
follow these meander loops are capable of transporting sand, then
the interpretation of current lineations from ancient sediments
should be expected to show greater diversity than is generally
shown.

SEDIMENTATION AND SEA LEVEL

Sea-level fluctuations can have pronounced effects on the
sediment supply to deep-water turbidite systems. Relative sea-
level changes can result from global-scale effects such as those
resulting from glacial/non-glacial transitions during the
Quaternary or from the effects of local or regional tectonic
activity. In general, periods of relative high stand of sea level
result in the trapping of terrigenous sediment in shelf/coastal
environments and a decrease in the amount of coarse clastics
delivered to deeper water basins. During relative low stands,
much of the detritus entering the marine environment can move
directly to the deep-sea (Stow et al., 1984; Shanmugam and Moiola,
1981).

For many submarine fans, the Holocene high-stand of sea level
has resulted in a blanket of hemipelagic or pelagic muds draping
the turbidites of the late Pleistocene. This relation does not
apply to all fans, however, and the Monterey Fan illustrates one
of the complexities that can develop. Monterey Fan receives
coarse clastic sediment regardless of relative sea level position;
what changes is the pathway. During lowered sea level, the
Ascension Canyon system can intercept sediment moving along the
coast in the littoral drift. During high stands, the canyon heads
on the outer shelf are bypassed by the coarse sediment moving in
the nearshore zone (Hess and Normark, 1976). Monterey Canyon
apparently can intercept coarse sediment at any position of sea
level; but during relative high stands, it becomes the dominant
pathway for sediment to the fan. Thus, the growth pattern for
Monterey Fan is complicated by repeated alternation of coarse
clastic input between Monterey and Ascension Canyons throughout
the Pleistocene.

Shifting of sediment source areas can be especially important
in areas where relative sea level changes result from tectonic
disruption within the source-transport areas (Stow et al., 1984).

DISCUSSION AND CONCLUSIONS

This limited review of the morphologic controls of turbidite sedimentation is intended to illustrate only a few of the complexities that might be encountered in provenance studies of ancient deposits. The major point is that we can no longer consider submarine fans to exhibit simple radial patterns of sediment dispersal nor can we assume uniform changes in sediment parameters, such as grain size, from fan apex to basin plain. The summary items that follow indicate a few of the problems that might be encountered in determining provenance characteristics for systems like the modern examples presented above.
1. Where sea level oscillations result in changes in sediment pathway to a fan, as for Monterey fan, sediment from different provenances might be mixed in the main depositional areas.
2. The change in local base level for Monterey Fan that resulted in extensive erosion within the channels of the upper and middle fan in the attempt to re-establish grade also leads to mixing of sediment types. In this case, the active transform offset along the San Andreas fault system during the late Neogene has probably resulted in a gradual shifting of sources contributing sediment to the fan. The extensive downcutting of the inner 200 km of the main channel can contribute sediment from "older" source areas to the most recent depositional areas on the fan.
3. Flow stripping, including levee overflow from non-curved channel segments, can result in compositional shifts of the sediment transported through the system. Hydraulic sorting of grains within a turbidity current can result in a variation of mineral/grain types vertically within the flow. Loss of the upper part of a turbidity current will then result in different compositions for deposits from the stripped flow than for deposits from the flow that remained channelized.
4. Navy Fan is not fed directly from a submarine canyon cutting into the continental shelf but lies in a basin that lies outboard of San Diego Trough, which is adjacent to the continental shelf. Thus, any turbidity current large enough to reach the deepest sill of San Diego Trough can contribute sediment to Navy Fan. At least four major canyon systems could possibly provide sediment to the fan, mixing provenance relationships.

The local morphologic features described for modern fans probably do not directly affect the provenance indicators of the sediment. They will, however, affect the sedimentologic interpretations for the fans. Headless channels, whether the small lobe-bounding type on Navy Fan or the large leveed valley on Monterey Fan, could be mistaken for major feeding or distributary channels if encountered in ancient deposits. The large scours (20 m to 1(+) km wide) seen on both Navy and Monterey Fans could be mistaken for channels if seen in section only (Normark et al., 1979). If filled with relatively coarse-grained sediment, all of these features could suggest more proximal settings than their actual position if observed in ancient sediment sequences.

These channel and scour features, as well as the sediment waves characteristic of large levee systems and the occurrence of major meander loops in other channel systems, provide valuable insight to the wide range of flow processes that operate on modern submarine fans. Turbidity currents probably come in a wide range of sizes and styles on most fans, as can be documented for Navy Fan. Until studies of ancient turbidite systems begin to document the sedimentologic effects of these local morphologic effects, we should be concerned about over-simplified interpretations of deep-water turbidites.

ACKNOWLEDGMENTS

This review draws heavily on several collaborative studies during the last decade. I especially acknowledge the ideas and stimuli generated by my work with D.J.W. Piper, G. R. Hess, D.A.V. Stow, A. J. Bowen, and C. E. Gutmacher. E. Mutti and F. Ricci Lucchi provided valuable opportunities to discuss these concepts during lectures to their students and colleagues at the Universities of Parma and Bologna. The manuscript benefited by critical reviews from P. R. Carlson and R. E. Hunter.

REFERENCES

Barnes, N.E., and Normark, W.R., 1984, Diagnostic parameters for comparing modern submarine fans and ancient turbidite systems: Geo-Marine Letters, v.3, Map.

Bouma, A.H., Coleman, J.M., and Leg 96 scientific party, 1984, Lithologic characteristics of Mississippi fan (abstract): Am. Assoc. Petroleum Geologists Bull., v.68, p.456.

Bowen, A.J., Normark, W.R., and Piper, D.J.W., 1984, Modelling of turbidity currents on Navy submarine fan, California Continental Borderland: Sedimentology, v.31, p.169-185.

Colella, A. and Normark, W.R., in press, High resolution side-scanning sonar survey of delta slope and inner fan channels of Crati submarine fan (Ionian Sea): Memorie della Società Geologica Italiana.

Damuth, J.E., Kolla, V., Flood, R.D., Kowsmann, R.O., Monteiro, M.C., Gorini, M.A., Palma, J.J.C., and Belderson, R.H., 1983, Distributary channel meandering and bifurcation patterns on Amazon deep-sea fan as revealed by long-range side-scan sonar (GLORIA): Geology, v.11, p.94-98.

Gardner, J.V., McCulloch, D.S., Eittreim, S.L., and Masson, D.G., in press, Long-range side-scan sonar studies of the central California EEZ (abstract): Geol. Soc. Amer. Annual Meeting, Reno, Nevada.

Garrison, L.E., Kenyon, N.H., and Bouma, A.H., 1982, Channel systems and lobe construction in the Mississippi Fan: Geo-Marine Letters, v.2, p.31-39.

Hess, G.R., and Normark, W.R., 1976, Holocene sedimentation history of the major fan valleys of Monterey Fan: Marine Geol., v.22, p.233-251.

Kastens, K., and Shor, A., in press, High resolution seismic and side-scan survey of a channel meander on the Mississippi Fan.

Menard, H.W., 1955, Deep-sea channels, topography, and sedimentation: Am. Assoc. Petroleum Geologists Bull., v.39, p.236-255.

Normark, W.R., 1970a, Channel piracy on Monterey deep-sea fan: Deep-Sea Research, v.17, p.837-846.

Normark, W.R., 1970b, Growth patterns of deep-sea fans: Am. Assoc. Petroleum Geologists Bull., v.54, p.2170-2195.

Normark, W.R., and Piper, D.J.W., 1972, Sediments and growth pattern of Navy deep-sea fan, San Clemente Basin, California Borderland: Jour. Geol., v.80, p.192-223.

Normark, W.R., Piper, D.J.W., and Hess, G.R., 1979, Distributary channels, sand lobes, and mesotopography of Navy submarine fan, California Borderland, with application to ancient fan sediments: Sedimentology, v.26, p.749-774.

Normark, W.R., Gutmacher, C.E., Chase, T.E., and Wilde, P., 1984, Monterey Fan: growth pattern control by basin morphology and changing sea levels: Geo-Marine Letters, v.3, p.93-99.

Piper, D.J.W., and Normark, W. R., 1983, Turbidite depositional patterns and flow characteristics, Navy submarine fan, California Borderland: Sedimentology, v.30, p.681-694.

Shanmugam, G., and Moiola, R.J., 1981, Eustatic control of turbidites and winnowed turbidites: Geology, v.10, p.231-235.

Shepard, F.P., 1966, Meander in valley crossing a deep-ocean fan: Science, v.154, p.385-386.

Shepard, F.P., and Buffington, E.C., 1968, La Jolla submarine fan-valley: Marine Geol., v.6, p.107-143.

Smith, D.L. and Normark, W.R., 1976, Deformation and patterns of sedimentation, South San Clemente Basin, California Borderland: Marine Geol., v.22, p.175-188.

Stow, D.A.V., Howell, D.G., and Nelson, C.H., 1984, Sedimentary, tectonic, and sea-level controls on submarine fan and slope-apron turbidite systems: Geo-Marine Letters, v.3, p.57-64.

TURBIDITE SYSTEMS AND THEIR RELATIONS TO DEPOSITIONAL SEQUENCES

Emiliano Mutti

Istituto di Geologia, Università di Parma,
Via Kennedy 4, 43100 Parma, Italy

ABSTRACT

Long-term global sea level variations and local tectonic control form the basic framework within which turbidite sediments develop as a response to breaks in the equilibrium between shelfal and basinal sedimentation. An understanding of the interaction of these processes and resulting types of turbidite deposition is still in its infancy and requires, as a preliminary approach, the precise framing of turbidite sediments within well defined depositional sequences.

All other things being equal, volume of gravity flows - which is largely a function of relative sea level position - determines the growth pattern of turbidite systems. Large flows deposit most of their sediment load as unchannelled sandstone lobes (type I systems). Particularly in elongate flysch basins, these sandstone lobes may develop as huge accumulations of virtually basinwide extent. Channels may form during the process but are essentially bypassed by most of the sand.

A decrease in the volume of gravity flows enhances the depositional character of the channels that progressively become the only site of sand deposition where small-volume and highly confined flows lose most of their fines through overbank processes (type II and III systems).

Within the same system, a decrease in the volume of gravity flows determines different stages of growth that are expressed by distinctive facies associations.

G. G. Zuffa (ed.), Provenance of Arenites, 65–93.

INTRODUCTION

The original depositional relations between shallow and deep marine sandstone facies have been investigated by many stratigraphers since 1950, when Kuenen and Migliorini published their fundamental paper on turbidity currents and graded beds. Most of these attempts were based on paleocurrent directions and sandstone composition, and on the implicit assumption that sands can be deposited in both shallow and deep marine environments at the same time. Several recent papers still portray such a relation as a simple process where active seaward progradation of marginal marine sands, particularly in deltaic settings, leads to shelfedge failures that trigger resedimentation mechanisms responsible for turbidite accumulation in adjacent deeper basins.

The problem is considerably more complex when attention is paid to the huge volumes of sand that are commonly involved in ancient turbidite sequences. Vail et al. (1977) have suggested that accumulation of major sequences of turbidite sands requires periods of relative lowering of sea level. During these periods of lowstand rivers bypass the shelf and deposit their sediment load directly onto the slope, whence sand would be funneled through submarine canyons into deep-sea fans. Although this model can certainly account for some types of ancient turbidite systems, it fails to explain other systems where volumes of individual sandstone beds or groups of beds require huge gravity flows that can hardly be produced by "normal" slumping seaward of active deltaic complexes.

A more "catastrophic" model to account for deposition of large-volume turbidite sandstone bodies is that recently discussed by Coleman et al. (1983) for the Pleistocene Mississippi canyon-fan system. The model involves huge volumes of virtually unconsolidated Pleistocene deltaic sediments being removed from the shelfedges of the Mississippi delta through catastrophic retrogressive slides that occurred in very short periods of time before the Holocene rise of sea level. A volume of sediment up to 8000 cubic km was removed in less than 5000 years and redeposited as turbidites in adjacent submarine fans as a result of increased slope instability related to lowering of sea level during Pleistocene (Coleman, written comunication 1982).

Retrogressive slides, produced by either tectonic uplift, eustatic fall, or a combination thereof, appear to be the basic factor in controlling the occurrence of large volumes of predominantly coarse-grained turbidite facies, i.e. sandstones, pebbly sandstones, and conglomerates deposited by highly concentrated gravity flows (Facies A through C of Mutti and Ricci Lucchi, 1972, 1975). The resedimentation process, however, requires previous deposition of at least equal volumes of shelfal sands and muds at sufficiently high rates of sedimentation to prevent lithification. Such conditions are apparently met only in actively prograding deltaic systems that form during sufficiently long periods of highstand to allow the shelf to be loaded with thick sequences of unconsolidated sands and muds. Consequently,large-volume turbidite sandstone bodies have to be derived from adjacent shelfal   areas that underwent phases of tectonic uplift or eustatic fall of sea level shortly after sufficiently long periods of highstand characterized by high rates of terrigenous sedimentation. Such settings are apparently those of many elongate flysch basins of thrust belts developed in relation to highly subsiding shelves fed by large interland drainage basins. These shelves may or may not be involved in the same structural setting of the associated turbidite basins. Frequent lowering of sea level, produced by tectonic uplift in an active structural setting leads to suitable conditions for turbidite deposition, but volumes of resedimented deposits will be less for lack of sufficient recharge time in adjacent shelves.

Relatively smaller volumes of resedimented sandstones occur in fault-bounded basins where either tectonic uplift or eustatic fall can only remove limited amounts of marginal marine sediments, chiefly coarse-grained fan-delta deposits, formed along irregular and narrow shelves fed by small interland drainage basins.Although particular settings of this type are not further discussed in this paper, it is likely that high rates of seaward progradation of marginal marine sand and gravel combined with periodic slump failures can result in contemporaneous deposition of substantial amounts of coarse-grained turbidite facies.

Predominantly fine-grained turbidites - mudstones and fine-grained current-laminated sandstones deposited by dilute turbidity currents (Facies D of Mutti and Ricci Lucchi, 1972, 1975) - occur in a variety of tectonic and paleogeographic settings and display a wide range of facies characteristics. Most commonly, these sediments record deep-water deposition during periods of highstand of sea level. Among such fine-grained turbidites, this paper focusses in particular on channel-levee complexes that form as a mere deep-water extension of actively prograding and mud-rich delta systems on adjacent shelves. As discussed more in detail later, the origin and geologic significance of fine-grained turbidites differ fundamentally from those of coarse-grained systems.

Long-term global sea level variations and local tectonic control form the basic framework within which turbidite sediments develop as a response to breaks in the equilibrium between shelfal and basinal sedimentation. An understanding of the interaction of these processes and resulting types of turbidite deposition is still in its infancy and requires, as a preliminary approach, the precise framing of turbidite sediments within well defined depositional sequences (in the sense of Vail et al., 1977).

This paper discusses some of the main types of ancient turbidite systems and their relations to depositional sequences in terms of relative sea level variations, basin size and configuration, tectonic setting, and type of sediment supplied. Most of the following discussion is based on data from ancient turbidite

systems in the Tertiary and Upper Cretaceous of the south-central
Pyrenees, Spain, and the Tertiary of the northwestern Italy. Part
of these data are summarized in papers by Cazzola et al. (1981,
1984), Cazzola and Rigazio (1983), Cazzola and Sgavetti (1984),
Mutti (1984a, 1984b), and Sgavetti et al.(1984). A more general
account on the Eocene Hecho Group turbidites of the south-central
Pyrenees that form the basis for most of the concepts developed
in this paper will be published on the occasion of the IAS
Regional Meeting to be held in Spain in 1985 (Mutti et al., in
preparation). The interpretation of data and the models presented
in this paper remain however the sole responsibility of this
author.

ANCIENT TURBIDITE SYSTEMS

    Ancient turbidites comprise depositional systems, i.e. rock
units whose different components are genetically interrelated,
that vary from small bodies of coarse-grained and poorly organized
deposits probably recording very short periods of geologic time,
to huge sedimentary prisms where regularly alternating sandstone
and shale accumulated over several millions of years.

MAIN TYPES OF TURBIDITE DEPOSITIONAL SYSTEMS

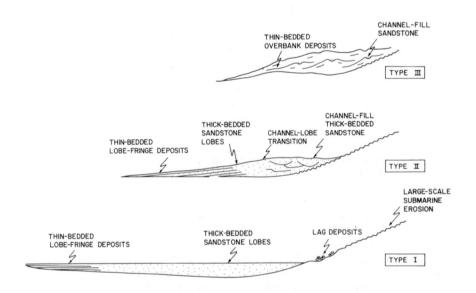

Figure 1. Main types of turbidite depositional systems. Systems
          differ from each other mainly in terms of where sand
          is predominantly deposited within the system.

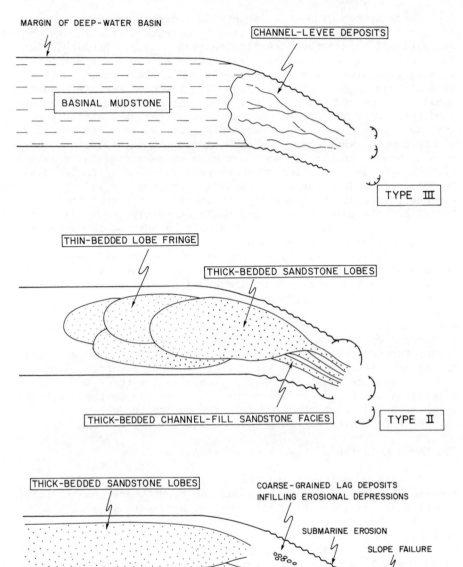

Figure 2. Main types of turbidite depositional systems. Deposition of sand within channels is progressively enhanced by the decrease in the volume of gravity flows.

These system display a variety of facies types and facies
associations that are often difficult to frame within the general
depositional models for turbidite sediments (e.g., Normark, 1970;
Mutti and Ricci Lucchi, 1972; Walker, 1978) currently accepted
in most literature. Some systems show sedimentary patterns that
clearly bear significant analogies with those of certain modern
submarine fans and basin plains; others are conversely characterized by overall geometries and patterns of facies distribution
that do not have any obvious modern analog, at least on the basis
of our present knowledge of Recent basins (Mutti, 1979; Normark
et al.,1984). Clearly the Holocene rise of sea level has drastically changed sedimentation processes and patterns of facies distribution in modern submarine turbidite systems (as well as in
the shelf areas). Unfortunately, Pleistocene sediments that are
buried beneath Holocene muds and could certainly provide a much
better analog for most ancient deep-water sandstone sequences are
still insufficiently known in detail.

Setting aside basin plain turbidites and exceptionally thick
resedimented units of basinwide extent (seismoturbidites of Mutti
et al., 1984) that will not be discussed herein, ancient turbidite
sequences most commonly comprise systems where channel-fill sediments are replaced in a downcurrent direction by nonchannelized
sandstone bodies, alternating sandstone and mudstone, or basinal
mudstone facies. Such patterns of facies distribution give rise
to point-source turbidite depositional systems fed through relatively long-lived channels not unlike, _mutatis mutandis_, most
modern submarine fans. This paper discusses the three main types
of systems which are the end-members of a continuum that, within
the same basin, is controlled by the progressive decrease of the
volume of turbidity currents and the increase of mud content.

The three main types of turbidite systems

As used hereafter, a turbidite system - not necessarily a
deep-sea fan system - denotes a body of rocks where channel-fill
deposits are replaced by nonchannelized sediments in a downcurrent
direction. Despite this common overall pattern, turbidite systems
differ considerably in terms of size, overall geometry, types of
facies and facies associations, and distribution and geometry of
sandstone bodies. As a result, these systems cannot be adequately
described by general models that do not take into account such
variability. This point has been discussed in previous papers
(Mutti, 1979; Nilsen, 1980; Mutti and Ricci Lucchi, 1981; Normark
et al., 1984).

Ancient turbidite systems differ from each other mainly in
terms of where the sand is deposited within the system. On this
basis, the three following main types of systems can be recognized
(Figs. 1 and 2).

Type I systems   are here defined as depositional systems
where the bulk of the sandstone occurs in nonchannelized and

elongate bodies, or lobes, in the outer region of the system.
These systems essentially fit the model of "highly efficient" fans
of Mutti (1979). The sandstone bodies are characterized by remarkable lateral continuity and roughly tabular geometry over distances up to several tens of kilometers parallel to the current
direction. Each lobe, commonly between 3-15 m thick, is character-
istically thick-bedded and grades in a downcurrent direction into
thinner bedded and finer grained deposits (lobe-fringe deposits
of Mutti and Ricci Lucchi, 1975; these sediments are discussed in
more detail later). In some systems, the aggregate thickness of
sandstone lobes and associated lobe-fringe deposits may reach
several hundreds of meters.

TURBIDITE FILL OF LARGE-SCALE
SUBMARINE EROSIONAL FEATURES

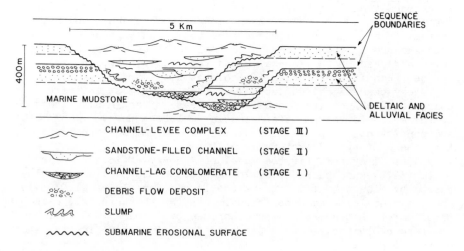

Figure 3. Turbidite fill of large-scale submarine erosional
features. Note different stages in the sedimentary fill.

Upsystem from these sediments, as observed in the Hecho Group,
are large-scale submarine erosions that cut into shelfal sediments
(Fig. 3). These extend basinward as erosional features as long as
10-15 km, incised into slope and basinal thin-bedded mudstone and
sandstone. In their upper reaches, these large-scale erosional
features are filled in with an abundance of chaotic deposits; downcurrent, the chaotic sediments become associated with and are
gradually replaced by facies of thin-bedded turbidite mudstone and
sandstone and lenticular bodies of sandstone and conglomerate.
These coarse-grained facies are the infill of minor channels

incised within the main erosional feature at different times
during the evolution of the system (see stages and sub-stages of
following sections). These minor channels are generally shallower
than 50 m and have widths up to 1-1.5 km. Because of the scale
involved, these large-scale erosional features can be detected
only through detailed photogeologic mapping in particularly well
exposed sequences.

The depositional setting of type I systems suggests most of
the sand bypassed the channel region and accumulated in the outer
and virtually flat basinal areas. This implies the sandstone lobes
of such systems are essentially physically detached from and not
time equivalent to the channel-fill sequences observed in the
inner portions of the same system (see following sections).

Type I systems are best observed in large and elongate flysch
basins of thrust belts. Superposed systems of this type may reach
aggregate volumes of turbidite sediments in excess of 20000 cubic
kilometers.

Type II systems   include all those depositional settings
where sandstone facies are predominantly deposited in the lower
reaches of channels and in the regions beyond channel mouths.These
systems form extensively channelized bodies that grade downcurrent
into sandstone lobes. Lobe and channel deposits are here physically attached, both laterally and vertically, through well-
defined transitional facies (Cazzola et al., 1981).

Very coarse-grained type II systems are almost entirely
composed of channelized deposits; decrease of grain size tends
to favour the development of associated lobes. However, the latter
are consistently less developed, in both volume and areal extent,
than those of type I systems.

Type II systems fit the model of "poorly efficient" fans of
Mutti (1979) and their depositional setting has been amply discussed by Nilsen (1980), Surlyk (1978),and Cazzola et al.(1981).
These systems display depositional patterns that strongly resemble
those of Recent suprafans as described by Normark (1970),although
the former obviously record periods of considerably higher rates
of sand supply.

In the Tertiary Piedmont Basin, northwestern Italy, type II
systems are extensively developed as the infill of Late Oligocene
and Early Miocene submarine depressions associated with synsedimentary faulting. These systems vary from very samll sandstone
bodies, less than 4 km across and with a thickness of 10-30 m, to
relatively large ones with width up to 20 km and thickness of
several hundreds of meters (Cazzola et al., 1981, 1984; Cazzola
and Rigazio, 1983; Cazzola and Sgavetti, 1984).

Type III systems   are characterized by small sandstone-filled
channels that are enclosed by and grade downcurrent into predominantly muddy sequences. The channelized sandstone facies do not
extend basinward and are therefore restricted to the inner portions
of the system. The channel-fill sequences are made up of fine- to
medium-grained sandstone in parallel-sided beds that characteris-

tically thin and pinch-out toward both edges of the channel. In
the studied examples, which include the Eocene Hecho Group, the
Tertiary Piedmont Basin, and the Jurassic Neuquen Basin in
Argentina (Mutti,unpublished data), channels are generally shallower than 10 m and have width up to 100 m. The channel-fill
deposits are enclosed by volumetrically predominant thin-bedded
mudstone, siltstone, and fine-grained sandstone that formed
through overbank processes. These thin bedded and essentially
muddy sediments are characterized by an abundance of features
produced by synsedimentary creeping, slumping, and shallow channeling that result in marked geometric unconformities at various
scales. Basinward, both channel-fill sequences and associated
overbank deposits grade into progressively thinner mudstone units
that are generally indistinguishable from "normal" basinal mudstone facies.

   Type III systems are apparently the ancient expression of
modern channel-levee complexes. However, the size of the studied
ancient systems is much smaller than that of similar and well-known
Recent complexes developed off major deltas(e.g.Damuth et al.,1983).

RELATIONS BETWEEN DEGREE OF SLOPE INSTABILITY AND VOLUME OF TURBIDITY CURRENTS

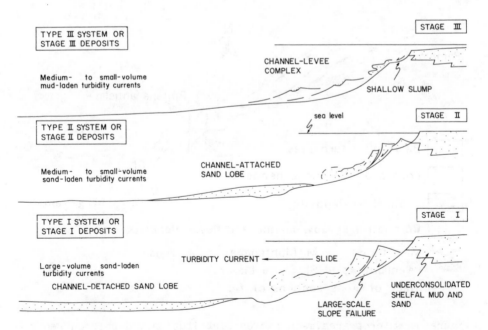

Figure 4. Relations between degree of slope instability and
          volume of turbidity currents.

Relations between types of turbidite systems and volume of
gravity flows

    Figure 4 shows the inferred relations between types of
turbidite systems and volume of gravity flows. The inference is
based on the volume of sediment which is involved in individual
beds or groups of beds within each system, as well as on the total
volume of the system itself. Examples of geometric reconstructions
of ancient turbidite systems by means of detailed correlation
patterns have been provided among others, by Enos (1969),Hirayama
and Nakajima (1977), Mutti et al.(1978), Casnedi et al.(1978),
Ricci Lucchi and Pignone (1978), Estrada Aliberas(1983), Remacha
(1983), Cazzola et al.(1984), Cazzola and Rigazio (1983), etc.
    In type I systems, individual beds and groups of beds that
occur in lobe sequences commonly reach volumes on the order of
1 and 10 cubic kilometers respectively. The latter volume is

OUTCROP AREA, MAIN FACIES DISTRIBUTION, AND PALEOCURRENTS OF THE EOCENE HECHO GROUP TURBIDITES, SOUTH-CENTRAL PYRENEES.

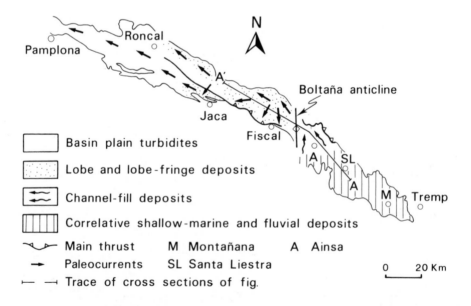

Figure 5. Outcrop area,main facies distribution,and paleocurrents
          of the Eocene Hecho Group turbidites,south-central
          Pyrenees. A-A' is trace of cross-sections of Fig.6.

INFERRED DEPOSITIONAL SETTING OF THE EOCENE HECHO GROUP TURBIDITES, SOUTH-CENTRAL PYRENEES

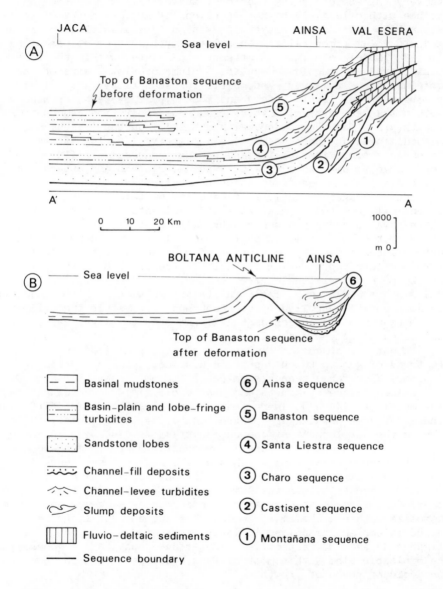

Figure 6. Inferred depositional setting of the Eocene Hecho Group turbidites, south-central Pyrenees. The basal Montanana sequence is Early Eocene in age. The uppermost Ainsa sequence is Bartonian in age.

comparable to the sediment volume of a modern channel-mouth bar
in a large delta complex. In type III systems, thin and laterally
impersistent beds of sandstone, and the abundant associated thin-
to very thin-bedded silty mudstone, suggest flows of rather small
volume with relatively high proportions of mud. Geometry of sandstone beds and bodies in type II systems indicates flows of intermediate volume, generally with relatively high proportions of sand.
   The model of Figure 4 attempts to relate, within the same
basin, types of turbidite systems, volume of flows, size of the
original slope failures responsible for such flows, and relative
position of sea level. Type 1 and most type 2 systems, or stages
within the same system, involve true resedimentation processes
for they are fed from older, although still unconsolidated marginal sediments. The process implies a relative lowering of sea level
to enhance slope instability and produce consequent failures of
very large volumes of sediment. Type III systems and stages involve small-scale resedimentation processes that are produced at
the seaward edges of actively prograding deltaic systems and are
essentially related to high rates of sedimentation.These processes,
which have been amply discussed by Coleman et al.(1983), lead to
the formation of a complex system of gullies and depositional
debris lobes. Downslope evolution of such debris lobes into relatively dilute and mud-rich turbidity currents is here thought to
be the main depositional process leading to the growth of type
III systems and stages. Facies distribution within these sediments
is apparently controlled by overbank and "flow-stripping" (Piper
and Normark, 1983) processes that force sand to be gradually
deposited along the channels.
   The three stages of Figure 4 may develop as independent
features of growth which result in distinct types of turbidite
systems. Therefore, each system results from a specific and
predominant stage of growth and can be classified according to
the model of Figure 1. These features may also represent different
phases of growth which are controlled by sea level variations
during the development of the same system. As discussed in more
detail later, the second case is apparently quite common and give
rise to composite systems which show ordered vertical sequences
and/or alternations of facies associations representing different
phases of growth within the same system. Depending on their scale,
these facies associations can be considered as stage or sub-stage
deposits within a system. Examples of stage and sub-stage facies
associations are discussed in later sections and are shown in
Figures 11 and 16, respectively. Sub-stage deposits are generally
recognizable also within most systems developed through one
predominant phase of growth.

TURBIDITE SYSTEMS AND DEPOSITIONAL SEQUENCES

   In the Eocene Hecho Group, turbidite sediments can be framed

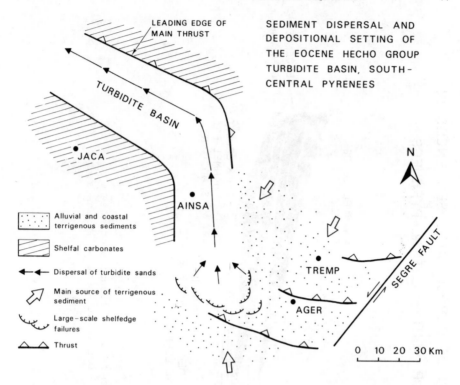

Figure 7. Sediment dispersal and depositional setting of the
Eocene Hecho Group turbidite basin, south-central Pyrenees.

within several depositional sequences (Figures 5 and 6). The basin
formed and evolved essentially as an elongate foredeep depression
parallel to the leading edge of southward moving thrust sheets.
As a result, the axis of the basin progressively shifted in the
same direction with time. Sediment was supplied from submarine
and subaerial erosion of shelfal and alluvial sediments accumulating in the east, along an imbricated thrust zone developed at
the convergence of two main zones of oblique wrenching (Figure 7).
Part of these shelfal and alluvial sediments are still preserved
in a broad and gentle syncline, the so-called Graus-Tremp basin,
north of the frontal thrust zone.

The Hecho Group turbidites are made up of several systems
that are correlative with marked stratigraphic unconformities
separating the shelfal and coastal sediments in the east. Most of
these unconformities are related to tectonic movements as indicated by angular relations between successive sequences even in
deep-water environments.

Figure 8 shows the model derived from this setting. From
base to top and within each sequence, individual turbidite systems

DEEP-WATER EXPRESSION OF DEPOSITIONAL SEQUENCES IN THE EOCENE HECHO GROUP, SOUTH-CENTRAL PYRENEES

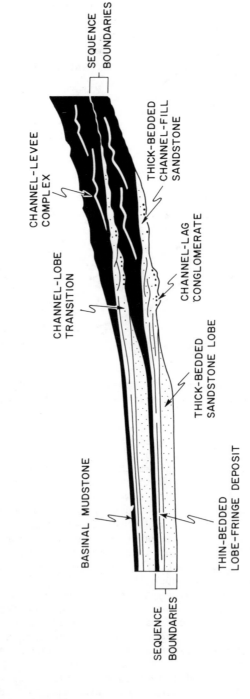

Figure 8. Deep-water expression of depositional sequences in the Eocene Hecho Group, south-central Pyrenees.

evolve as composite features from stage I into stage II and eventually into stage III. As clearly shown by lateral stratigraphic relations directly observed in the field, stage I and II deposits are associated with periods of lowstand of sea level; stage III deposits are conversely the mere deep-water expression of actively prograding delta systems on the adjacent shelfal zone. Channel-levee complexes are therefore a type of turbidite deposition that develops during periods of highstand provided adjacent shelves are characterized by sufficiently high rates of deltaic progradation. Vertical evolution from stage I to stage II deposits within the same system is interpreted as the result of progressive decrease in the volume of turbidity currents following slope retreat and rise of sea level. Stage III deposits develop on top of stage I and stage II sandstone facies when progressive rise of sea level has allowed the onset of active deltaic progradation on adjacent shelves. Basically these sequences of different turbidite stages within the same system indicate a rapid lowering of sea level followed by a gradual restablishement of shelfal deposition.

The uppermost depositional sequence, the Ainsa, is a composite stratigraphic unit made up of several minor sequences only partly represented in the diagram of Figure 6. The main sequence, as well as the smaller ones that develop within it, result from a major phase of tectonic deformation of the basin that occurred during late Middle Eocene - a global highstand period in the Vail chart (Vail et al., 1977). As indicated by the cross section, a series of tectonic pulses within a relatively short period of time produced an overall thinning-upward in the resulting turbidite systems because of the progressive shortening of recharge time in the adjacent shelves. The uppermost and smaller sequences, only partly depicted in the diagram, are virtually devoid of basal turbidite sediments. These are progressively replaced by slumped slope mudstones. The example provides a model for entirely tectonically controlled turbidite systems within a global period of highstand of sea level.

Strong tectonic confinement of the Hecho Group basin results in the development of markedly elongate turbidite systems.This is true in particular for the stage I sandstone lobes of the Charo sequence (Figure 6) that may form bodies with length in excess of 50 km parallel to the basin axis (Remacha, 1983).

Figures 9 and 10 refer to turbidite-bearing basins in the Upper Cretaceous Aren Sandstone, south-central Pyrenees, and the Tertiary Piedmont Basin,northwesternmost Italy,respectively. In both cases, although in quite different tectonic settings, turbidite sediments occur mainly as relatively small bodies within predominantly mudstone facies.

The Aren turbidites were deposited through several tectonic episodes associated with the synsedimentary uplift of an anticline (Sgavetti et al.,1984). Due to the high frequency of pulses of tectonic uplift, about 10-15 depositional sequences developed over a total sedimentary thickness of only some 500 m and encompassing

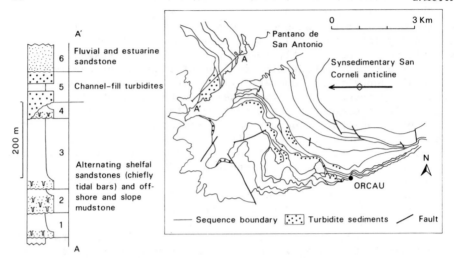

A'

6 Fluvial and estuarine sandstone

5 Channel-fill turbidites

4

3

Alternating shelfal sandstones (chiefly tidal bars) and off-shore and slope mudstone

2

1

A

200 m

Pantano de San Antonio

Synsedimentary San Corneli anticline

N

ORCAU

0   3 Km

——— Sequence boundary   [∴∴] Turbidite sediments   ╱ Fault

Sequences 1 through 5 formed as a result of phases of tectonic uplift during an overall period of sea level rise

Sequence 6, which is overlain by continental red beds formed as a result of combined tectonic uplift and lowering of sea level.

SEQUENCE BOUNDARIES AND TURBIDITE OCCURRENCE WITHIN THE TECTONICALLY CONTROLLED AND OVERALL REGRESSIVE AREN SANDSTONE SUCCESSION (UPPER CRETACEOUS, SOUTH-CENTRAL PYRENEES)

(Map slightly modified after SGAVETTI et al., 1984)

Figure 9. Sequence boundaries and turbidite occurrence within the tectonically controlled and overall regressive Aren Sandstone succession (Upper Cretaceous, south-central Pyrenees).

a time interval of approximately 2 million years. Most of the depositional sequences developed very poorly organized type II systems of very small volume and characterized primarily by highly chaotic sediments. The paucity of resedimented sandstone facies can be ascribed to insufficient recharge time of the adjacent shelves.

The Oligocene and Miocene turbidites of the Tertiary Piedmont Basin (Figure 10) consist of sand-rich, type II systems that were deposited within fault-related depressions. These systems are separated by laterally extensive mudstone units. Widening and deepening of the basin resulted from structural subsidence within an overall extensional pattern. As discussed by Cazzola & Sgavetti (1984), each turbidite system, or group of laterally equivalent systems, forms the lower stratigraphic portion of a depositional sequence. The upper portion is expressed by slope and basinal mudstone deposited during periods of relative highstand. Sequence boundaries are commonly characterized by marked onlap terminations of sandstone units onto the highstand mudstones of the underlying sequence.

As indicated by Figure 10, the turbidite bodies which are lower in the stratigraphic succession are characterized by highly lenticular geometry, extensive channeling, and very coarse texture. They formed as the infill of narrow fault-bounded troughs supplied from the resedimentation of the fan-delta deposits of the Molare Fm, originally deposited along narrow and irregular shelf zones, following a period of tectonic uplift.

Higher in the succession, turbidite sandstone bodies become progressively finer grained and have more tabular geometries, with the tendency to extend all across the basin floor. This suggests progressively more extensive feeder shelves with sufficient re-charge times to provide increasingly larger volumes of sand for resedimentation processes (Cazzola et al., 1981).

Based on the above discussion, Figures 11 and 12 show conceptual models for the relations between turbidite systems and depositional sequences in two common tectonic settings.

DIAGRAMMATIC CROSS-SECTION OF UPPER OLIGOCENE AND LOWER MIOCENE TURBIDITE SEDIMENTS IN THE TERTIARY PIEDEMONT BASIN (NORTHWESTERN ITALY).
(Sligthly modified after Cazzola & Sgavetti, 1984)

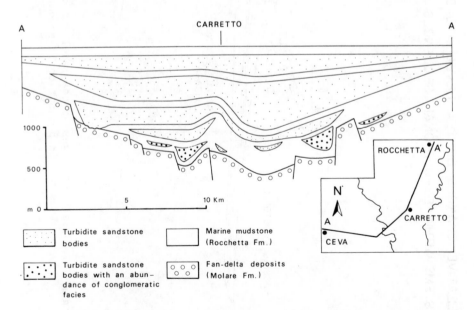

Figure 10. Diagrammatic cross-section of the Upper Oligocene and Lower Miocene turbidite sediments in the Tertiary Piedmont Basin (northvestern Italy).

RELATIVE CHANGES OF SEA LEVEL AND DIFFERENT STAGES OF GROWTH OF TURBIDITE SYSTEMS WITHIN DEPOSITIONAL SEQUENCES DEVELOPED IN ELONGATE FLYSCH BASINS

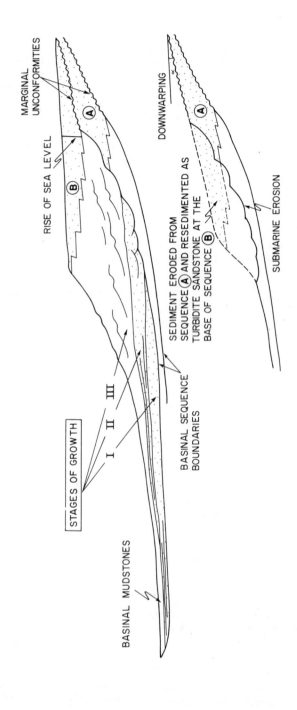

Figure 11. Relative changes of sea level and different stages of growth of turbidite systems within depositional sequences developed in elongate flysch basins.

Figure 11 depicts deposition in elongate flysch basins filled from one end through resedimentation of thick successions of essentially unconsolidated deltaic sediments. Tectonic confinement and ponding play a major role in these basins in that turbidite sandstone tend to develop elongate and unusually thick bodies. The abundance of fines related to deltaic sedimentation on adjacent shelves enhances the mobility of turbidity currents that can spread their sand load virtually all across the basin. In addition, both overbank and lobe fringe thin-bedded and fine-grained turbidite sandstone may develop in considerably large volumes. Except for tectonic confinement and consequent anomalous thickness of turbidite sandstone bodies, these settings should resemble those of divergent continental margins.

Figure 12 depicts smaller settings in areas of strong tectonic control. The lack of extensive shelves and deltaic complexes results in sand-rich and small-volume turbidite systems that are fed by fan-deltas or other types of coarse-grained alluvial and nearshore deposits. The superposition of several turbidite systems of this type separated by relatively thick mudstone units indicates a global rising of sea level since the combination of tectonic uplift and a global falling of sea level would obviously lead to the rapid infilling of the basin with alluvial deposits.

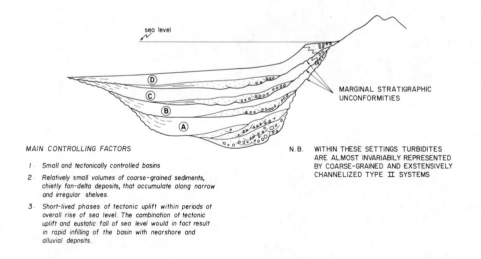

TURBIDITE-BEARING DEPOSITIONAL SEQUENCES IN BASINS
WITH STRONG MARGINAL TECTONIC UPLIFT

sea level

(D)
(C)
(B)
(A)

MARGINAL STRATIGRAPHIC
UNCONFORMITIES

MAIN CONTROLLING FACTORS

1. Small and tectonically controlled basins

2. Relatively small volumes of coarse-grained sediments, chiefly fan-delta deposits, that accumulate along narrow and irregular shelves.

3. Short-lived phases of tectonic uplift within periods of overall rise of sea level. The combination of tectonic uplift and eustatic fall of sea level would in fact result in rapid infilling of the basin with nearshore and alluvial deposits.

N.B. WITHIN THESE SETTINGS TURBIDITES ARE ALMOST INVARIABLY REPRESENTED BY COARSE-GRAINED AND EXSTENSIVELY CHANNELIZED TYPE II SYSTEMS

Figure 12. Turbidite-bearing depositional sequences in basins with strong marginal tectonic uplift.

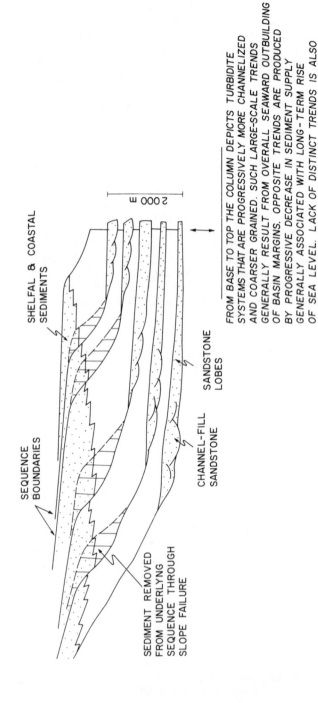

FROM BASE TO TOP THE COLUMN DEPICTS TURBIDITE
SYSTEMS THAT ARE PROGRESSIVELY MORE CHANNELIZED
AND COARSER GRAINED. SUCH LARGE-SCALE TRENDS
GENERALLY RESULT FROM OVERALL SEAWARD OUTBUILDING
OF BASIN MARGINS. OPPOSITE TRENDS ARE PRODUCED
BY PROGRESSIVE DECREASE IN SEDIMENT SUPPLY
GENERALLY ASSOCIATED WITH LONG-TERM RISE
OF SEA LEVEL. LACK OF DISTINCT TRENDS IS ALSO
COMMON.

Figure 13. Relations between depositional sequences and vertical
sequential arrangement of turbidite depositional systems.

VERTICAL CYCLICITY AND SEA LEVEL VARIATIONS

The concept of depositional sequences as related to sea level variations (Vail et al., 1977) provides a powerful tool to a better understanding of the sequential arrangement so obviously displayed by most ancient turbidite sediments at the scale of hundreds or thousands of metres ("suites" of Ricci Lucchi, 1975). In reality, as indicated by Figure 13, sequences of this order of magnitude do not bear any relation to progradation or recession within individual turbidite systems as originally thought (Mutti & Ricci Lucchi, 1972), but simply result from long-term outbuilding or recession of basin margins. At this scale, an almost endless series of possible sequential arrangements can be envisaged on the basis of basin evolution in terms of global variations of sea level, local tectonic setting, and amounts of sediment supplied. As an example, Figure 14 shows the basic difference between progradation within a turbidite system and progradation related to the stacking of different turbidite systems each of which is part of a distinct depositional sequence.

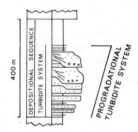

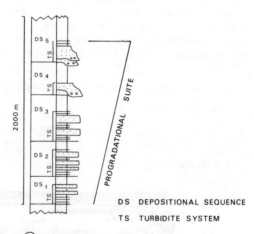

DS DEPOSITIONAL SEQUENCE
TS TURBIDITE SYSTEM

① Within the same system progradation is expressed by the superposition of coarse-grained channel-fill deposits on nonchannelized and finer grained sandstone lobes. Basinward progradation is common in type II systems developed in relatively small basins and over-supplied with sand.

② The overall thickening-and coarsening-upward trend of turbidite systems simply reflects long-term progradation of basin margins. Within each depositional sequence, turbidite systems develop as prograding, receding, or vertically accreting features depending upon local conditions.

Figure 14. Progradation within a turbidite system (left) and progradation related to the stacking of different turbidite systems each of which is part of a distinct depositional sequence (right).

    Considerably more interesting are the implications of shorter
term sea level variations as a possible cause of medium- and small-
scale vertical cyclicity observed in turbidite sediments   at   the
scale of meters and tens of meters, i.e. the classic sequences or
cycles depicted by vertical superposition of different facies.
    These cycles, first recognized by Mutti and Ricci Lucchi
(1972) and subsequently used in most literature, although with
considerable misuse and controversies (see Hiscott, 1981), are
essentially restricted to two types: (1) thinning- and fining-
upward cycles, thought to be characteristic of channel-fill sequences and suggesting phases of channel activity followed by
abandonment; and (2) thickening- and coarsening-upward cycles
thought to be diagnostic of prograding sand lobes.
    As discussed in more recent literature (e.g. Mutti et al.,
1978; Ricci Lucchi and Pignone, 1978; Walker, 1980 in Nilsen;
Hiscott, 1981; Mutti and Sonnino, 1981; Mutti, 1983; Remacha,
1983), interpreting turbidite facies sequences is considerably
more complex particularly as to type I systems, where sandstone
lobes are primarily the product of vertical accretion processes
rather than progradation. However, facies cycles are nicely depicted by most ancient turbidite sequences and the problem
remains whether these cycles are produced by autocyclic or allocyclic mechanisms or a combination thereof. Part of these cycles
are certainly related to autocyclic mechanisms, i.e. produced by

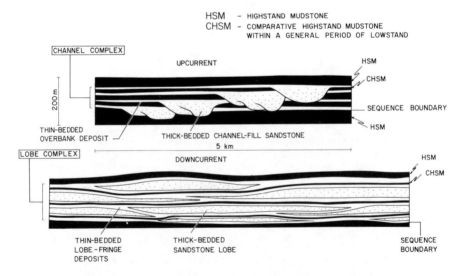

Figure 15.   Turbidite depositional system comprised by
             channel-fill and lobe sediments.

SEA LEVEL CHANGES AND RESULTING FACIES CYCLICITY IN A TURBIDITE
SYSTEM COMPRISED BY CHANNEL-FILL AND LOBE SEDIMENTS

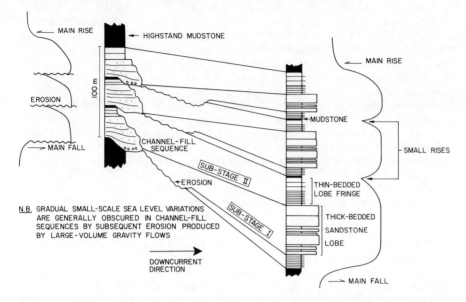

Figure 16. Sea level changes and resulting facies cyclicity in a
turbidite system comprised by channel-fill and lobe
sediments. Sub-stages are conceptually similar to
stages but develop at a smaller scale. Sub-stages are
characterized by an alternation of packages of coarse-
grained and fine-grained turbidite facies that form
cycles generally between 3-15 m thick in both channel-
fill and lobe deposits. Care should be exerted however
in not confusing such facies cycles with those related
to autocyclic mechanisms such as channel shifting and
compensation that develop within a sub-stage deposit.
Virtually all turbidite systems display substages.

processes such as lateral shifting of channel axis with time, basinward progradation of sand lobes in type II systems, or compensation of depositional relief. Others, however, are apparently
much better explained through allocyclic mechanisms and particularly by small-scale sea level variations that affect the volume
of turbidity currents with time.
     Figures 15 and 16 show facies cyclicity in type I systems
as related to small-scale fluctuations of sea level. The model,
which is mainly derived from the Hecho Group, indicates these
systems form as an actual alternation of type I and II sub-stages.
Sub-stage I is related to the main relative lowstand and develops
through channel erosion or re-incision and deposition of sandstone

lobe facies. During this sub-stage coarse-grained lag deposits
may accumulate in the channel axis. Sub-stage II is related to
periods of comparative highstand that are produced by small rises
of sea level during the main lowstand. This results in the gradual infilling of channels with thick-bedded sandstone facies and
deposition of thin-bedded sandstone and mudstone in the lobe
region. The model is supported by detailed correlation patterns
in several ancient basins (Casnedi et al., 1978; Mutti et al.,
1978; Remacha, 1983; Cazzola et al.,1984; G.Ghibaudo, personal
communication, 1984), where thick-bedded sandstone lobes and
associated lobe-fringe deposits can be traced over distances up
to tens of kilometers. As indicated by these correlation patterns,
thick-bedded sandstone lobe facies eventually grade peripherally
into thinner bedded fringe sediments; however, considerable proportions of the thin-bedded facies which are vertically associated
with the sandstone lobes cannot be explained in the same way.
Actually, these sediments form packages that can be traced laterally as far as the vertically associated thick-bedded lobes. For
this reason, such packages are here interpreted as sub-stage II
deposits, equivalent to channel-fill sequences and channel-lobe
transition deposits located farther upsystem. Small scale fluctuations of sea level appear to occur as "background noise" throughout the development of submarine turbidite systems and probably
account for a large part of the medium- to small-scale cyclicity
observed in ancient turbidite lobe and channel-fill deposits.
These fluctuations are thought to be of the same order of magnitude as those which produce the classic nearshore facies sequences
in both terrigenous and carbonate sediments.These cyclic facies
patterns are poorly developed or absent only in those cases where
the vertical accretion rate largely exceeds the rate of small-
scale sea level variations. Such conditions are commonly associated with phases of rapid tectonic uplift and abundance of
sediment supply.

SOME CONCLUDING REMARKS

     Deposition of turbidite sediments occur in a variety of
sedimentary and tectonic settings each of which leaves a print
on the resulting type of facies,facies associations,and geometry
of sandstone bodies. Setting aside basin plain sediments and
seismoturbidites, most ancient turbidite systems form through
patterns where channel-fill facies are replaced in a downcurrent
direction by nonchannelized sediments. All other things being
equal, volume of gravity flows determines where sand will
be deposited within a system. Large flows are related to huge
submarine erosional features and deposit most of their sediment
load as virtually basinwide and unchannelled sandstone lobes.
Channels may form during the process but are essentially bypassed
by most of the sand. Decrease of volume of turbidity currents

enhances the depositional character of the channels that progressively become the only site of sand deposition when highly confined and small-volume flows lose most of their fines through overbank processes.

Substantial amounts of turbidite sandstone facies are associated only with type I and II systems. Such systems are characteristic of periods of lowstand of sea level - although some type II systems can probably develop also during periods of highstand under conditions of very high rates of seaward progradation of fan-deltas across narrow shelves.

Turbidite sandstone facies of type I and II systems are derived from still poorly consolidated shelfal sands and muds that undergo failures and mass movement triggered by relative lowering of sea level. In these cases, compositional and textural characteristics of turbidite sandstones are therefore inherited from those of shelfal and coastal sediments deposited during preceding periods of highstand. Lithification of shelfal sediments plays a major role in determining how much older and deeper in the section the source for resedimented turbidite sands can be. Lithified sediments will not in fact be directly involved in normal resedimentation processes and can be fragmented and displaced only by exceptional events related to tectonic activity (olistholiths, seismoturbidites, megabreccias, etc).

Type III systems, and probably most sequences of essentially fine-grained and thin-bedded turbidites, are deposited during periods of highstand of sea level and are related to "normal" conditions of shelf-edge instability such as produced by high rates of seaward progradation of deltas or other types of nearshore systems, major storms, and earthquakes. Turbidite sandstones, chiefly fine-grained deposits, are directly derived in these cases from contemporaneous or nearly contemporaneous shelfal sediments.

A Pleistocene example of turbidite deposition with rates of sedimentation up to 6 m/y has been recently documented by Coleman et al. (1983) for the Mississippi canyon. Large-scale turbidite sandstone bodies, particularly in type I systems which are derived from catastrophic slope failures, could actually record very short periods of geologic time, possibly beyond the biostratigraphic resolution. Careful biostratigraphic and sedimentological investigations are therefore required in ancient turbidite sandstone sequences in order to correctly evaluate rates of sedimentation when applied to this type of deposit. Results of this work could help considerably in discriminating between lowstand and highstand turbidite sediments.

ACKNOWLEDGMENTS

The help of M. Sgavetti and R.Valloni with various aspects of this study is greatly appreciated. The writer also thanks R.M. Lowe and P.H. Stauffer for reviewing and commenting on the manuscript. The financial support for the final phase of this study was provided by the Italian Ministry of Education (MPI).

REFERENCES

Casnedi, R., Moruzzi, G., and Mutti, E., 1978, Correlazioni elettriche dei lobi deposizionali torbiditici nel Pliocene del sottosuolo abruzzese: Memorie Società Geologica Italiana, v. 18, p. 23-30.

Cazzola, C., Fonnesu, F., Mutti, E., Rampone, G., Sonnino, M., and Vigna, B., 1981, Geometry and facies of small, fault-controlled deep-sea fan systems in a transgressive deposi-tional setting (Tertiary Piedmont Basin, Northwestern Italy), in Ricci Lucchi, F., ed., Excursion Guidebook: 2nd IAS Reg. Meeting, Bologna, p.5-56.

Cazzola, C., Mutti, E., and Vigna, B., 1984, The Cengio Submarine Turbidite System of the Tertiary Piedmont Basin, Northwestern Italy: Geo-Marine Letters, v.3, p.173-177.

Cazzola, C., and Rigazio, G., 1983, Caratteri sedimentologici dei corpi torbiditici di Valla e Mioglia, Formazione di Rocchetta (Oligocene-Miocene) del Bacino Terziario Piemontese: Giornale di Geologia, v. 45, p. 87-100.

Cazzola, C., and Sgavetti, M., 1984, Geometria dei depositi torbiditici della Formazione di Rocchetta e Monesiglio (Oligocene superiore - Miocene inferiore) nell'area compresa tra Spigno e Ceva: Giornale di Geologia, Bologna, (in press).

Coleman, J. M., Prior, D. B., and Lindsay, J.F., 1983, Deltaic influences on shelfedge instability processes: SEPM Special Publication 33, p. 121-137.

Damuth, J. E., Kowsmann, R. O., Flood, R. D., Belderson, R. H., and Gorini, M. A., 1983, Age relationships of distributary channels on Amazon deep-sea fan: Implications for fan growth pattern: Geology, v. 11, P. 470-473.

Enos, P., 1969, Anatomy of a flysch: J. Sed. Petrology, v. 39, p. 680-723.

Estrada Aliberas, M. R., 1983, Lobulos deposicionales de la parte
    superior del grupo de Hecho entre el anticlinal de Boltana y
    el Rio Aragon (Huesca): Resumen de Tesis Doctoral; Publicac.
    de la Univers. Autonoma de Barcelona Bellaterra, Facultad
    de Ciencias, 24 p.

Hirayama, J., and Nakajima, T., 1977, Analitical Study of Turbid-
    ites, Otadai Formation, Boso Peninsula, Japan: Sedimentology,
    v. 24, p. 747-779.

Hiscott, R. N., 1981, Deep-Sea Fan Deposits in the Macigno
    Formation (Middle-Upper Oligocene) of the Gordana Valley,
    Northern Apennines, Italy - Discussion: J. Sed. Petrology,
    v. 51, p. 1015-1033.

Kuenen, Ph. H., and Migliorini, C. I., 1950, Turbidity currents
    as a cause of graded bedding: Jour. Geology, v. 58, p. 91-127.

Mutti, E., 1979, Turbidites et cones sous-marins profonds, in
    Homewood P., ed., Sedimentation detritique (fluviatile,
    littorale et marine): Institut de Géologie, Université de
    Fribourg, Suisse, p. 353-419.

Mutti, E., 1983, Facies Cycles and Related Main Depositional
    Environments in Ancient Turbidite Systems: Abstracts of the
    AAPG Annual Convention, Dallas, p. 135.

Mutti, E., 1984 a, Turbidite facies and sea-level variations in
    the Eocene Hecho Group, South Central Pyrenees, Spain:
    Abstracts of the 5th European Regional Meeting IAS,
    Marseille, p. 310-311.

Mutti, E., 1984 b, The Hecho Eocene Submarine Fan System, South-
    Central Pyrenees, Spain: Geo-Marine Letters, v. 3, p. 199-202.

Mutti, E., Nilsen, T. H., and Ricci Lucchi, F., 1978, Outer fan
    deposiitonal lobes of the Laga Formation (upper Miocene and
    lower Pliocene),east-central Italy, in Stanley, D. J., and
    Kelling, G., eds, Sedimentation in Submarine Canyons,
    Trenches and Fans: Dowden, Hutchinson and Ross, Stroudsburg,
    Pennsylvania, p. 210-223.

Mutti, E., and Ricci Lucchi, F., 1972, Le torbiditi dell'Appennino
    settentrionale: Introduzione all'analisi di facies: Memorie
    Società Geologica Italiana, v. 11, p. 161-199.

Mutti, E., and Ricci Lucchi, F., 1975, Turbidite facies and facies
    associations, in Examples of Turbidite Facies and Facies
    Associations from Selected Formations - Northern Apennines:
    9th Int.Congr.Sedim.,Nice,Guidebook Field Trip 11, p. 21-36.

Mutti, E., and Ricci Lucchi, F., 1981, Introduction to the excursions on siliciclastic turbidites: Abstracts of the 2nd European Regional Meeting IAS, Bologna, p. 1-3.

Mutti, E., Ricci Lucchi, F., Seguret, M., and Zanzucchi, G., 1984, Seismoturbidites: a new group of resedimented deposits: Marine Geology, v. 55, p. 103-116.

Mutti, E., and Sonnino, M., 1981, Compensation cycles: a diagnostic feature of turbidite sandstone lobes: Abstracts of the 2nd European Regional Meeting IAS, Bologna, p. 120-123.

Nilsen, T. H., 1980, Modern and ancient submarine fans: Discussion of papers by R. G. Walker and W. R. Normark: AAPG Bulletin, v. 64, p. 1094-1101.

Normark, W. R., 1970, Growth patterns of deep-sea fans: AAPG Bulletin, v. 54, p. 2170-2195.

Normark, W. R., Mutti, E., and Bouma, A. H., 1984, Problems in Turbidite Research: Geo-Marine Letters, v. 3, p. 53-56.

Piper, D. J. W., and Normark, W. R., 1983, Turbidite depositional patterns and flow characteristics, Navy Submarine Fan, California Borderland: Sedimentology, v. 30, p. 681-694.

Remacha, E., 1983, "Sand Tongues" de la Unidad de Broto (Grupo de Hecho), entre el anticlinal de Boltaña y el Rio Osia (Prov. de Huesca): Tesis Doctoral; Publicac. de la Univers. Autonoma de Barcelona Bellaterra, Facultad de Ciencias, 163 p.

Ricci Lucchi, F., 1975, Depositional cycles in two turbidite formations of northern Apennines: J. Sed. Petrology, v. 45, p. 3-43.

Ricci Lucchi, F., and Pignone, R., 1978, Ricostruzione geometrica parziale di un lobo di conoide sottomarina: Memorie Società Geologica Italiana, v. 18, p. 125-133.

Sgavetti, M., Mutti, E., Rosell, J., and Legarreta, L., 1984, Tectonically controlled depositional sequences in the Upper Cretaceous Aren Sandstone, South Central Pyrenees, Spain: Abstracts of the 5th European Regional Meeting IAS, Marseille, p. 406-407.

Surlyk, F., 1978, Submarine fan sedimentation along fault scarps on tilted fault blocks (Jurassic-Cretaceous boundary, East Greenland): Grønlands Geologiske Undersøgelse, 128, 108 p.

Walker, R. G., 1978, Deep-water sandstone facies and ancient
    submarine fans: Model for stratigraphic traps: AAPG Bulletin,
    v. 62, p.932-966.

Vail, P. R., Mitchum, R. M., and Thompson, III, S., 1977, Seismic
    stratigraphy and global changes of sea level. Part 4:
    Global cycles of relative changes of sea level, in Payton,
    C. E., ed., Seismic Stratigraphy - Application to Hydrocarbon
    Exploration: AAPG Memoir 26, p. 83-97.

DIAGENETIC PROCESSES THAT AFFECT PROVENANCE DETERMINATIONS IN
SANDSTONE

Earle F. McBride

Department of Geological Sciences
University of Texas at Austin
Austin, Texas   78713-7909   U.S.A.

ABSTRACT

     Diagenetic processes that alter the depositional composition
of sands must be considered when making provenance interpretations.
These modifying processes operate from the zone of weathering to
the deep subsurface where diagenesis grades into metamorphism.
Of greatest importance is the total dissolution of grains of
feldspar, rock fragments, and heavy minerals by corrosive meteoric
or subsurface waters to severely modify original sand composition.
Some "diagenetic quartzarenites" have been produced by essentially
wholesale dissolution of non-quartz grains under severe weathering.
Even partial dissolution of some rock fragments can alter them
beyond identification. Of great importance also is the replacement of detrital grains by authigenic carbonates, clays, zeolites,
and many other minerals. Although the identity of pseudomorphous
grains of distinctive size and shape can logically be deduced for
some sandstones, the identity of most totally replaced grains cannot
be determined with confidence. The albitization of plagioclase
more calcic than An_{10} and of K-feldspar should be expected for all
deeply buried sandstones. In the Texas Gulf Coast basin, albitization is important for sandstones that reach burial temperatures
in excess of 120° C.
     There are several additional processes that can complicate
provenance determinations. In environments where detrital
ferromagnesian minerals, magnetite and rutile are undergoing oxidation to produce the pigment of secondary red beds, hematite,
and its precursors can invade micropores in rock fragments
sufficiently to prevent their identification. Of lesser importance, but locally troublesome, is the loss of primary textures

G. G. Zuffa (ed.), Provenance of Arenites, 95–113.

of carbonate and other rock fragments that undergo recrystallization or severe compaction.

INTRODUCTION

Petrographers have long been aware that sands undergo a
variety of textural and mineralogical changes during burial
and subsidence in a sedimentary basin, during structural deformation of buried sequences, and during outcrop weathering.
Several of these post-depositional changes modify the original
detrital mineralogy of a sand and must be considered when any
interpretation is made concerning the provenance of an ancient
sandstone. This paper reviews post-depositional (diagenetic)
changes that sands undergo during burial and which are of concern
to workers in sandstone provenance. Also considered are the
effects of outcrop weathering on provenance determinations,
because they can be equally important as burial diagenetic events.
Of principal concern in this review is the loss of detrital
framework grains by dissolution, the alteration of grains by
replacement or recrystallization, and the loss of identity of
certain ductile grains during compaction which can generate
pseudomatrix. Except for severe ductile grain deformation,
which requires fairly deep burial, the other processes can occur
during outcrop weathering as well as at all burial depth ranges.
Specific processes, such as the albitization of feldspar, are in
part a function of temperature and, thus, indirectly of burial
depth. Common reactions of detrital grain types during diagenesis
are listed in Table 1.

DISSOLUTION

Nearly all detrital grain types can undergo some degree of
dissolution by meteoric water in outcrop and at shallow to
intermediate depths in the subsurface or by modified connate
("formation") water in the deeper subsurface. Partial dissolution of a detrital grain can sufficiently modify its texture
and composition so that its identity becomes questionable (Fig. 1),
whereas totally dissolved grains present an absolute loss of
provenance data (Figs. 2, 3, 5). The only exception to the latter
statement are grain molds of distinctive shape that are surrounded
by cement (Fig. 3). Such molds can provide a clue to the composition of the now-lost grains.
Table 2 is a list of detrital grains that are reported in the
literature or that I have seen that show evidence of in situ
dissolution. About the only grains that show no or trivial signs
of dissolution are the common ultrastable heavy minerals
zircon, tourmaline, and rutile. Even quartz grains undergo
dissolution along strained zones and microcrystal boundaries to
develop dissolution channels (Fig. 6) in intensely weathered

Table 1.   Common Diagenetic Processes That Affect Detrital Grain
Types.
Processes are listed in sequence of most common (first) to less
common.

K-feldspar: calcitized
              zeolitized
              dissolved
              kaolinized
              albitized

Plagioclase: calcitized
              dissolved
              albitized
              zeolitized

Carbonate rock fragments: dissolved
                            dolomitized
                            recrystallized

Volcanic rock fragments: calcitized
                           dissolved
                           chloritized
                           zeolitized

Micaceous metamorphic rock fragments: mashed
                                        calcitized

Chert rock fragments: dissolved
                        calcitized

Claystone/shale/siltstone rock fragments: mashed

Muscovite: kaolinized
             dissolved

Biotite: dissolved
           argillized

Unstable heavy minerals: dissolved
                           calcitized

Figure 1. Rock fragments that have undergone selective dissolution of some components to sufficiently obscure their identity.
Grain B is possibly chert and A is possibly a volcanic rock
fragment. Pore space is blue. Depth = 1500 m. Pennsylvanian
sandstone, Texas, U.S.A.

Figure 2. Outcrop sample that is a "diagenetic" quartzarenite
owing to complete dissolution of feldspars, rock fragments,
and most heavy minerals. A few oversize pores (O) formed where
detrital grains have been lost by dissolution. Grains are lined
by brown clay cutans of infiltered clay. Eocene sandstone,
Texas, U.S.A.

Figure 3. The nearly complete dissolution of a rock fragment
that is partly surrounded by calcite cement (C) produces a
large secondary pore (blue). Eocene, subsurface, Texas, U.S.A.

Figure 4. Sandstone in which all feldspar and possibly some
volcanic rock fragments have been replaced without a change in
size or shape by kaolinite (K). Pore space is blue; brown clay
of pedogenic origin coats most grains. Eocene sandstone from
outcrop, Texas, U.S.A.

sandstones. It is not always possible to distinguish in quartz
(and other detrital grains) which channels are the result of
direct dissolution of quartz versus the channels that are the
result of dissolution of calcite that replaced part of the quartz
grain during the cementation event. Although quartz grains can
be totally replaced by carbonate and pyrite, I have not seen a
sandstone in which I believe entire quartz grains have dissolved.
However, the presence in South America of karst developed on
quartz-rich sandstone bedrock (V. Schmidt, 1984, personal comm.)
suggests that total quartz dissolution in the soil zone is
possible.

    More than 40 years ago heavy mineral workers showed the
serious problem that exists in using heavy minerals to determine
provenance because of the tendency for less stable mineral species
to be selectively dissolved by ground water. Bramlette's (1941)
study of heavy minerals in concretions versus porous host sandstones remains a classic (Table 2). Hornblende was largely
removed from the porous beds of the Hambre Formation and epidote
was largely removed from porous beds of the Modelo Formation.
These species are much more abundant in the concretions, which
are tightly cemented and impermeable. Pettijohn (1941) noted the
strong tendency for some heavy minerals to be absent in older
rocks; for younger rocks to have a more varied suite than older
rocks; and for certain authigenic minerals (tourmaline, zircon)
to be much more common in old than in young sandstones. Pettijohn

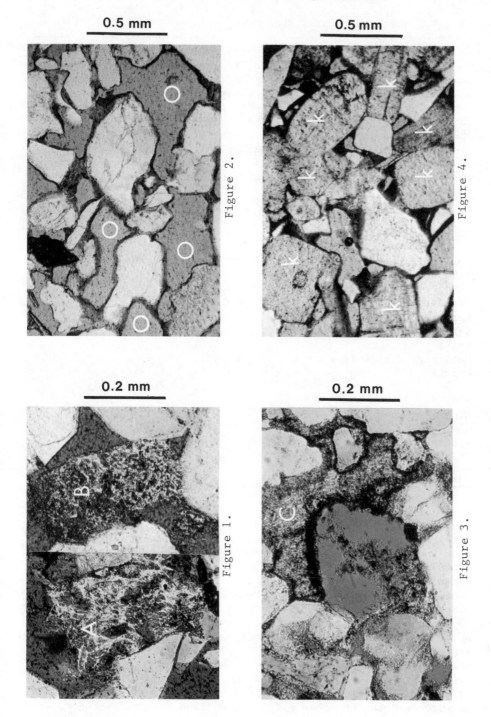

Figure 1.

Figure 2.

Figure 3.

Figure 4.

Table 2.   Detrital Grains That Dissolve.

Grains are listed in approximate sequence of greatest susceptibility (top) to lowest susceptibility. Carbonate rock fragments
are readily dissolved in outcrop and soils, but can be more stable
than plagioclase in deeply buried sandstones with carbonate
cements. Sources for light minerals/grains: Pettijohn (1957);
Hayes (1979); McBride (unpublished). The heavy mineral list is
Morton's (1984, Table 3) order of stability for sandstones undergoing burial-related intrastratal solution and differs from the
stability order in soils and acid-weathering environments. The
dissolution shown by rutile, tourmaline, zircon and quartz is
trivial.

Light Minerals/Grains   Heavy Minerals

Carbonate rock fragments   Olivine, pyroxene
Volcanic glass and fine-grained   Andalusite, sillimanite
  volcanic rock fragments   Amphibole
Plagioclase (>10% An)   Epidote/Zoisite
Perthite   Sphene
K-feldspar   Kyanite
Chert   Staurolite
Metamorphic rock fragments   Garnet
Quartz   Apatite, chloritoid, spinel
                                   Rutile, tourmaline, zircon

Micas

Biotite
Chlorite
Muscovite

(1941) attributed the scarcity of less stable heavy minerals in
older sandstones to their selective loss through dissolution
after deposition. Although there are other factors than dissolution that can contribute to producing the trend documented by
Pettijohn (e.g., van Andel, 1959), the studies of Yurkova (1970),
Scavnicar (1979), and Morton (1979; 1984; this volume) clearly
document the serious effects of the dissolution of some heavy
minerals in the subsurface. Yurkova showed a progressive decrease
in abundance and the ultimate loss of epidote with increasing
burial depth in some Miocene sandstones in the USSR. Epidote is
absent at depth in water-bearing sandstones, but is present in
oil-saturated sandstones. It is inferred that epidote was preserved in the oil-saturated sandstones where migratory water was
excluded after oil entrapment, thereby preventing solution loss
of more epidote, but that epidote dissolved from adjacent

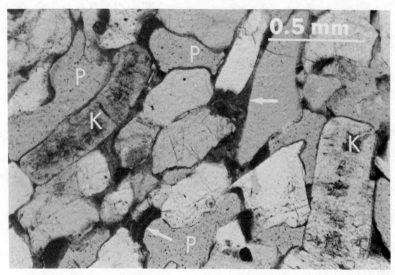

Figure 5. Outcrop sandstone sample with giant authigenic kaolinite grains (K) that replaced feldspars. P = secondary pores formed where detrital grains have been completely dissolved. Arrows indicate clay cutans of infiltered clay. Eocene sandstone, Texas, U.S.A.

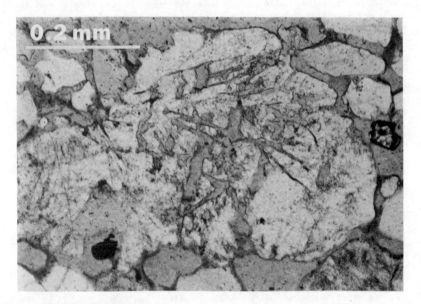

Figure 6. Monocrystalline quartz grain that has undergone selective dissolution along internal strain boundaries. Dissolution occurred during Holocene weathering of an Eocene sandstone. Gray areas are pores. Eocene sandstone, Texas, U.S.A.

Table 3. Comparison of Heavy Mineral Abundance in Calcareous
Concretions Versus Host Sandstone

Epidote was dissolved from the Modelo host rock and hornblende
from the Hambre host rock. Mineral abundance in the concretions
is presumed to be close to the original abundance in the sand.
Not all minerals present are listed (after Bramlette, 1941;
Pettijohn, 1957, p. 676). Values are percent of total heavy
minerals.

	Zircon	Garnet	Sphene	Epidote	Hornblende
Hambre Sandstone	12	3	10	37	5
Hambre Concretion	5	3	6	17	44
Modelo Sandstone	20	15	22	2	--
Modelo Concretion	12	5	10	53	--

sandstones that were open to migrating formation water. Morton's
(1979; 1984; this volume) detailed work on Paleocene sandstones
in the subsurface of the North Sea shows that at least six heavy
mineral species undergo major dissolution loss with increasing
burial depth. Amphibole, epidote, kyanite, sphene, and staurolite,
all present at shallow burial depths, are absent in the same age
sandstones at depths approaching 3000 m, and garnet was markedly
reduced in abundance relative to stable grains. Another example
of the loss of certain heavy minerals with depth of burial of
Plio-Pleistocene sandstones in the northwestern Gulf of Mexico
basin is shown in Figure 7.

     Table 2 shows the relative stability of heavy minerals
deduced by Morton (1984) based upon the depth sequence of loss
by dissolution in the rocks that he studied plus the work of
others. Morton (1984) concludes that the relative order of
stability is controlled by the composition of the interstitial
waters, whereas the limits of persistence depend on pore-fluid
temperature, rate of water migration, and geologic time. He
also stresses the point that the stability of heavy minerals in
soil zones and in acid, near-surface waters is somewhat different
from their stability during burial diagenesis.

     Another example of a depth-dependent dissolution reaction
is the loss of K-feldspar with depth in Tertiary sandstones in
the Texas part of the Gulf Coast basin. Several workers (L.S.
Land, 1981, oral communication; Fisher, 1982) have noted that
K-feldspar begins to decrease in abundance in the Wilcox (Eocene)
and Frio (Oligocene) formations at 2700 to 3000 m and is lost
entirely at depths of 3200 to 4000 m (Fig. 8). Although some

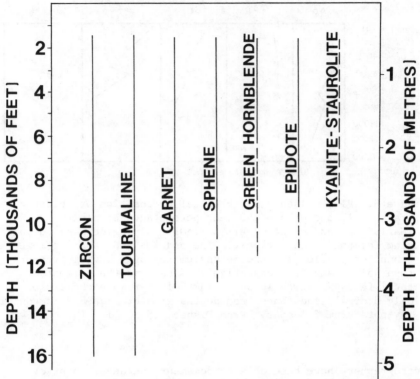

Figure 7. Depth plot of the distribution of seven heavy mineral species in Plio-Pleistocene sandstones of the northern Gulf of Mexico basin. Solid line = abundant; dashed line = rare. Unpublished data of K. L. Milliken.

loss of K-feldspar is due to calcitization and albitization, much of it apparently dissolves directly to yield oversize pores, some of which are subsequently filled by kaolinite or other cements. The dissolution loss of K-feldspar is obviously a temperature-dependent process in the Gulf Coast basin, but it did not take place in all areas. Sandstones of the deeply buried (>4000 m) Norphlet Formation (Jurassic) have pristine detrital K-feldspar grains, many with extensive K-feldspar overgrowths (McBride, 1981). However, the Norphlet sandstones overlie salt beds of the Louann Formation (Jurassic) and were bathed in pore waters that were rich in Na and K ions which stabilized the feldspars.

   Since the use of dyed-epoxy impregnated thin sections,

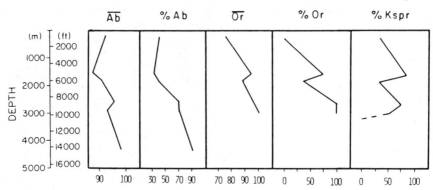

Figure 8. Plot of depth vs. chemical composition of feldspar
grains. Ab and Or are the mean percentages of albite
(NaAlSi₃O₈) and orthoclase (KAlSi₃O₈) in plagioclase and potassium feldspar, respectively. %Ab and %Or indicate the mean
amount of plagioclase and potassium feldspar having greater
than 95% Ab and Or, respectively. %Kspr is the percent of
total feldspar which is potassium feldspar in each sample.
Only 3 of 41 sandstones from depths greater than 3.2 km contain
any potassium feldspar. From Fisher, 1982, Fig. 14.

petrographers have been able to document the development of
secondary porosity where detrital grains and cements are dissolved
from sandstones in the subsurface. Although considerable uncer-
tainty exists about how to distinguish primary from secondary
intergranular pores in sandstones, intragranular and oversize
pores are clearly secondary. The once-held idea that dissolution
takes place only in outcrop has been discarded with the recognition that many detrital grains undergo partial to complete loss
through dissolution at depths greater than 500 m (Table 2). Dissolution of compositionally zoned grains, such as plagioclase
(Figs. 9, 10), and grains that vary in internal grain size, such
as chert (Fig. 1), can be remarkably selective as shown by the
lacy, fritted, or other delicate remnants of such grains. Dissolution of limestone grains is commonly also texture-selective,
and grain size alone is generally not the controlling factor.
Selective dissolution is shown also in sandstones where limestone
grains dissolve but calcite cement does not.

REPLACEMENT

The most common replacement events that hamper provenance
studies are the replacement of grains of carbonate

Figure 9.   Selective dissolution of plagioclase that is compositionally zoned.   The grain has overgrowths of K-feldspar in
places.   Scale bar is 10 μm.   Permian, depth = 1520 m, New
Mexico, U.S.A.   Photo by C. R. Williamson.

Figure 10.   Selective dissolution of detrital plagioclase.
Jurassic, depth = 5350 m, Mississippi, U.S.A.

Figure 13. Biotite (B) and muscovite (M) grains that are partly
  altered to kaolinite and which have expanded greatly along
  their cleavages. Biotite has dissolved in places (arrows).
  Polars crossed. Eocene sandstone, outcrop, Texas, U.S.A.

Figure 14. Oversized patch of poikilotopic calcite that occupies
  the site of a replaced detrital grain plus adjacent pores.
  Polars crossed. Unknown sample.

Figure 15. Sandstone that has undergone extensive argillation,
  where feldspar and some rock fragments have been replaced by
  clay and where pores are filled by clay cement. Distinction
  between altered grains and cement is obscured because of compaction. All grains in this photo, other than quartz, have
  been argillized. Polars crossed. Second Frontier Sandstone,
  Cretaceous, depth = 3667 m, Wyoming, U.S.A.

Figure 16. Strongly compacted litharenite composed chiefly of
  shale, argillite, claystone and siltstone grains. Poorly
  consolidated grains have been mashed sufficiently to lose their
  identity as discrete grains and are now pseudomatrix. All pores
  in the sand have been plugged by pseudomatrix; there is no pore-
  filling cement. Ordovician Martinsburg Formation, Virginia,
  U.S.A.

or clay minerals and the albitization of plagioclase and K-
feldspar. However, there are many additional replacement minerals
that locally are important (e.g., Hayes, 1979, Table 1), such
as the zeolitization of feldspar and volcanic rock fragments in
volcanogenic sandstones and the replacement of feldspar by sul-
fate and other evaporite minerals in sandstones associated with
evaporites (e.g., Walker, 1984). If replacement of a detrital
grain is only partial (Figs. 11-13), its composition generally
can be determined. However, if replacement is complete (Figs. 14,
15), generally its identity is lost. The precursor of pseudomorphous grains sometimes can be inferred on the basis of grain
shape (Figs. 4, 5, 11), but this task is much more difficult if
replacement involves a change in shape or volume (Figs. 5, 12).
The kaolinization of some feldspar (Fig. 5) and mica (Figs. 12,
13) commonly is accompanied by large increases in volume. The
cementation of sands by carbonate cements is generally accompanied
by partial (Fig. 10) to total (Fig. 14) replacement of some grains
by the carbonate. Although it is not possible to distinguish
between carbonate cement that fills detrital grain molds from
that which replaces grains, the presence of abundant oversize
patches of carbonate cement in subsurface sandstones that have not
been exposed to shallow meteoric water is presumptive evidence
of replacement (Fig. 14).

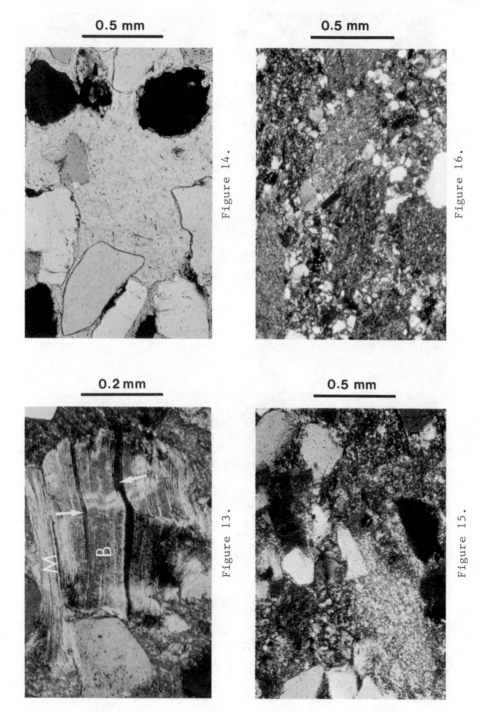

0.5 mm

Figure 14.

0.5 mm

Figure 16.

0.2 mm

Figure 13.

0.5 mm

Figure 15.

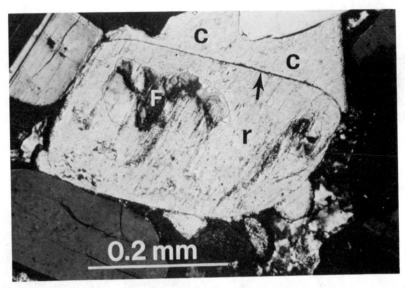

Figure 11. Feldspar grain that has been 70 percent replaced by calcite. The detrital boundary of the feldspar pseudomorph is clearly visible (arrow). C = pore-filling calcite; R = replacement calcite; F = unreplaced feldspar. Polars crossed. Oligocene. outcrop, Texas, U.S.A.

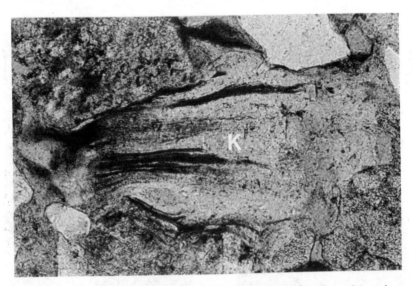

Figure 12. Kaolinization of a biotite grain (dark relicts) resulted in a great increase in volume. K = kaolinite. Eocene, outcrop, Texas, U.S.A. Same scale as Figure 11.

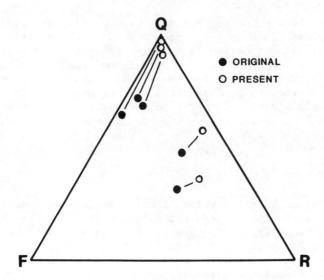

Figure 17.   QFR diagram showing the present composition of five
sandstones and the reconstructed composition assuming 15 percent of the grains now occupied by oversize pores were rock
fragments and 85 percent were feldspar.

Extensive alteration of detrital grains by clay minerals
takes place in many environments, including soils in both humid
and arid climates, and in the deep subsurface. Although the
kaolinization of feldspars (Figs. 4, 5) is the commonest type of
clay alteration, extensive to pervasive argillation of rock fragments and ferromagnesian minerals (Figs. 12, 15) by kaolinite
and other clay species can take place. Red beds (cf. Walker,
1967) and volcanogenic sandstones (cf. Carrigy and Mellon, 1964)
are among the most common sands to undergo pervasive argillation,
but some deeply buried non-volcanogenic sandstones are subject
to extensive argillation also.
    The albitization of detrital plagioclase in sandstones that
undergo substantial burial has been documented by numerous workers
(e.g., Dickinson et al., 1969; Iijima and Utada, 1972; Boles, 1982;
Fisher, 1982). Calcium-bearing plagioclase generally undergoes
replacement to albite with Ab content greater than 95 percent.
The temperature of albitization ranges from 105° to 150° (Milliken
et al., 1981; Boles, 1982). In the southern Texas Gulf Coast,
albitization of the Oligocene Frio Formation takes place at
depths between 2500 and 4000 m, whereas the Eocene Wilcox Formation
in central Texas Gulf Coast shows a gradual albitization between
depths of 1500 and 4200 m (Fig. 8). Albitization of K-feldspar,
commonly leading to the formation of "chessboard" texture, has
been documented by Walker (1984).

Because few petrographers have used particular species of detrital plagioclase as clues to provenance, the albitization of plagioclase has not seriously hampered provenance interpretations to date. This loss of identity of original plagioclase species will be felt when more workers begin using the electron microprobe and other devices to analyze detrital grains in detail. In addition, the albitization of K-feldspar presents obvious serious problems. Albitization is a function of temperature, pore water composition, and time.

MASHING OF DUCTILE GRAINS

Mashing of ductile grains is not a serious problem for provenance determination unless the deformation is severe enough to alter significantly the texture of the grains. Severe deformation is possible for some clayey rock fragments that are poorly consolidated. Ductile flowage of clayey grains can make it impossible to distinguish extrabasinal rock fragments from intrabasinal ripup clasts, and makes it difficult to distinguish detrital matrix, deposited as discrete clay floccules, from pseudomatrix, severely deformed clasts (Fig. 16). Any recrystallization that accompanies ductile grain deformation makes provenance determinations less reliable.

OTHER PROCESSES

Other diagenetic processes that can be of concern include oxidation of detrital grains and recrystallization. Oxidation products can obscure the original texture and composition of grains. Iron oxides produced by the oxidation of ferromagnesian minerals within volcanic rock fragments can obscure grains severely enough that it is impossible to recognize them as of volcanic parentage. The oxidation of discrete detrital ferromagnesian and magnetite grains such that the oxidation products invade and coat adjacent detrital grains can also prevent proper grain identification. Although significant oxidation is most common at the surface, oxidizing meteoric water can reach depths of several thousand metres in some basins.

Recrystallization is not a serious problem in detrital silicate grains, but it can sufficiently modify the texture of limestone grains to make provenance interpretations difficult. Replacement of limestone grains by dolomite or other carbonate minerals is more common a problem than is recrystallization.

COMPENSATING FOR DIAGENESIS AND WEATHERING PROCESSES

In provenance analyses one hopes to deduce the relative

amounts of detritus in a sand that came from specific igneous, metamorphic or sedimentary sources, and, through paleocurrent data or other stratigraphic and paleontologic data, to determine the location of the source area(s). For the first goal above one must infer what the composition was of a sand at the time it was deposited. All post-depositional changes that modified detrital composition must be taken into account in order to deduce the original composition. Although competent petrographers are aware that this "restoration of composition" must be done (and also aware of the difficulties of doing so), few workers have addressed the problem. Most workers present QFR diagrams or point-count data of sandstones that have been modified to some degree by weathering or burial diagenesis, whereas it is the original composition that is needed for detailed provenance interpretations. The importance of dealing with restored compositions is illustrated in Figure 17. In these Tertiary outcrop samples, all feldspar and most rock fragments have been leached to yield sandstones that are composed chiefly of monocrystalline and polycrystalline quartz grains. Figure 14 shows the present composition of several samples on a QFR diagram in addition to the original composition, assuming that 85 percent of the oversize pores were feldspar and 15 percent were rock fragments. Of course, even if the original detrital composition of a sand can be deduced, many problems remain in determining its provenance. Most of the problems are addressed by other articles in this volume.

ACKNOWLEDGEMENTS

Kitty Lou Milliken and Charles R. Williamson kindly supplied unpublished data. I benefited from discussions with many participants of the conference on "Reading Provenance From Arenites" convened by Prof. Gian Gaspare Zuffa in June 1984. The Owen-Coates Fund of the Geology Foundation of the University of Texas at Austin contributed toward publication costs of this article.

REFERENCES

Boles, J.R. 1982, Active albitization of plagioclase, Gulf Coast Tertiary: Amer. Jour. Science 282, pp. 165-180.

Bramlette, M.N. 1941, The stability of minerals in sandstone: Jour. Sedimentary Petrology 11, pp. 32-26.

Carrigy, M.A., and Mellon, G. B. 1964, Authigenic clay mineral cements in Cretaceous and Tertiary sandstones of Alberta: Jour. Sedimentary Petrology 34, pp. 461-467.

Dickinson, W.R., Ojakangas, R.W., and Steward, R.J.   1969, Burial metamorphism of the Late Mesozoic Great Valley Sequence, Cache Creek, California: Geol. Soc. America Bull. 80, pp. 519-526

Fisher, R.S. 1982, Diagenetic history of Eocene Wilcox Sandstones and associated formation waters, south-central Texas: Ph.D. dissertation, Univ. Texas at Austin, 185 p.

Hayes, J.B. 1979, Sandstone diagenesis - the hole truth, p. 127-140 in P.A. Scholle and P.R. Schluger (eds.), Aspects of diagenesis, Soc. Economic Paleontologists and Mineralogists Spec. Publ. No. 26, 443 p.

Iijima, A., and Utada, M.   1972, A critical review of the occurrence of zeolites in sedimentary rocks in Japan: Japanese Jour. Geology - Geography 42, pp. 61-83.

McBride, E.F. 1981, Diagenetic history of Norphlet Formation (Upper Jurassic), Rankin County, Mississippi: Gulf Coast Assoc. Geological Societies Trans. 31, pp. 347-351.

Milliken, K.L., Land, L.S., and Loucks, R. G. 1981, History of burial diagenesis determined from isotopic geochemistry, Frio Formation, Brazoria County, Texas: American Assoc. Petroleum Geologists Bull. 65-8 pp. 1397-1413.

Morton, A.C.   1979, Depth control of intrastratal solution of heavy minerals from the Paleocene of the North Sea: Jour. Sedimentary Petrology 49, pp. 281-286.

_____ 1984, Stability of detrital heavy minerals in Tertiary sandstones from the North Sea basin: Clay Minerals 19.

Pettijohn, F.J. 1941, Persistence of heavy minerals and geologic age: Jour. Geology 49, pp. 610-625.

_____ 1957, Sedimentary rocks: 2nd Ed., Harper and Row, 718 p.

_____, Potter, P.E., and Siever, Raymond   1973, Sand and sandstone, Springer-Verlag, New York, 618 p.

Scavnicar, B.   1979, Pjescenjaci Pliocenai Miocena savaske potoline. Zbornik Radova, Serija A 6/2, pp. 351-382.

Van Andel, T.J.   1959, Reflections on the interpretation of heavy mineral analyses: Jour. Sedimentary Petrology 29, pp. 153-163.

Yurkova, R.M.   1970, Comparison of post-sedimentary alteration of oil-, gas-, and water-bearing rocks: Sedimentology 15, pp. 53-68.

Walker, T.R.   1967, Formation of red beds in modern and ancient deserts: Bull. Geol. Soc. America 78, pp. 353-368.

_____   1984, Diagenetic albitization of potassium feldspar in arkosic sandstones: Jour. Sedimentary Petrology 54, pp. 3-16.

TYPES OF POROSITY IN SANDSTONES
AND THEIR SIGNIFICANCE IN INTERPRETING PROVENANCE

G. Shanmugam

Mobil Research and Development Corporation
P. O. Box 819047
Dallas, Texas 75381

ABSTRACT

Recognition of pore types in sandstones is important in
interpreting provenance because primary porosity is indicative
of depositional framework, whereas secondary porosity is
suggestive of postdepositional, primarily diagenetic, framework. A classification of porosity in sandstones is proposed
to recognize primary and secondary porosity. Secondary porosity caused by dissolution of framework grains is very common
in sandstones. Dissolution of both carbonate and silicate
minerals has been attributed to release of CO_2 either during
organic maturation or during inorganic clay-carbonate
reaction, and to flushing of CO_2 charged waters related to
erosional unconformity. Because sandstone composition is
often controlled by the dissolution of unstable framework
grains such as feldspar and rock fragments (including chert),
it is imperative to estimate the amount of grain dissolution
in determining the original composition. Failure to recognize
the original composition of sandstones will lead to incorrect
interpretation of provenance and paleogeography.

INTRODUCTION

Types, recognition, magnitude, and general significance
of secondary porosity in sandstone have been discussed elsewhere (Schmidt and McDonald, 1979a, 1979b; Shanmugam, 1983,
1984). Secondary porosity plays an important role not only in
improving reservoir quality but also in modifying original
composition of sandstones by grain dissolution. Unfortu-

115

G. G. Zuffa (ed.), Provenance of Arenites, 115–137.
© *1985 by D. Reidel Publishing Company.*

nately, conventional techniques of point counting (Galehouse, 1971) recognize only existing framework grains, and they do not estimate the dissolved grains. Consequently, petrographers estimate sandstone composition that is modified by diagenetic processes such as dissolution and replacement. This "diagenetic composition" is not useful in interpreting provenance which requires an understanding of sandstone composition at the time of deposition, i.e., "depositional composition." The purpose of this paper is to emphasize the significance of recognizing pore types in sandstones and their relevance to interpreting provenance.

Types of porosity are illustrated by photomicrographs in plain light (Figures 1, 4, 5, 6, 7, 8, and 9). Blue areas in color photomicrographs represent porosity as samples are impregnated with blue epoxy resin.

CLASSIFICATION OF POROSITY IN SANDSTONES

In the present study, a classification of porosity of sandstones (Table 1) is proposed based on the genetic and physical characteristics of pore types. The timing of pore genesis is the factor that separates primary porosity from secondary porosity (Choquette and Pray, 1970; Aguilera, 1980). By definition, primary porosity originates either at the time of final deposition or even prior to final deposition as in the case of some transported rock fragments; whereas secondary porosity always originates after deposition. On the basis of their position, pores are classified as intergranular (between grains) or intragranular (within grains) in the case of pores of both primary and secondary origin.

An understanding of processes responsible for creating porosity is important in appreciating the distribution of pore types in sandstones. Primary intergranular pores are invariably caused by sediment sorting and packing; they are common. In well-sorted sands, primary intergranular pores tend to be distributed rather uniformly when compared to poorly sorted sands. Pore geometry is controlled by degree of roundness of grains, compaction, and cementation. Primary intragranular pores have various origins such as vesicles in volcanic glass fragments and morphological cavities in fossils; they are very rare.

Four types of secondary pores are grain fractures, rock fractures, intergranular and intragranular pores (Shanmugam, 1983). Grain fracture is confined to an individual grain, and it does not extend to adjacent grains. Grain fractures are commonly formed by overburden stresses in conjunction with

Table 1. Classification of porosity in sandstones

PORE TYPES	DISTRIBUTION	POSITIVE EFFECTS ON RESERVOIR QUALITY
Primary intergranular	Common and isolated	Small to large
Primary intragranular	Very rare and isolated	Very small
Secondary intergranular	Very common and isolated to continuous	Moderate to large
Secondary intragranular	Rare to common and isolated	Small to moderate
Grain fracture	Rare to common and isolated	Very small
Rock fracture (open)	Rare to common and pervasive	Moderate to large

diagenetic processes such as dissolution and compaction;
however, a tectonic origin for grain fractures is not uncommon. Fractures that cut through several grains or through the
whole rock unit are termed rock fractures; they are caused by
tectonic stresses. Fractures are important pathways for fluid
migration and are commonly responsible for generation of
dissolution porosity. Dissolution plays a major role in
creating both intergranular and intragranular secondary porosity. Intergranular and intragranular porosity are created by
dissolution of cement and grain, respectively. Complete
dissolution of grains, however, often results in the development of intergranular porosity from intragranular type.
Therefore secondary intergranular porosity is the most common
type in sandstones (Table 1). Shrinkage of grains, matrix,
and cement (Schmidt and McDonald, 1979b) does not appear to be
an important mechanism in generating secondary porosity.

The term microporosity, i.e., pores with pore-aperture
radii less than 0.5 microns (Pittman, 1979), may be used for
both primary and secondary types in the proposed classification. This is because microporosity can be either primary
(among clay) or secondary (microporous chert caused by dissolution) in origin.

The classification system shown in Table 1 has the
following advantages: (1) nomenclature is indicative of
genetic and physical characteristics of pore types, and
(2) positive effects of pore types on reservoir quality may be
inferred. Basic types of porosity are shown in Figure 1.

CONCEPT OF SECONDARY POROSITY

Organic Maturation

The concept of relating development of secondary porosity
to the arrival of oil was first developed by Chepikov et al
(1961). They suggested that the arrival of oil in the reservoir leads to dissolution of cements and causes an improvement
in reservoir properties. The concept of dissolution of
carbonate minerals by CO_2 released from maturing organic
matter was introduced by later workers (Schmidt and McDonald,
1979a). The idea of thermal degradation of kerogen and
related calcite dissolution can be written as:

Kerogen   ---thermal exposure---> CO_2 + petroleum + H_2O + N_2.

$$CaCO_3 + CO_2 + H_2O \rightleftharpoons Ca^{2+} + 2HCO_3^-$$
calcite

TYPES OF POROSITY IN SANDSTONES

PRIMARY INTERGRANULAR

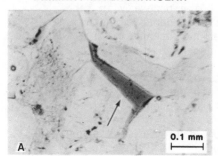

PRIMARY INTRAGRANULAR

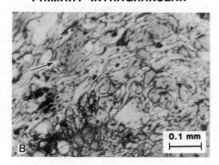

SECONDARY INTERGRANULAR

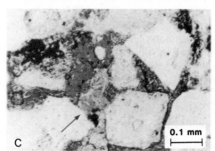

SECONDARY INTRAGRANULAR

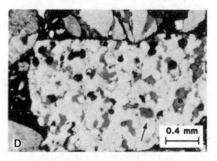

GRAIN FRACTURE

ROCK FRACTURE

FIG. 1. A. Depositional porosity reduced by quartz overgrowth. B. Depositional porosity in a volcanic glass fragment. C. Dissolution of calcite (red) cement. D. Internal dissolution of a siltstone fragment. E. Fracturing of a quartz grain. F. Rock fracture partially filled with calcite (red).

Secondary porosity is an attractive concept in hydro-
carbon exploration because it allows generation of porosity by
dissolution of minerals at depths traditionally considered
unsuitable for reservoir development. In this regard,
Yermolova and Orlova (1962) noted, "High porosity in sandstones can therefore be produced at considerable depths; the
main process in the cases considered has been removal at depth
of authigenic minerals that act as cement."

Savkevich (1969) was the first to illustrate that porosity could actually increase with depth due to dissolution of
early formed cement. Later workers subsequently arrived at
similar curves for distribution of porosity with depth
(Shanmugam, 1984). The distribution of primary and secondary
porosity with increasing depth of burial is illustrated by a
conceptual model in Figure 2.

Clay-Carbonate Reaction

The reaction of clay minerals with carbonates during deep
burial diagenesis provides a powerful mechanism for generating
large quantities of CO_2 by inorganic origin (Hutcheon et al,
1980). The disappearance of kaolinite and the intimate
association of chlorite with corroded dolomite indicates:

$$5 \ CaMg(CO_3)_2 + Al_2Si_2O_5(OH)_4 + SiO_2 + 2H_2O =$$
dolomite   kaolinite   quartz

$$Mg_5Al_2Si_3O_{10}(OH)_8 + 5CaCO_3 + 5CO_2$$
chlorite   calcite

Hutcheon et al (1980) estimated that if 10% of a given
sandstone were comprised of kaolinite, dolomite, and quartz in
the same ratio as in the above reaction, one cubic meter of
sandstone would produce approximately 25,800 litres of CO_2 at
25°C at 1 atm. Thus, secondary porosity produced by CO_2
charged waters in the subsurface may be related to both
organic (Schmidt and McDonald, 1979a) and inorganic (Hutcheon
et al, 1980) mechanisms.

Importance of Unconformities

The creation of secondary porosity related to erosional
unconformities was first suggested by Krynine (1947), who
noted that carbonate cement in sandstone undergoes dissolution
during uplift and weathering. In the North Sea, dissolution
of feldspar in the Jurassic sandstone has been explained by
the circulation of meteoric waters related to the Cimmerian
uplifts (Sommer, 1978; Hancock and Taylor, 1978; Bjørlykke,
1982). In the Brent Formation, dissolution of mineral

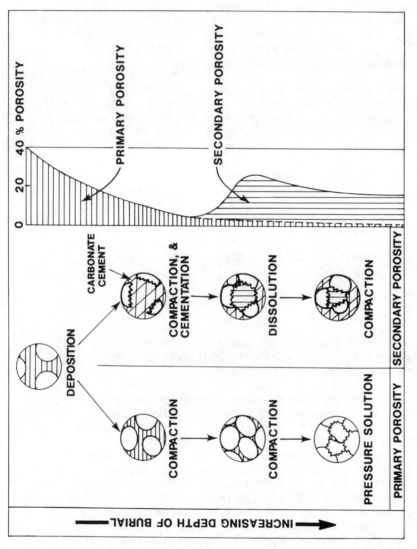

FIG. 2. A conceptual model showing distribution of primary and secondary porosity with depth (partly based on Savkevich, 1969).

constituents, chiefly feldspars, has been alluded to the
large-scale removal of ions such as K^+, Ca^{++}, Fe^{++}, Mg^{++},
and even Si^{4+} during fresh water circulation (Sommer, 1978).
Apparently large-scale ionic migration and water circulation
opened up the reservoir by increasing dissolution porosity.
Percolating meteoric waters, however, will gradually lose
their capacity to generate dissolution porosity as they move
downward, away from the unconformity, because these waters
will approach saturation with respect to dissolved
constituents.

Chemical breakdown of feldspars, particularly
plagioclase, during acid attack by CO_2-charged waters related
to the unconformity is a major cause for creating secondary
porosity. Chemical weathering of feldspars to kaolinite
(Garrels and MacKenzie, 1971) can be expressed by:

1) $2NaAlSi_3O_8 + 2CO_2 + 11H_2O =$
       albite

$Al_2Si_2O_5(OH)_4 + 2Na^+ + 2HCO_3^- + 4H_4SiO_4$
kaolinite

2) $4KAlSi_3O_8 + 4CO_2 + 22H_2O =$
       k-feldspar

$Al_4Si_4O_{10}(OH)_8 + 4K^+ + 4HCO_3^- + 8H_4SiO_4$
kaolinite

These reactions explain the common association of kaolinite
with unconformities (Al-Gailani, 1981) where feldspars are
usually exposed to chemical attack.

Similar to feldspars, chert also undergoes weathering in
proximity to an unconformity (Krumbein, 1942). Swanson (1981)
considers weathered chert as a definite indication of an
unconformity. The process of weathering chert known as
tripolitization has been attributed to a low-pH subaerial
environment (Christopher, 1980). The occurrence of tripolitic
chert in Missouri and Oklahoma has also been explained by
weathering (Keller, 1978). The presence of minute liquid-
filled spherical bubbles in chert is apparently responsible
for its rapid dissolution and for its sponge-like character
(Folk, 1968). Detrital chert is less stable than quartz
because the fine grain size and bubble content present a large
surface area to chemical attack (Folk, 1968).

Weathered chert commonly develops a whitish outer rim
(Folk, 1968). Such tripolitic chert has been observed in the

Prudhoe Bay reservoir (Sadlerochit), Alaska, where dissolution of chert may be attributed to flushing of CO_2-charged meteoric waters originating at the Neocomian unconformity.

RECOGNITION OF SECONDARY POROSITY

Chepikov et al (1961) introduced the first set of criteria for recognizing secondary porosity in sandstones. Their criteria include: (1) corroded grains, (2) adjacent porous and less-porous zones, (3) sinuous pores, (4) remnant cement and (5) inhomogeneity of packing.

Recognition of secondary porosity should be based on multiple evidence as non-fractured secondary pores tend to mimic primary pores. Secondary pores are generally larger in size, more irregular in shape and more random in distribution than primary pores (Schmidt and McDonald, 1979b). An understanding of the origin of pores is critical in recognizing secondary porosity. Schmidt and McDonald (1979b) classified secondary porosity based on origin and on pore texture. Their petrographic criteria for recognition of secondary porosity include: (1) partial dissolution, (2) molds, (3) inhomogeneity of packing, (4) oversized pores, (5) elongate pores, (6) corroded grains (7) intra-constituent pores and (8) fractured grains.

In this paper, a comprehensive set of twenty criteria is proposed for recognition of secondary porosity (Figure 3). Selected examples of grain dissolution are shown by photomicrographs in Figures 4 through 9.

EFFECTS OF GRAIN DISSOLUTION ON SANDSTONE POROSITY AND TEXTURE

Complete dissolution of unstable rock fragments such as chert creates oversized secondary pores as shown by the remnant clay rims (Figure 10). During subsequent compaction, undissolved quartz grains tend to readjust to partially fill the newly developed secondary porosity and to cause a reduction in secondary porosity. These pores often mimic primary pores. Failure to recognize such pores as secondary in origin would lead to underestimation of reservoir quality because extensive dissolution of framework grains often develops "dissolution channels" by connecting adjacent pores which might increase effective porosity.

Complete dissolution of certain framework grains also alters depositional texture. For example, a pebbly sandstone

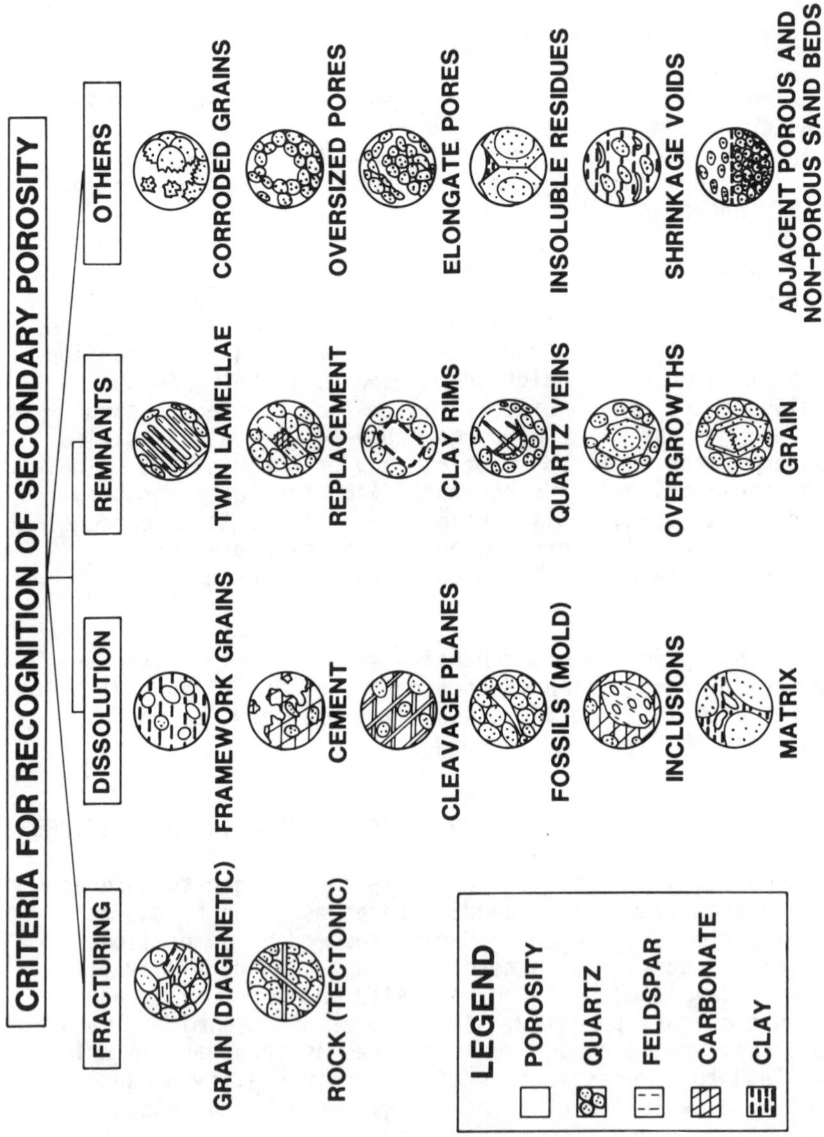

CRITERIA FOR RECOGNITION OF SECONDARY POROSITY

FRACTURING
- GRAIN (DIAGENETIC)
- ROCK (TECTONIC)

DISSOLUTION
- FRAMEWORK GRAINS
- CEMENT
- CLEAVAGE PLANES
- FOSSILS (MOLD)
- INCLUSIONS
- MATRIX

REMNANTS
- TWIN LAMELLAE
- REPLACEMENT
- CLAY RIMS
- QUARTZ VEINS
- OVERGROWTHS
- GRAIN

OTHERS
- CORRODED GRAINS
- OVERSIZED PORES
- ELONGATE PORES
- INSOLUBLE RESIDUES
- SHRINKAGE VOIDS
- ADJACENT POROUS AND NON-POROUS SAND BEDS

LEGEND
- POROSITY
- QUARTZ
- FELDSPAR
- CARBONATE
- CLAY

FIGURE 3.

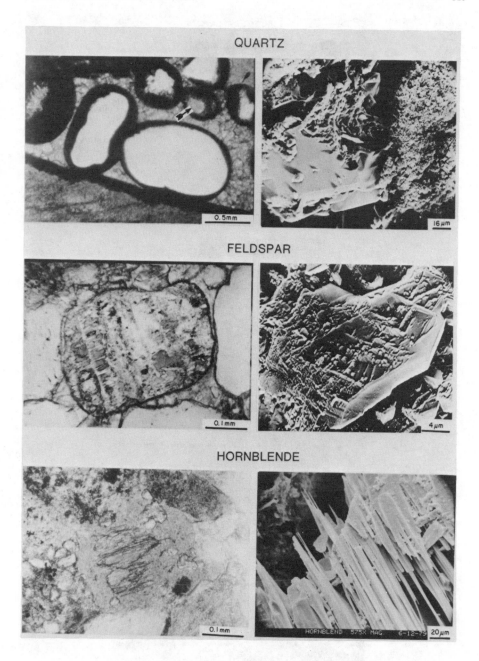

FIG. 4. Dissolution of grains shown by photomicrographs (left) and by SEM (right).

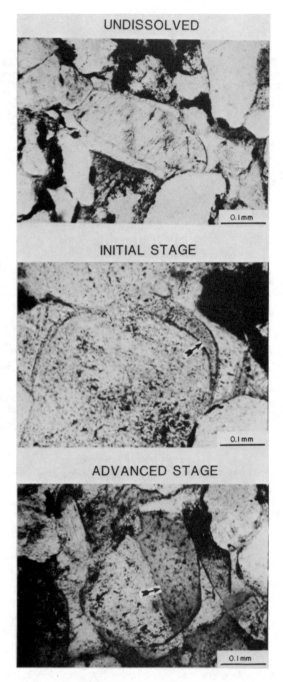

FIG. 5. Dissolution of k-feldspar; arrow shows dissolved area
between the grain and its undissolved overgrowth.

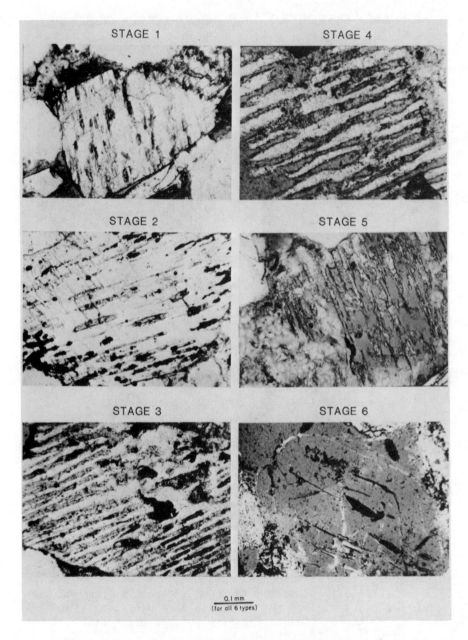

FIG. 6. Progressive stages of feldspar dissolution; stage 6 is an extreme case where the entire grain is dissolved; North Sea reservoirs.

PARTIALLY DISSOLVED
FELDSPAR

COMPLETELY DISSOLVED
FELDSPAR

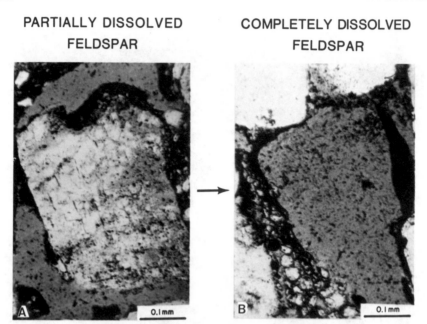

MULTIPLE CLAY RIMS

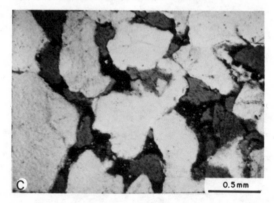

FIG. 7. Recognition of shape and composition of dissolved grains.
A. Partially dissolved feldspar with a clay rim. B. Completely dissolved feldspar with a remnant clay rim.
C. Multiple remnant clay rims suggest dissolution of
several feldspar grains.

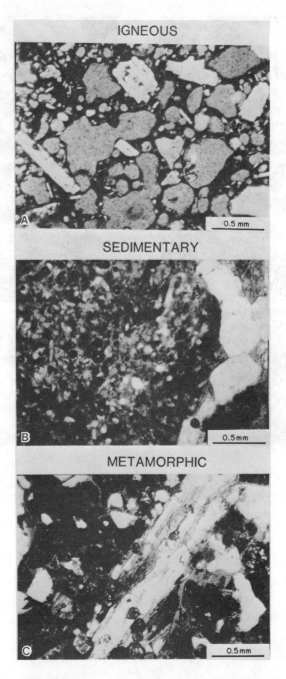

FIG. 8. Dissolution of rock fragments. A. Volcanic glass. B. Siltstone. C. Mica-garnet schist.

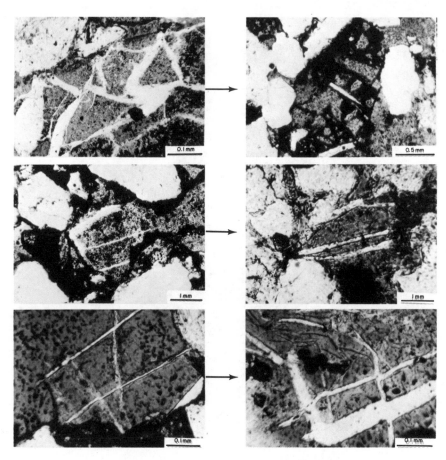

FIG. 9. Dissolution of detrital chert grains; arrows indicate
progressive stages of dissolution; notice undissolved
quartz veins in all cases; compaction of quartz veins
results in their collapse (bottom right); Prudhoe Bay
reservoir, Alaska.

can be modified to a sandstone due to dissolution of unstable
chert pebbles (Figure 10). Such post-depositional modifications of texture (e.g., Mousinho de Meis and Amador, 1974)
could lead to erroneous interpretations regarding hydro-
dynamics of depositional processes, distance of transport, and
source area. Size analyses of sandstones may have an inherent
flaw inasmuch as such analyses do not take dissolution of
framework grains into consideration.

EFFECTS OF GRAIN DISSOLUTION ON SANDSTONE COMPOSITION

Successful interpretation of provenance depends primarily
on our ability to decipher original composition of sandstone
based on framework grains. Quartz, feldspar, and rock fragments (including chert) are the three major types of framework
grains used in classifying sandstones (Folk, 1968). Conventional techniques of point counting framework composition are
accurate only for sandstones in which framework grains are
unaffected by dissolution. In sandstones with dissolved
grains, the following steps are suggested to account for the
effects of dissolution.

(1)   Recognize secondary porosity caused by grain
      dissolution.

(2)   Reconstruct original grain boundaries by using
      criteria such as remnant clay rims.

(3)   Include the dissolved portion of a framework grain
      as a grain, rather than as porosity, while point
      counting.

In sandstones with completely dissolved framework grains,
it is difficult to establish grain composition. Some general
inferences, however, may be made regarding dissolved constituents on the basis of the overall diagenetic history of a
formation.

The effects of grain dissolution on sandstone composition
are hypothetically illustrated in three diagenetic stages in
Figure 11. In the first stage, prior to dissolution of framework grains, the sandstone has more chert than feldspar, and
therefore it should be termed as a lithic sandstone. In the
second stage, when most of the chert grains have been dissolved, the relative abundance of feldspar increases with
respect to chert. The sandstone would now be classified as a
feldspathic sandstone. In the third stage, when most chert
and feldspar grains have been dissolved, the sandstone becomes
enriched in quartz, and it would now be classified as a

EFFECTS OF GRAIN DISSOLUTION ON SANDSTONE POROSITY AND TEXTURE

STAGE	POROSITY	TEXTURE
DEPOSITION 	**PRIMARY**	**PEBBLY SANDSTONE**
DISSOLUTION  OVERSIZED PORE CLAY RIM	**PRIMARY AND SECONDARY**	**SANDSTONE**
COMPACTION 	**PRIMARY AND REDUCED SECONDARY**	**SANDSTONE**
⊙ **QUARTZ**	▦ **CHERT (ROCK FRAGMENT)**	

FIGURE 10.

EFFECTS OF GRAIN DISSOLUTION ON SANDSTONE COMPOSITION

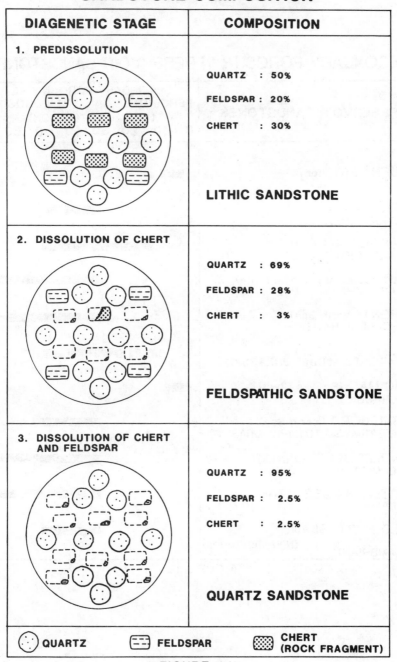

DIAGENETIC STAGE	COMPOSITION
1. PREDISSOLUTION	QUARTZ : 50% FELDSPAR : 20% CHERT : 30% **LITHIC SANDSTONE**
2. DISSOLUTION OF CHERT	QUARTZ : 69% FELDSPAR : 28% CHERT : 3% **FELDSPATHIC SANDSTONE**
3. DISSOLUTION OF CHERT AND FELDSPAR	QUARTZ : 95% FELDSPAR : 2.5% CHERT : 2.5% **QUARTZ SANDSTONE**

QUARTZ   FELDSPAR   CHERT (ROCK FRAGMENT)

FIGURE 11.

SECONDARY POROSITY IN RESERVOIR SANDSTONES

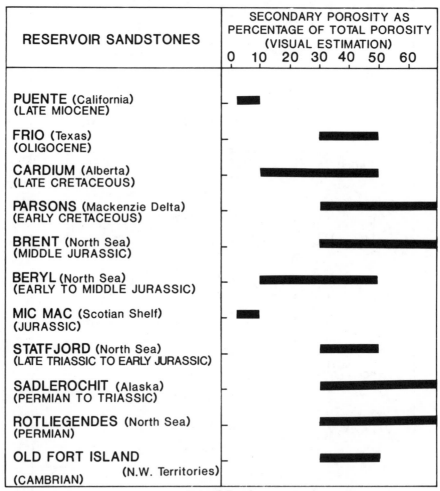

FIGURE 12.

quartz sandstone. Such a proportional increase in stable grains due to early dissolution of unstable grains has been observed by Walker et al (1978) and by Mathisen (1984). It is clear that increasing degree of grain dissolution can drastically modify the "depositional composition" of sandstones. Failure to recognize the diagenetic effects of grain dissolution during petrographic examination of sandstones may result in estimating the "diagenetic composition" of sandstones.

Secondary porosity is the dominant type of porosity in major reservoir sandstones worldwide (Figure 12). In 33 sandstone units, secondary porosity is created primarily by dissolution of framework grains (Shanmugam, 1985). Because grain dissolution appears to control the composition of many sandstones, it is critical to recognize framework dissolution and to estimate the amounts and types of dissolved grains in determining both the "depositional composition" and the provenance of sandstones.

ACKNOWLEDGMENTS

I thank Prof. G. G. Zuffa for inviting me to present this paper at the NATO Conference on "Reading Provenance from Arenites" which was held in Cosenza, Italy, June 3-11, 1984. I also thank Prof. Emiliano Mutti who was primarily responsible for my participation in the Conference. I am grateful to Mobil Research and Development Corporation for permission to publish this work. Drs. D. W. Kirkland and R. J. Moiola kindly reviewed the manuscript. D. L. Guidici typed the manuscript.

REFERENCES

Aguilera, R., 1980, Naturally fractured reservoirs. Tulsa, Oklahoma, Petroleum Publishing Company, 703 p.

Al-Gailani, M. B., 1981, Authigenic mineralizations at unconformities: implication for reservoir characteristics. Sed. Geology, 29, pp. 89-115.

Bjørlykke, K., 1982, Formation of secondary porosity in sandstones: How important is it and what are controlling factors? (abs.). AAPG Bull., 66, pp. 549-550.

Chepikov, K. P., et al., 1961, Corrosion of quartz grains and examples of the possible effect of oil on the reservoir properties of sandy rocks. Doklady of the Academy of Sciences of the U.S.S.R., Earth Science Sections, 140, pp. 1111-1113, (in English).

Choquette, P. W., and L. C. Pray, 1970, Geologic nomenclature
    and classification of porosity in sedimentary carbonates.
    AAPG Bull., 54, pp. 207-250.
Christopher, J. E., 1980, The Lower Cretaceous Mannville Group
    of Saskatchewan - a tectonic over-view. Sask. Geol. Soc.
    Spec. Pub. 5, pp. 3-32.
Folk, R. L., 1968, Petrology of sedimentary rocks. Austin,
    Texas, Hemphill's, 170 p.
Galehouse, J. S., 1971, Point counting in R. E. Carver, ed.,
    Procedures in sedimentary petrology. New York, Wiley -
    Interscience, pp. 385-407.
Garrels, R. M., and F. T. MacKenzie, 1971, Evolution of
    sedimentary rocks. New York, W. W. Norton & Co., Inc.,
    397 p.
Hancock, N. J., and A. M. Taylor, 1978, Clay mineral
    diagenesis and oil migration in the Middle Jurassic Brent
    Sand Formation. Jour. Geol. Soc. London, 135, pp. 69-72.
Hutcheon, I., A. Oldershaw, and E. D. Ghent, 1980, Diagenesis
    of Cretaceous sandstones of the Kootenay Formation at
    Elk Valley (Southeastern British Columbia) and Mt. Allen
    (Southwestern Alberta). Geochim. et Cosmochim. Acta,
    44, pp. 1425-1435.
Keller, W. D., 1978, Textures of tripoli illustrated by
    scanning electron micrographs. Eco. Geology, 73,
    pp. 442-446.
Krumbein, W. C., 1942, Criteria for subsurface recognition of
    unconformities. AAPG Bull., 26, pp. 36-62.
Krynine, P. D., 1947, Petrologic aspects of prospecting for
    deep oil horizons in Pennsylvania. 11th Technical
    Conference on Petroleum Production, Pennsylvania, State
    College, pp. 81-95.
Mathisen, M. E., 1984, Diagenesis of Plio-Pleistocene
    nonmarine sandstones, Cagayan basin, Philippines: Early
    development of secondary porosity in volcanic sandstones.
    AAPG Memoir 37 (in press).
Mousinho de Meis, M. R., and E. S. Amador, 1974, Note on
    weathered arkosic beds. Jour. Sed. Petrology, 44,
    pp. 727-737.
Pittman, E. D., 1979, Porosity, diagenesis and productive
    capability of sandstone reservoirs. SEPM Spec. Pub. 26,
    pp. 159-173.
Savkevich, S. S., 1969, Variation in sandstone porosity in
    lithogenesis (as related to the prediction of secondary
    porous oil and gas reservoirs). Doklady of the Academy
    of Sciences of the U.S.S.R., Earth Science Sections, 184,
    pp. 161-163, (in English).
Schmidt, V., and D. A. McDonald, 1979a, The role of secondary
    porosity in the course of sandstone diagenesis. SEPM
    Spec. Pub. 26, pp. 175-207.

_____ 1979b, Texture and recognition of
  secondary porosity in sandstones. SEPM Spec. Pub. 26,
  pp. 209-225.
Shanmugam, G., 1983, Secondary porosity in sandstones (abs.).
  AAPG Bull., 67, pp. 54.
_____ 1984, Secondary porosity in sandstones: Basic
  contributions of Chepikov and Savkevich. AAPG Bull., 68,
  pp. 106-107.
_____ 1985, Significance of secondary porosity in
  interpreting sandstone composition. AAPG Bull., 69 (in
  press).
Sommer, F., 1978, Diagenesis of Jurassic Sandstones in the
  Viking Graben. Jour. Geol. Soc. London, 135, pp. 63-67.
Swanson, R. G., 1981, Sample examination manual. AAPG Methods
  in Exploration Series, IV-45 p.
Walker, T. R., B. Waugh, and A. J. Grone, 1978, Diagenesis in
  first-cycle desert alluvium of Cenozoic age, southwestern
  United States and northwestern Mexico. Geol. Soc.
  America Bull., 89, pp. 19-32.
Yermolova, Y. P., and Orlova, N. A., 1962, Variation in
  porosity of sandy rocks with depth. Doklady of the
  Academy of Sciences of the U.S.S.R., Earth Science
  Sections, 144, pp. 55-56, (in English).

PROVENANCE OF FELDSPATHIC SANDSTONES - THE EFFECT OF DIAGENESIS
ON PROVENANCE INTERPRETATIONS: A REVIEW

Kenneth P. Helmold

Cities Service Oil and Gas Corporation, Box 3908, Tulsa,
Oklahoma 74102, USA

ABSTRACT

   Chemical composition, zoning, twinning, and structural state
are four intrinsic properties of feldspar that are useful in
deciphering the provenance of sandstones. The utility of feldspar
as a provenance indicator is moderated by its chemical and
mechanical instability in the sedimentary environment. During
weathering, transportation, and burial the processes of mechanical
abrasion, replacement, dissolution, and albitization may effectively modify or remove feldspar from the detritus thereby altering its composition. These modifications must be recognized in
order to correctly decipher the provenance of a sandstone.

INTRODUCTION

   Feldspar comprises 10 to 15 percent of the "average" sandstone and, next to quartz, is the most abundant mineral in sandstones. Largely as a result of this ubiquity, feldspar has long
been used as a provenance indicator for sandstones. Detrital
feldspar, which originally formed under high temperature and
pressure conditions in igneous and metamorphic rocks, is chemically and mechanically unstable in the sedimentary environment.
Consequently, feldspar may be selectively modified or removed from
the detritus during weathering, transportation, and burial, resulting in a decrease of its effectiveness as a provenance indicator. Because the effects of transportation and diagenesis on
feldspar are often subtle and easily overlooked or misinterpreted,
misleading or erroneous conclusions regarding the source area may
result.

G. G. Zuffa (ed.), Provenance of Arenites, 139–163.
© *1985 by D. Reidel Publishing Company.*

The purpose of this paper is twofold. The first is to review those chemical and physical properties of feldspar that make it useful as a provenance indicator. Chemical composition is given particular emphasis in light of recent electron microprobe studies (Trevena and Nash, 1981; Putnam and Pedskalny, 1983) which have proven the utility of microanalyses in the delineation of possible source rocks. The second purpose is to stress the effect of diagenesis on the interpretation of feldspathic-sandstone provenance. Understanding the possible diagenetic modifications of detrital feldspar is important in making viable interpretations regarding provenance of feldspathic sandstones.

FELDSPAR AS A PROVENANCE INDICATOR

Feldspar is a useful provenance indicator because it is abundant in many sedimentary basins and because the variations in its physical and chemical properties have genetic implications. Feldspar which formed in the plutonic regime, for example, has inherent physical and chemical properties which are different from those of feldspar which formed in volcanic and metamorphic environments. Documentation of the physical and chemical properties of feldspar from particular geologic environments is voluminous (Van der Plas, 1966; Barth, 1969; Smith, 1974) and is beyond the scope of this paper. Emphasis in this paper is given to those properties which are most useful to the sedimentary petrologist interested in deciphering the provenance of sandstones. Because the detrital feldspar grains in a sandstone may have been derived from a number of different sources, the properties of an individual feldspar grain are useful only in determining the provenance of that particular grain. This results in the determination of feldspar provenance in sandstones being made on a grain-by-grain basis. The diagnostic properties must, therefore, be easily determined in order to be useful. They include chemical composition, zoning, twinning, and structural state.

Chemical Composition of Feldspar

In the past, feldspar compositions were determined by tedious wet chemical methods that were too time consuming to be of practical use to the sedimentary petrologist. The application of the electron microprobe in the 1960's made possible the rapid and precise chemical analysis of individual feldspar grains. Trevena and Nash (1981) have compiled over 5000 microprobe analyses of feldspar from crystalline rocks. Those analyses indicate that potassium content is a useful parameter for determining the origin of both plagioclase and alkali feldspars. There is a gradual decrease in the maximum potassium content of plagioclase from volcanic to plutonic to metamorphic environments (Fig. 1).

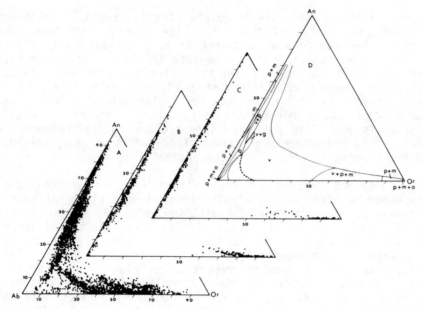

Figure 1. Composition of feldspar (weight %) in igneous and metamorphic rocks as determined by electron microprobe analysis. A) Composition of feldspar from volcanic rocks. B) Composition of feldspar from plutonic rocks. C) Composition of feldspar from metamorphic rocks. D) Compositional range of eight provenance groups of feldspar. v = volcanic; p = plutonic; m = metamorphic; v+g = volcanic or granophyre; v+p = volcanic or plutonic; p+m = plutonic or metamorphic; v+p+m = volcanic, plutonic, or metamorphic; p+m+a = plutonic, metamorphic, or authigenic (from Trevena and Nash, 1981).

Due to the subtle nature of this trend, however, the fields of high-K volcanic and plutonic plagioclase partially overlap, and those of low-K metamorphic and plutonic plagioclase are also similar. Volcanic alkali feldspar varies widely in potassium content with its composition ranging from $Ab_{74}An_{13}Or_{13}$ to $Ab_{12}An_1Or_{87}$, while alkali feldspar from both plutonic and metamorphic rocks are potassium rich (Trevena and Nash, 1981). Plutonic alkali feldspar ranges in composition from $Ab_{43}An_1Or_{56}$ to $Ab_2An_0Or_{98}$ and alkali feldspar from metamorphic rocks ranges from $Ab_{47}An_2Or_{51}$ to $Ab_2An_0Or_{98}$ (Trevena and Nash, 1981).

Trevena and Nash (1981) have delineated eight provenance groups in the albite-anorthite-orthoclase ternary diagram which define the compositional range of feldspar from the various crystalline environments (Fig. 1). These groups can be used as a guide for determining the source of detrital feldspar from sandstones. This is particularly helpful for feldspathic sandstones of mixed provenance. For example, the Chuska Sandstone from northwest New Mexico and northeast Arizona contains both high-K and low-K plagioclase, and intermediate (50 to 85% Or) and potassium-rich (>90% Or) alkali feldspar (Trevena and Nash, 1981). Trevena and Nash (1981) interpret the high-K plagioclase to be mostly of volcanic origin and the low-K plagioclase to be derived from plutonic and/or metamorphic rocks. The potassium-rich alkali feldspar is most likely derived from plutonic and/or metamorphic rocks, while the intermediate alkali feldspar is probably volcanic and/or plutonic in origin (Trevena and Nash, 1981).

Sibley and Pentony (1978) utilized feldspar compositions, determined by microprobe analysis, to delineate different source terranes for sediment from DSDP drilling sites 299 and 301 in the Sea of Japan. Sibley and Pentony (1978) concluded that feldspar from site 299 was derived from the Japanese volcanic arc and was transported to the depositional site via submarine canyons. This detritus contains a greater proportion of plagioclase than that from site 301. In addition, the plagioclase from site 299 has a higher An content (40% vs. 29%) which is a direct reflection of the composition of the andesites comprising the arc (Sibley and Pentony, 1978). Sediment from site 301 is a mixture of detritus shed from andesites of the arc and granitic rocks on the Asian continent. This detritus contains a greater proportion of quartz and K-feldspar and is indicative of an acidic plutonic source (Sibley and Pentony, 1978).

Feldspar Zoning

Chemical zoning is well documented in both plagioclase and alkali feldspars (Smith, 1974); however, only zoning in plagioclase has received much attention as a possible provenance indicator. Though not always definitive, it has proved to be a useful property. Pittman (1963) examined 90 plutonic, volcanic, hypabyssal, and metamorphic rocks in order to determine the relative abundance of progressively-zoned, oscillatory-zoned, and unzoned plagioclase in each. The results of this study (Fig. 2) show that the presence of any type of zoning is a strong indicator of an igneous origin for plagioclase. In particular, oscillatory zoning suggests either a volcanic or hypabyssal source. Conversely, the absence of oscillatory zoning does not rule out a volcanic or hypabyssal source because unzoned plagioclase is abundant in some volcanic and hypabyssal rocks (Pittman, 1963). Progressive zoning is equally abundant in all the igneous rocks examined, and therefore is of no use in distinguishing plutonic from volcanic and hypabyssal plagioclase.

Metamorphic rocks are characterized by the near absence of zoned plagioclase (Pittman, 1963). Care must be exercised when

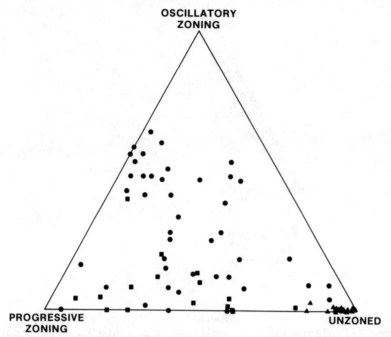

Figure 2. Proportion of progressively-zoned, oscillatory-zoned, and unzoned plagioclase in volcanic and hypabyssal rocks (dots), plutonic rocks (squares), and metamorphic rocks (triangles); (from Pittman, 1963).

interpreting the origin of unzoned plagioclase in a sedimentary rock, particularly a fine-grained one, because detrital plagioclase grains are commonly only fragments of the original crystals. A coarsely-zoned igneous plagioclase may, upon transportation and abrasion, yield several unzoned plagioclase grains which may be misinterpreted as metamorphic in origin.

Trevena and Nash (1981) obtained similar results in an electron microprobe study of detrital plagioclase in sandstones. Most moderately- to strongly-zoned plagioclase belongs to the high-K volcanic-plutonic trend, while plagioclase of the low-K metamorphic-plutonic trend is either weakly zoned or unzoned. All oscillatory-zoned plagioclase plots on the high-K volcanic-plutonic trend, with most in the field of volcanic plagioclase. Plagioclase in the low-K metamorphic-plutonic field is either unzoned or progressively zoned (Trevena and Nash, 1981).

Feldspar Twinning

Twinning is the physical property perhaps most commonly used

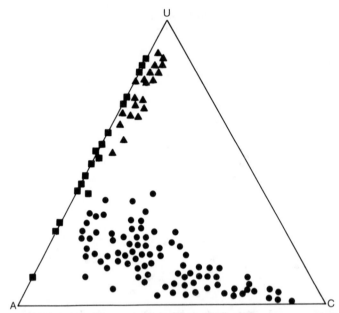

Figure 3.   Proportion of untwinned (U), C-twinned (C), and A-twinned (A) plagioclase in volcanic and plutonic rocks (dots), schists and gneisses (squares), and hornfels (triangles); (from Gorai, 1951).

to identify feldspars.   Correlation of twinning with host rock lithology has been attempted by many workers (Gorai, 1951; Turner, 1951; Tobi, 1962).   Most of these studies have dealt exclusively with the plagioclase series. Gorai (1951) determined the twinning laws of over 1500 plagioclase twins from various plutonic, volcanic (including hypabyssal), and metamorphic (including schists, gneisses, and hornfels) rocks. He divided the twins into A-twins, which include the albite, pericline, and acline laws, and C-twins, which essentially include all the other laws. A-twins occur in all lithologies, but are usually less abundant in hornfelsic rocks. C-twins are restricted to volcanic and plutonic rocks, except for hornfelses in which they occur in only minor amounts (Fig. 3).

The relative frequency of twinned and untwinned plagioclase in both plutonic and volcanic rocks varies with composition. C-twins are more abundant in calcium-rich plagioclase, particularly that more calcic than An_{50} (Gorai, 1951). Both A-twins and untwinned crystals are more abundant in sodic plagioclase. Smith (1974) points out there is essentially no difference in twin frequencies between volcanic and plutonic rocks.

Table 1. Plagioclase twin and zone types in selected sample from Mohawk Lake beds (Pleistocene),
northern Sierra Nevada Range, California (from
Pittman, 1970).

	5% Albite	76% Oligo- clase and Andesine	1% Plagio- clase $>An_{50}$
A-twins	28%	29%	30%
C-twins	20	25	23
Untwinned	48	43	47
Untwinned-unzoned	4	3	0
	100%	100%	100%
Progressive zoning	79%	68%	70%
Oscillatory zoning	21	32	30
	100%	100%	100%

Similar trends do not appear to exist in metamorphic plagioclase. Gorai (1951) did not detect any variation in the frequency
of twinned plagioclase with composition in schists and gneisses.
However, both Gorai (1951) and Turner (1951) did note the increase
in frequency of twinned plagioclase with increasing grain size.

In a provenance study of the Mohawk Lake beds (Pleistocene)
of northern California, Pittman (1970) recorded the relative
frequency of both zoned and twinned plagioclase. This unit was
chosen for study because the source terrane for the sediments
consists of andesite and minor rhyolite, basalt, metamorphic, and
granitic rocks. The results of his study are shown in Table 1.
The occurrence of C-twinned plagioclase in the sands is consistent
with an igneous source. Oscillatory zoning in plagioclase is
also indicative of volcanic or hypabyssal source rocks. Pittman
(1970) noted that the near absence of untwinned-unzoned plagioclase rules out a major metamorphic source.

Twinning of K-feldspar has been most often utilized in
provenance studies to distinguish microcline-bearing rocks from
those rich in orthoclase or sanidine. Microcline is almost
always twinned according to the albite or pericline laws, and most
often exhibits both types of twinning to form the well-known
cross-hatch twins (Deer et al., 1962). This is best seen in thin
sections cut roughly parallel to both composition planes, while
in other crystal orientations only albite or pericline twinning
may be observed. Cross-hatch twinning is an unequivocal indicator

of microcline, but its absence does not necessarily imply that the K-feldspar is not microcline. In both orthoclase and sanidine, Carlsbad twins occur most frequently, although Baveno and Manebach twins may occasionally be found (Deer et al., 1962).

Plymate and Suttner (1983) used the frequency of cross-hatch twinning in K-feldspar to determine the provenance of Holocene stream alluvium from southwest Montana. The Radar Creek Pluton, a granodiorite body within the Boulder Batholith, was chosen as the study site because the aerial distribution of microcline- and orthoclase-bearing rocks is known (Tilling, 1968). Additional sampling of the basement rocks by Plymate and Suttner (1983) confirmed that distribution (Fig. 4A). Sand samples from a stream draining the basement rocks were analyzed petrographically to determine the frequency of cross-hatch twinning in the detrital K-feldspar. The distribution of cross-hatch twinned K-feldspar in the alluvium (Fig. 4B) very clearly reflects the distribution of microcline-rich rocks in the source terrane (Plymate and Suttner, 1983). Local fluctuations in the overall trend demonstrate the influence that tributaries upstream from sampling sites have on concentrating or diluting the microcline population. Plymate and Suttner (1983) concluded that optical determination of the frequency of cross-hatch twinning is a quick and efficient method for correlating K-feldspar-bearing sandstones with their source rocks.

Structural State of Feldspar

Structural state determination of plagioclase is of little practical use to the sedimentary petrologist as a provenance indicator because of the difficulty in obtaining data. The optical orientation of plagioclase may yield information as to structural state of albite and oligoclase, but is of little value for more calcic plagioclase (Hutchison, 1974). Standard X-ray diffraction techniques alone are not sufficient to determine structural state. Knowledge of composition in conjunction with X-ray diffraction data may yield structural state determinations (Bambauer et al., 1967), but chemical analysis is time consuming unless electron microprobe techniques are employed. No study utilizing structural state of plagioclase as a provenance indicator is known to the author.

The structural state of alkali feldspar has received more attention as an indicator of provenance, largely because these data are easily acquired using standard X-ray diffraction techniques (Wright, 1968; Suttner and Basu, 1977). The degree of Al-Si ordering in the tetrahedral (T) sites controls the structural state of the feldspar. If Al is randomly distributed among the four possible T sites [$t_1(o)$, $t_1(m)$, $t_2(o)$, $t_2(m)$], the feldspar is totally disordered (e.g., sanidine and high-albite;

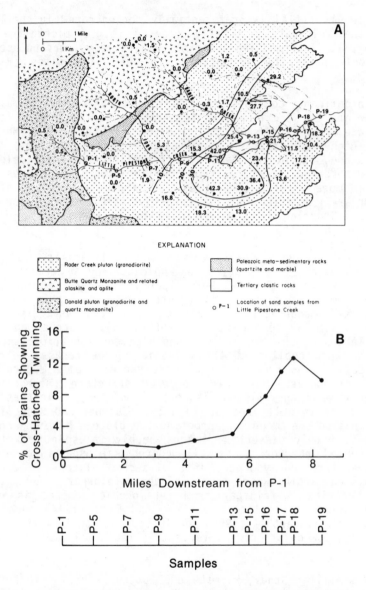

Figure 4. Use of cross-hatch twinning in K-feldspar as a prove-nance indicator. A) K-feldspar variation in the Rader Creek Pluton and adjacent intrusive bodies determined optically. Values shown are the percentages of K-feldspar in each sample with cross-hatch twinning. B) K-feldspar variation in Little Pipestone Creek (shown in A) determined optically. Values shown are the percentages of K-feldspar (35-80 mesh size) in each sample with cross-hatch twinning (from Plymate and Suttner, 1983).

Barth, 1969). If Al is preferentially concentrated in the $T_1(o)$ sites, the feldspar is totally ordered (e.g., microcline and low-albite). Intermediate ordering results (e.g., orthoclase) if Al is only partially concentrated in the $T_1(o)$ sites (Barth, 1969).

The dominant controls on the structural state of alkali feldspar are the equilibration temperature of the atomic structure, rate of subsolidus cooling, and the activity of H_2O (Martin, 1974). High equilibration temperatures result in more Al-Si disorder and, therefore, favor the formation of alkali feldspar with a disordered structure. Low equilibration temperatures will result in more ordered structures. Slow rate of cooling allows time for Al-Si ordering and, consequently, results in well-ordered feldspar. Ordering is also enhanced by an adequate supply of water (Martin, 1974).

Suttner and Basu (1977) tested the applicability of using structural data as a provenance indicator by determining the structural state of detrital K-feldspar in Holocene fluvial sands derived from known plutonic, volcanic, and metamorphic sources. Detrital K-feldspar derived from volcanic sources has a disordered structure, similar to high sanidine (Fig. 5). Suttner and Basu (1977) interpret this to reflect the high equilibration temperature and rapid cooling of the volcanic source rocks. Detrital K-feldspar derived from plutonic rocks varies from a moderately-ordered (orthoclase) to a well-ordered structure (microcline). Feldspar from younger plutons is more disordered, while that from older plutons is well ordered (Fig. 5). Suttner and Basu (1977) relate this to the amount of unroofing the plutons have undergone. The older, highly dissected plutons supply feldspar from their slowly cooled interiors where well-ordered K-feldspar is abundant. K-feldspar from the younger, less dissected plutons is derived from chilled margins where more disordered feldspar is prevalant. All the detrital K-feldspar from metamorphic sources is well-ordered maximum microcline. Suttner and Basu (1977) note that this agrees with Barth's (1969) conclusion that maximum microcline is the most common structural variant in regionally metamorphosed rocks.

In a similar study, Plymate and Suttner (1983) compared the structural state of detrital K-feldspar in Holocene alluvium with the known aerial distribution of plutonic source rocks (Fig. 6A). They observed a downstream increase in obliquity (degree of ordering) of the detrital K-feldspar that reflects the transition from orthoclase-rich to microcline-rich rocks in the source area (Fig. 6B). They concluded that structural data for locally derived detrital K-feldspar could be used to detect variations in source rock lithology. Plymate and Suttner (1983) also noted that, even though structural information obtained via X-ray dif-

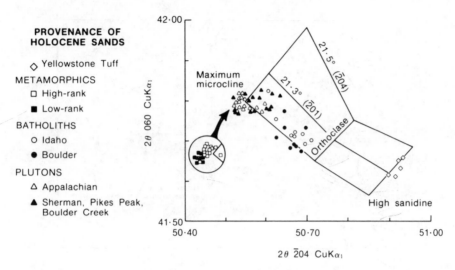

Figure 5. Structural state of K-feldspar in Holocene sands derived from known igneous and metamorphic rocks (from Suttner and
Basu, 1977).

fraction techniques is more precise, optical determination of
structural state is more time efficient.

DIAGENETIC MODIFICATION OF FELDSPAR

Implicit in all provenance interpretations of sandstones is
the assumption that detrital modes accurately reflect the composition of the source rocks. Any physical or chemical process that
alters the composition of the detritus during weathering, transportation, or burial can result in misleading or erroneous provenance interpretations. Many, but not all, diagenetic processes
active in sandstones result in increased mineralogic maturity
(quartz content) of the detritus. Selective removal of labile
components (including feldspar) by processes such as weathering,
granular disintegration, and subsurface dissolution makes accurate
assessment of provenance increasingly difficult. The ultimate
product of these processes is the creation of quartz-rich sandstones. Provenance interpretations for such sandstones can be
very difficult and somewhat controversial (Blatt and Christie,
1963; Basu et al., 1975). Other diagenetic processes (e.g.,
albitization) may not increase mineralogic maturity, but may
still hinder determination of the original source rocks. The
diagenetic processes that are particularly effective in modifying
feldspar include mechanical abrasion, replacement, dissolution,
and albitization.

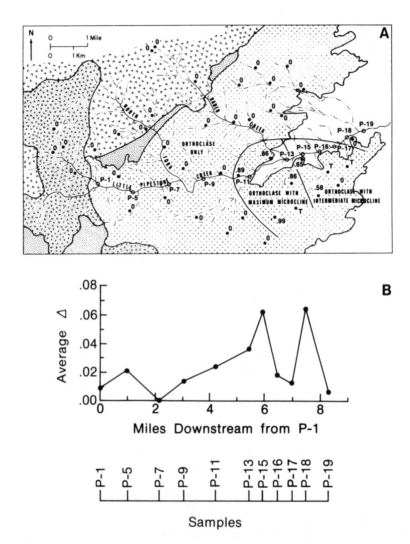

Figure 6. Use of the structural state of K-feldspar as a provenance indicator. A) K-feldspar variation in the Rader Creek
Pluton and adjacent intrusive bodies determined by X-ray diffraction methods. Numbers shown are the Δ value (obliquity) of
the triclinic phase. 0 = orthoclase; T = orthoclase with a trace
of microcline. See Fig. 4A for explanation of rock types. B)
K-feldspar variation in Little Pipestone Creek (shown in A)
determined by X-ray diffraction methods. Numbers shown are the
average Δ value (obliquity) for 30 individual K-feldspar grains
(5-16 mesh size) from each sample (from Plymate and Suttner,
1983).

Mechanical Abrasion of Feldspar

Mechanical abrasion can reduce the grain size and, hence, alter the frequency of feldspar grains that exhibit certain types of twinning or zoning. For example, Pittman (1969) documented the destruction of plagioclase twins by stream transport. Samples of Holocene alluvium from the Merced River, California, were systematically collected downstream from granodiorite outcrops of the Sierra Nevada Batholith. The relative frequencies of A-twinned, C-twinned, and untwinned plagioclase were determined for the fine-, medium-, and coarse-sand fractions. The data show an increase in both the A-twin/C-twin and untwinned/twinned ratios with distance of transport for all size fractions (Pittman, 1969). Pittman thought that the selective destruction of C-twins was the main cause for the downstream increase of the A-twin/C-twin ratio. He proposed that plagioclase grains twinned according to the Albite-Carlsbad A law might break along the 010 cleavage which is parallel to the composition plane of the Carlsbad twin. This would produce two daughter A-twins from the parent C-twin. Similarly, Pittman (1969) concluded that the downstream increase in the untwinned/twinned ratio resulted from breakage of A- and C-twins along their composition planes, thereby yielding untwinned grains.

The increase in the A-twin/C-twin and untwinned/twinned ratios resulting from transportation can influence provenance interpretation that are based on twin types. For example, detritus derived from igneous (volcanic and/or plutonic) sources should contain both A- and C-twinned plagioclase (Fig. 3). If transportation is sufficient to selectively destroy all, or most, of the C-twins, the remaining detritus would be rich in A-twinned and untwinned plagioclase. Due to the absence of C-twins, the source terrane of the transported detritus would probably be interpreted to consist of metamorphic rocks (Fig. 3). A high untwinned/ twinned ratio would falsely suggest hornfelsic rocks as the source.

Although not documented in the literature, one might expect that the ratio of unzoned/zoned plagioclase would show a similar increase with distance of transport. Mechanical abrasion of coarsely-zoned plagioclase should concentrate unzoned grains in the finer size fractions. This would have the same effect on the evaluation of provenance as the destruction of plagioclase twins. Plagioclase in igneous (volcanic and plutonic) rocks are characterized by an abundance of zoning. If stream transport selectively removes zoned plagioclase, the resulting detritus will be enriched in unzoned plagioclase. Based on the absence of zoned plagioclase, the source terrane of the detritus would probably be interpreted to consist of metamorphic rocks.

Table 2. Distribution of altered feldspar and diagenetic
clay in siliciclastic sandstones (from Wilson and Tillman,
1974).

	Number of Samples	Number (percent) Containing Diagenetically Altered Feldspar	Number (percent) Containing Diagenetic Clay
Tertiary	225	142 (63.2)	201 (89.4)
Mesozoic	329	237 (72.1)	310 (94.3)
Paleozoic	231	176 (76.2)	205 (88.8)
Total	785	555 (70.7)	716 (91.3)

Replacement and Dissolution of Feldspar

        Calcic plagioclase and K-feldspar are chemically unstable in
the sedimentary environment, and are, therefore, susceptable to
diagenetic modifications. Although the results of these modifications have long been recognized in sandstones (Gilbert, 1949;
Coombs, 1954), their widespread occurrence has only recently
been appreciated. Wilson and Tillman (1974) documented the
relative abundance of altered feldspar and diagenetic clay in 785
Phanerozoic sandstones. An average of 70 percent of the sandstones contain altered feldspar, with a tendency for altered
feldspar to be more common in the older sandstones (Table 2).
Approximately 90 percent of the sandstones contain at least minor
amounts of authigenic clay. Although most of this clay lines or
fills intergranular pores, some of it occurs as a replacement of
detrital feldspar.

        Wilson and Tillman (1974) distinguished feldspar that is
replaced by authigenic clay from that which has undergone intrastratal dissolution. They noted that both types of alteration
may affect the classification and, hence, interpretation of
provenance for the sandstones. An example of this is provided by
the Norphlet Formation (Jurassic) of southern Mississippi (Wilson
and Pittman, 1977). The Norphlet sandstones from one well (well
B in Fig. 7A) contain 5 to 10 percent unaltered K-feldspar, but
no plagioclase. Angular clasts that consist mainly of illite
comprise roughly 7 percent of the framework grains. They were
interpreted as completely altered plagioclase grains based on the
alignment of illite parallel to presumed cleavage, angular nature
of the clasts, and the occurrence of quartz-K-feldspar-illite
clasts that resemble plutonic rock fragments (Wilson and Pittman,

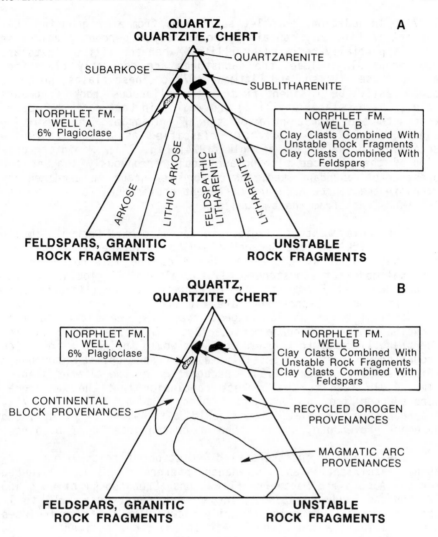

Figure 7. Replacement of detrital plagioclase can affect both
the classification and inferred provenance of a sandstone. An
example of plagioclase replacement by clay is from the Norphlet
Formation of southern Mississippi. (A) Classification of sandstone changes from subarkose to sublitharenite if the altered
plagioclase is not correctly identified (modified from Wilson and
Tillman, 1974). (B) Interpretation of tectonic setting changes
from a continental block provenance to a recycled orogen provenance if the altered plagioclase is not correctly identified
(provenance fields from Dickinson and Suczek, 1979; data from
Wilson and Tillman, 1974).

1977). In addition, Norphlet sandstones from a neighboring well (well A in Fig. 7A) contain approximately 6 percent plagioclase that is partially replaced by illite. When the illite clasts are counted as plagioclase, the sandstones are correctly classified as subarkoses (Wilson and Pittman, 1977). These clasts, however, could easily be misidentified as argillaceous rock fragments (i.e., shale, slate, phyllite) and would, in that case, be apportioned to the lithic pole of the QFL diagram. The resulting sandstone would be incorrectly classified as a sublitharenite according to the scheme of Folk (1974; Fig. 7A). Misinterpretation of the replaced plagioclase could erroneously suggest the presence of abundant sedimentary or low-grade metasedimentary rocks in the source area and different provenances for the Norphlet sandstones from the two wells.

The replacement of plagioclase can also have tectonic implications. Dickinson and Suczeck (1979) demonstrated that the tectonic setting of a sedimentary basin can be inferred from the detrital modes of sandstones. If the altered plagioclase grains in the Norphlet Formation are interpreted as argillaceous rock fragments and the sandstones are classified as sublitharenites, the detrital modes would suggest that they were derived from recycled orogenic terrane (Fig. 7B). In particular, the low chert/argillaceous rock fragment ratio would indicate a collision orogen provenance, or possibly a foreland uplift provenance (Dickinson and Suczek, 1979). Recognition of the altered plagioclase would lead to the correct conclusion that the sandstones were sourced from a continental block (Badon, 1975; Fig. 7B). The abundance of plagioclase would also suggest an uplifted basement provenance, rather than a craton interior provenance.

The effect of intrastratal dissolution of feldspar on sandstone classification and provenance interpretations is documented for the Almond and Mesaverde Formations (Cretaceous) of Colorado (Wilson and Tillman, 1974). These sandstones contain 10 to 17 percent feldspar which is primarily microcline, but also includes minor plagioclase and orthoclase. In addition, sedimentary rock fragments, primarily chert and dolomite, occur in variable amounts. These sandstones are classified as subarkoses according to the scheme of Folk (1974). Approximately one-third of the feldspar is relatively fresh, with the remainder having undergone fairly extensive dissolution (Wilson and Tillman, 1974). The partially dissolved feldspar might be misidentified as silt-sized quartz grains. Completely dissolved grains might simply be counted as porosity. In either case, the detrital modes would be enriched in quartz. While the Mesaverde sandstones would still be classified as subarkoses, they would plot much closer to the Q-pole of the QFL diagram (Fig. 8A). The Almond sandstones would be reclassified primarily as sublitharenites and would also plot closer to the Q-pole (Fig. 8A).

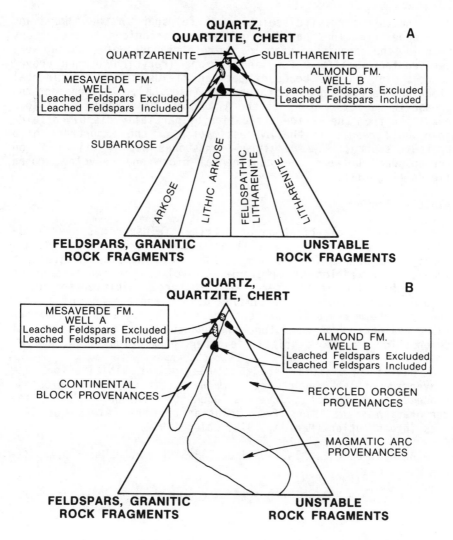

Figure 8. Dissolution of detrital plagioclase can affect both the classification and inferred provenance of a sandstone. An example is from the Cretaceous Mesaverde and Almond Formations of Colorado. (A) Classification of the Almond sandstones changes from subarkoses to sublitharenites if the altered plagioclase is not correctly identified (modified from Wilson and Tillman, 1974). (B) Interpretation of tectonic setting for the Almond sandstones changes from a continental block provenance to a recycled orogen provenance if the altered plagioclase is not correctly identified (provenance fields from Dickinson and Suczek, 1979; data from Wilson and Tillman, 1974).

The intrastratal dissolution of feldspar in the Almond and Mesaverde Formations has less effect on tectonic interpretations than in the previous example. These sandstones, which were derived from uplifted foreland terrane (recycled orogen provenance), plot on the boundary between fields of the continental block and recycled orogen provenances (Fig. 8B). The apparent increase in quartz content of the Almond and Mesa Verde Formations resulting from the misidentification of partially dissolved feldspar would result in the overestimation of the importance of a cratonic source. The resulting sandstones would still plot on the boundary between the continental block and recycled orogen fields (Fig. 8B).

Albitization

Diagenetic albitization of detrital feldspar has been documented for both plagioclase (Land and Milliken, 1981; Boles, 1982; Helmold and van de Kamp, 1984) and K-feldspar (Walker, 1984). Albitization of plagioclase involves the replacement of Ca^{+2} by Na^+ ions in the feldspar structure. Albitization of K-feldspar may proceed by a similar direct replacement or it may involve intermediate replacement steps (Walker, 1984). In either case, evaluation of sandstone provenance based on the chemical composition of the detrital feldspar population is hindered.

During the initial stage of plagioclase albitization, the conversion of the anorthite component of calcic plagioclase proceeds as an equal volume replacement reaction with the incorporation of Na^+ ions from solution and the release of Ca^{+2} ions into solution (Merino, 1975; Boles, 1982):

$$0.8NaAlSi_3O_8 \cdot 0.2CaAl_2Si_2O_8 + 0.203H_4SiO_4 + 0.201Na^+ + 0.796H^+ =$$
(oligoclase)

$$1.001NaAlSi_3O_8 + 0.2Ca^{+2} + 0.199Al^{+3} + 0.804H_2O \quad (1)$$
(albite)

In some cases the reaction proceeds to completion in this manner, but more commonly it results in other mineral by-products such as laumontite, calcite, and kaolinite.

In those sandstones which contain laumontite, calcium-bearing plagioclase is partly albitized and partially replaced by laumontite according to the reaction (Boles and Coombs, 1977):

$$NaAlSi_3O_8 \cdot CaAl_2Si_2O_8 + 2SiO_2 + 4H_2O =$$
(plagioclase)   (quartz)

$$NaAlSi_3O_8 + CaAl_2Si_4O_{12} \cdot 4H_2O \quad (2)$$
(albite)   (laumontite)

Boles and Coombs (1977) calculated that for a given volume of plagioclase altered, an equal volume of albite-laumontite plus additional laumontite will be produced to replace grains and fill pores. They proposed an additional reaction, especially for large volumes of sandstones in which albitization is complete, that necessitates the supply of Na^+ from solution and the release of Ca^{+2} ions into solution:

$$NaAlSi_3O_8 \cdot CaAl_2Si_2O_8 + 3SiO_2 + 2H_2O + Na^+ =$$
$$\text{(plagioclase)} \qquad \text{(quartz)}$$

$$2NaAlSi_3O_8 + 0.5CaAl_2Si_4O_{12} \cdot 4H_2O + 0.5Ca^{+2} \qquad (3)$$
$$\text{(albite)} \qquad \text{(laumontite)}$$

Boles (1982) noted the presence of authigenic calcite within many albitized grains from the Frio Formation of the Gulf Coast, and speculated that it formed from Ca^{+2} released during albitization in combination with carbonate ions from pore fluids according to the reaction:

$$NaAlSi_3O_8 \cdot CaAl_2Si_2O_8 + 4SiO_2 + 2Na^+ + CO_3^{-2} =$$
$$\text{(plagioclase)} \qquad \text{(quartz)}$$

$$3NaAlSi_3O_8 + CaCO_3 \qquad (4)$$
$$\text{(albite)} \quad \text{(calcite)}$$

Authigenic kaolinite is also apparently associated with albitization in the Frio Sandstone (Lindquist, 1977; Boles, 1982). Boles (1982) pointed out that kaolinite probably is a sink for aluminum released during the albitization process and suggested the reaction:

$$2SiO_2 + 0.5H_2O + H^+ + Na^+ + CaAl_2Si_2O_8 =$$
$$\text{(quartz)} \qquad\qquad \text{(anorthite)}$$

$$NaAlSi_3O_8 + 0.5Al_2Si_2O_5(OH)_4 + Ca^{+2} \qquad (5)$$
$$\text{(albite)} \qquad \text{(kaolinite)}$$

The diagenetic albitization of K-feldspar may proceed by either direct replacement or by one of several intermediate replacement steps (Walker, 1984). In the simplest case, K-feldspar is replaced by albite with the receipt of Na^+ ions from solution and the release of K^+ ions into solution (Land and Milliken, 1981):

$$KAlSi_3O_8 + Na^+ = NaAlSi_3O_8 + K^+ \qquad\qquad (6)$$
$$\text{(K-feldspar)} \qquad \text{(albite)}$$

This type of replacement produces chemically pure albite grains that commonly display "chessboard" twinning and that are pseudo-morphs of the original K-feldspar grain (Walker, 1984).

Walker (1984) documented the albitization of K-feldspar through an intermediate replacement step. Initially the K-feld-spar grain is almost completely replaced by either anhydrite, calcite, or dolomite. Later this replacement product is in turn replaced by albite. Alignment of the new albite crystals sug-gests that nucleation occurred on remnants of K-feldspar that survived the initial replacement. The resulting albite grain may also exhibit characteristic "chessboard" twinning.

The albitization reaction is important not only because of the exchange of the major cations (i.e., Ca^{+2}, Na^+, K^+), but also because of the increase in chemical purity of the resultant al-bite. Electron microprobe studies of albitized plagioclase from the Frio Formation (Land and Milliken, 1981; Boles, 1982) and Eocene sandstones from the Santa Ynez Mountains, California (Helmold and van de Kamp, 1984), show that diagenetic albite is very pure, usually containing less than one mole percent each of anorthite and orthoclase. Analyses of albitized K-feldspar grains show them to be equally pure albite (Walker, 1984). The increased chemical purity that results from albitization, in particular the loss of calcium and potassium, precludes the use of albitized feldspar as a provenance indicator in the manner described by Trevena and Nash (1981). Compositionally, authigenic albite plots in the same field as some plutonic and metamorphic plagioclase (Fig. 1). Failure to recognize the existance of albitized feldspar in a sandstone may result in interpretation of the source area to consist of acidic plutonic and albite-bearing metamorphic rocks.

The Charney Formation (Cambrian) from near Quebec, Canada, is a good example of how albitization of detrital K-feldspar has complicated the interpretation of provenance. Middleton (1972) reported that feldspar in the Charney sandstones consisted en-tirely of sodic plagioclase (albite and oligoclase) to the ex-clusion of K-feldspar. Based on remnant film perthite micro-structures, Middleton concluded that the sodic plagioclase was originally perthitic K-feldspar that was diagenetically albitized. He interpreted the source area to consist of high-rank metamorphic and plutonic rocks (granites and gneisses) rich in K-feldspar. Lajoie (1973) rejected the possibility of a diagenetic origin for the albite in favor of source rock control over feldspar composi-tion. He thought the source terrane consisted dominantly of low-grade metamorphic rocks (quartzo-feldspathic schists) rich in untwinned albite. Recent electron microprobe and X-ray diffrac-tion analyses by Ogunyomi et al. (1981) confirm the diagenetic origin of the albite. They concluded that the feldspars were originally antiperthites derived from anorthositic massifs of the Grenville Province.

Walker (1984) encountered a similar problem in a study of the Fountain Formation (Pennsylvanian) from the Denver area, Colorado. Model analyses of sandstones from the fluvial facies of the Fountain Formation indicate that the feldspar in many of the samples consists almost exclusively of albite, with little or no K-feldspar. Sandstones from the marine facies of the Fountain Formation and Holocene alluvium derived from the same source terrane as the Fountain Formation have different detrital modes. They contain roughly equal proportions of plagioclase and K-feldspar, and in general are richer in feldspar than sandstones of the fluvial facies. Walker (1984) concluded that these differences are the result of greater diagenetic modification to the fluvial facies sandstones. In these fluvial sandstones both albitization of K-feldspar and replacement of plagioclase by clay are important diagenetic reactions. Similar modifications in the marine facies sandstones were prevented by early calcite cementation (Walker, 1984). Failure to recognize the greater diagenetic modifications of the fluvial sandstones could possibly result in the suggestion of different source areas for sandstones from the two facies.

SUMMARY

Four key intrinsic properties have been identified which enable feldspars to be used as provenance indicators for sandstones. These are chemical composition, twinning, zoning, and structural state. Chemical composition and nature of twinning are two properties which are equally useful for both plagioclase and K-feldspar. Although zoning occurs in minerals of both feldspar series, it has received much more attention as an indicator of provenance for plagioclase. Structural state determinations can be made for all feldspars, but because of the difficulty of acquiring these data for plagioclase, sedimentary petrologist have used it as a provenance indicator only for K-feldspar.

Because of its chemical and mechanical instability in the sedimentary environment, feldspar undergoes modification during erosion, transportation, and burial. This modification includes mechanical abrasion, replacement, dissolution, and albitization. Fracturing and breakage of detrital feldspar due to mechanical abrasion have pronounced effects on the frequencies of twinning and zoning. Prolonged transport in high energy environments may sufficiently alter the relative frequencies of the various types of twinning and zoning so that interpretations regarding the source rocks will be affected.

The diagenetic replacement of feldspar, particularly by clay, and its intrastratal dissolution can result in detrital modes which do not accurately reflect the composition of the detritus deposited in the basin. Not only can the affected sandstone be incorrectly classified, but evaluation of the tectonic setting of the source area can also be hampered. This type of modification is so pervasive that it can affect all four of the properties routinely used to evaluate provenance.

Albitization can dramatically alter the chemical composition of detrital feldspar, and can also effect the degree of Al-Si ordering in the lattice. The reaction may take place in the source area as a deutric or hydrothermal alteration of crystalline basement rock, or it may proceed as a diagenetic alteration in deeply buried sediments. Determination of the exact timing of albitization can be very difficult, and adds further to the problem of identifying source rocks. An understanding of the possible diagenetic modifications that feldspar can undergo is imperative for a correct interpretation of the provenance of feldspathic sandstones.

REFERENCES

Badon, C. L., 1975, Stratigraphy and petrology of Jurassic Norphlet Formation, Clarke County, Mississippi: Am. Assoc. Petroleum Geol. Bull. v. 59, p. 377-392.

Bambauer, H. U., M. Corlett, E. Eberhard and K. Viswanathan, 1967, Diagrams for the determination of plagioclases using X-ray powder methods: Schweiz. Mineral. Petrog. Mitt., v. 47, p. 333-349.

Barth, T. R. W., 1969, Feldspars, Wiley Interscience, New York, 261 p.

Basu, A., S. W. Young, L. J. Suttner, W. C. James and G. H. Mack, 1975, Re-evaluation of the use of undulatory extinction and polycrystallinity in detrital quartz for provenance interpretations: Jour. Sed. Petrology, v. 45, p. 873-882.

Blatt, H. and J. M. Christie, 1963, Undulatory extinction in quartz of igneous and metamorphic rocks and its significance in provenance studies of sedimentary rocks: Jour. Sed. Petrology, v. 33, p. 559-579.

Boles, J. R., 1982, Active albitization of plagioclase, Gulf Coast Tertiary: Am. Jour. Sci., v. 282, p. 165-180.

Boles, J. R. and D. S. Coombs, 1977, Zeolite facies alteration of sandstones in the Southland Syncline, New Zealand: Am. Jour. Sci., v. 277, p. 982-1012.

Coombs, D. S., 1954, The nature and alteration of some Triassic sediments from Southland, New Zealand: Royal Soc. New Zealand Trans., v. 82, p. 65-109.

Deer, W. A., R. A. Howie and J. Zussman, 1962, Rock-forming minerals, v. 4, Framework silicates: Longmans, Green and Co., London, 435 p.

Dickinson, W. R. and C. A. Suczek, 1979, Plate tectonics and sandstone compositions: Am. Assoc. Petroleum Geol. Bull., v. 63, p. 2164-2182.

Folk, R. L., 1974, Petrology of sedimentary rocks: Hemphill Publishing Co., Austin, Texas, 182 p.

Gilbert, C. M., 1949, Cementation of some California Tertiary reservoir sands: Jour. Geology, v. 57, p. 1-17.

Gorai, M., 1951, Petrological studies on plagioclase twins: Am. Mineralogist, v. 36, p. 884-901.

Helmold, K. P. and P. C. van de Kamp, 1984, Diagenetic mineralogy and controls on albitization and laumontite formation in Paleogene arkoses, Santa Ynez Mountains, California, in McDonald, D. A. and R. C. Surdam (eds.), Clastic Diagenesis: Am. Assoc. Petroleum Geol. Memoir 37, p. 239-276.

Hutchison, C. S., 1974, Laboratory handbook of petrographic techniques: Wiley-Interscience, New York, 527 p.

Lajoie, J., 1973, Albite of secondary origin in Charny sandstones, Quebec: Discussion: Jour. Sed. Petrology, v. 43, p. 575-576.

Land, L. S. and K. L. Milliken, 1981, Feldspar diagenesis in the Frio Formation, Brazoria County, Texas Gulf Coast: Geology, v. 9, p. 314-318.

Lindquist, S. J., 1977, Secondary porosity development and subsequent reduction, overpressured Frio Formation sandstone (Oligocene), south Texas: Gulf Coast Assoc. Geol. Soc. Trans., v. 27, p. 99-107.

Martin, R. F., 1974, Controls of ordering and subsolidus phase relations in the alkali feldspars, in MacKenzie, W. S. and J. Zussman (eds.), The Feldspars: Manchester University Press, Manchester, p. 313-336.

Merino, E., 1975, Diagenesis in Tertiary sandstones from Kettleman North Dome, California I. Diagenetic mineralogy: Jour. Sed. Petrology, v. 45, p. 320-336.

Middleton, G. V., 1972, Albite of secondary origin in Charny sandstones, Quebec: Jour. Sed. Petrology, v. 42, p. 341-349.

Ogunyomi, O., R. F. Martin and R. Hesse, 1981, Albite of secondary origin in Charny sandstones, Quebec: a re-evaluation: Jour. Sed. Petrology, v. 51, p. 597-606.

Pittman, E. D., 1963, Use of zoned plagioclase as an indicator of provenance: Jour. Sed. Petrology, v. 33, p. 380-386.

Pittman, E. D., 1969, Destruction of plagioclase twins by stream transport: Jour. Sed. Petrology, v. 39, p. 1432-1437.

Pittman, E. D., 1970, Plagioclase feldspar as an indicator of provenance in sedimentary rocks: Jour. Sed. Petrology, v. 40, p. 591-598.

Plymate, T. G. and L. J. Suttner, 1983, Evaluation of optical and X-ray techniques for detecting source-rock-controlled variation in detrital potassium feldspar: Jour. Sed. Petrology, v. 53, p. 509-519.

Putnam, P. E. and M. A. Pedskalny, 1983, Provenance of Clearwater Formation reservoir sandstones, Cold Lake, Alberta, with comments on feldspar composition: Can. Petroleum Geol. Bull., v. 31, p. 148-160.

Sibley, D. F. and K. J. Pentony, 1978, Provenance variation in turbidite sediments, Sea of Japan: Jour. Sed. Petrology, v. 48, p. 1241-1248.

Smith, J. V., 1974, Feldspar minerals: Springer-Verlag, New York, v. 1, 627 p.; v. 2, 690 p.

Suttner, L. J. and A. Basu, 1977, Structural state of detrital alkali feldspars: Sedimentology, v. 24, p. 63-74.

Tilling, R. I., 1968, Zonal distribution of variations in structural state of alkali feldspar within the Rader Creek Pluton, Boulder Batholith Montana: Jour. Petrology, v. 9, p. 331-357.

Tobi, A. C., 1962, Characteristic patterns of plagioclase twinning: Norsk Geologisk Tidskrifft, v. 42, Proceedings NATO Study Institute on Feldspars, p. 264-271.

Trevena, A. S. and Nash, W. P., 1981, An electron microprobe study of detrital feldspar: Jour. Sed. Petrology, v. 51, p. 137-150.

Turner, F. J., 1951, Observations of plagioclase in metamorphic rocks: Am. Mineralogist, v. 36, p. 581-589.

Van der Plas, L., 1966, The identification of detrital feldspars, Elsevier Publishing Co., Amsterdam, 305 p.

Walker, T. R., 1984, Diagenetic albitization of potassium feldspar in arkosic sandstones: Jour. Sed. Petrology, v. 54, p. 3-16.

Wilson, M. D. and E. D. Pittman, 1977, Authigenic clays in sandstones: Recognition and influence on reservoir properties and paleoenvironmental analysis: Jour. Sed. Petrology, v. 47, p. 3-31.

Wilson, M. D. and R. W. Tillman, 1974, Diagenetic destruction of feldspar and genesis of clay: their influence on sandstone classification and grain size analysis: Geol. Soc. Amer. Abstracts with Programs, v. 5, p. 130-131.

Wright, T. L., 1968, X-ray and optical study of alkali feldspars: II. An X-ray method for determining the composition and structural state from measurement of 2θ values for three reflections: Am. Mineralogist, v. 53, p. 88-104.

OPTICAL ANALYSES OF ARENITES: INFLUENCE OF METHODOLOGY ON COMPOSI-
TIONAL RESULTS

Gian Gaspare Zuffa
Dipartimento di Scienze della Terra
Università della Calabria
87030 Castiglione Cosentino Stazione, Cosenza, ITALY

ABSTRACT

This paper illustrates the problems involved in selecting the
petrographical classes to be used in quantitative studies of arenites and proposes criteria whereby the dependence of rock composition on grain size is minimized and a satisfactory separation
is obtained between arenaceous components which are either coeval
with the deposit (mainly intrabasinal) or else significantly older
(mainly extrabasinal).

INTRODUCTION

The usefulness of arenites to paleogeographic and paleotectonic reconstructions is due both to their abundance and to the
wealth of information they can provide, particularly from microscopic analyses. The term arenite is used in this paper in a textural sense (Pettijohn, 1949, p. 226) to mean sandy grains, 0.0625
to 2 millimeters in diameter, regardless of composition and genesis (Grabau, 1904, p. 242).
The factors controlling the composition of arenites are extremely difficult to evaluate. Limitations and errors are often
introduced into paleogeographic and paleotectonic reconstructions
·by incomplete or incorrect data concerning climate and relief of
the source area, mechanical transport, depositional and post-depositional processes. Additionally, results can be greatly influenced by the particular methodology adopted for optical analyses.
In order to avoid errors in paleorecostrutions it is essential
that all these factors be taken into consideration.
The gross composition of arenites given by the standard QFL

165

G. G. Zuffa (ed.), Provenance of Arenites, 165–189.

(or QFR) triangular diagrams can be strongly influenced by grain
size. Since fine and coarse-grained samples of the same provenance
contain different quantities of coarse-grained rock fragments, they
plot in quite separate areas of the triangular diagrams. This leads
to two fundamental drawbacks: (1) classifications of arenites unaccompanied by grain size data are seldom comparable; (2) in the
standard QFL diagrams, the compositional trends of samples within
the same sedimentary sequence may not be due to provenance evolution but rather to variation in the grain size of the samples. This
paper revises the counting technique, independently proposed by
Gazzi (1966), Dickinson (1970), later discussed by Zuffa (1980),
and recently tested by Ingersoll et al. (1984), which minimizes
the dependence of rock composition on grain size. A general scheme
modified from Zuffa (1980) is presented.

Misinterpretation of modes of arenites containing framework
grains of both intrabasinal and extrabasinal origin may lead to
unreliable reconstructions of paleobasins and paleosource areas
(Zuffa, 1980; Mack, 1984). The distinction between sand grains
which can be either coeval with the deposit (mainly intrabasinal)
or else significantly older (mainly extrabasinal) mainly concerns
carbonate and volcanic particles. This paper proposes some criteria to help overcome the practical problems which arise.

DISTINCTION BETWEEN EXTRABASINAL AND INTRABASINAL CARBONATE PAR-
TICLES IN A SANDY FRAMEWORK

Extrabasinal dolostone and limestone rock fragments provide
information on the source area. Carbonate allochems (Folk, 1959),
formed within the sedimentary basin, provide information on the
physical, chemical and biological conditions of the depositional
basin (Fig. 1).

The abundance of terrigenous calcareous components in arenites has been underestimated or ignored for three main reasons: 1)
environments capable of producing carbonate debris are generally
thought to be very rare, because of the chemical instability and
physical weakness of carbonate rocks (Wilson, 1975, p. 7), 2) such
debris can lose its terrigenous features during diagenesis, and
3) the debris found in many shallow-marine arenites (impure chemical rocks: Folk, 1974, p. 168; hybrid arenites, Zuffa, 1980) tends
to be confused with particles of intrabasinal origin (e.g., peloids,
intraclasts, etc.). Evidence that terrigenous carbonate debris was
an important detrital component during both the Upper Cretaceous
and Paleogene times (e.g., Gazzi and Zuffa, 1970; Sestini, 1970;
De Rosa and Zuffa, 1979; Gandolfi et al.,1983), and in modern sediments (Gazzi et al., 1973; Gandolfi and Paganelli, 1975), suggests
that more attention should be paid to this largely ignored component. Though end members of extrabasinal carbonate arenites (calclithite, Folk, 1959) are commonly recognized in the literature,
problems arise with shallow- and deep-marine arenites containing

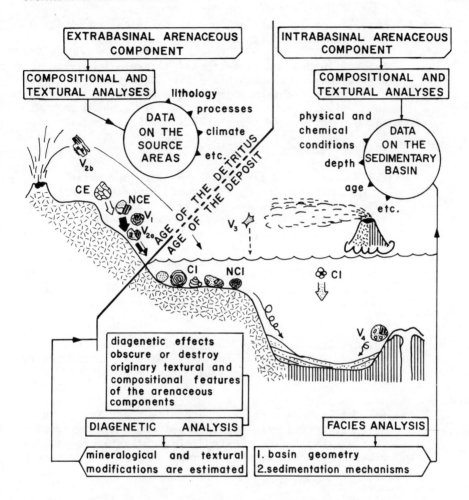

Figure 1.   Schematic relationship between arenaceous components
and data provided by the compositional, textural and
facies analysis of marine arenites.
NCE: noncarbonate extrabasinal grains; CE: carbonate
extrabasinal grains; NCI: noncarbonate intrabasinal
grains; CI: carbonate intrabasinal grains; V: volcanic
grains (for numbers 1, 2a, 2b, 3, and 4, refer to Table 2). Modified from Zuffa, 1980.

Figure 2. Examples of intrabasinal and extrabasinal carbonate and
           noncarbonate sand grains.

           (a) NCE: noncarbonate extrabasinal grain (phyllite);
           CE: carbonate extrabasinal grain (dolostone); NCI: non-
           carbonate intrabasinal grain (glauconite); CI: carbonate
           intrabasinal grain (bioclast; size is about 5 mm). Cros-
           sed nicols. Hybrid Arenite (Zuffa, 1980; see also Fig.
           11), Ager Valley, Pyrenees, Eocene (De Rosa and Zuffa,
           1979).

           (b) Sands carried from Southern Alps (northwestern Italy)
           to the Adriatic Sea by the Brenta River (Gazzi et al.,
           1973). CE: micritic and fossiliferous terrigenous lime-
           stone grains; NCE: terrigenous coarse (metamorphic) and
           fine-grained (chert, 1 mm) lithic grains. Crossed nicols.

           (c) CE: terrigenous micritic limestone (0,6 mm); ar-
           rows indicate Tintinnid fauna of Upper Jurassic - Lower
           Cretaceous age; CI: bioclasts. Crossed nicols. Paludi
           Formation, Eocene, eastern Calabria (Zuffa and De Rosa,
           1978).

           (d) CE: Terrigenous rounded micritic limestone with
           veins (0,4 mm); CI: bioclast, Paludi Formation, Eocene,
           eastern Calabria, Italy (Zuffa and De Rosa, 1978).

           (e) Terrigenous carbonate particles (Laga Formation,
           Miocene, Central Italy). CE_1: dolostone (0,8 mm);
           CE_2: microspatitic limestone; CE_3: slightly-foliated
           micritic limestone; CE_4: micritic limestone. Crossed
           nicols.

           (f) Calclithite (Messanagros Sandstone, Rodi, Oligocene).
           CE: various types of carbonate extrabasinal grains;
           siliciclastic grains include altered diabase (NCE_1) and
           chert (NCE_2) rock fragments. Courtesy of E. Mutti.

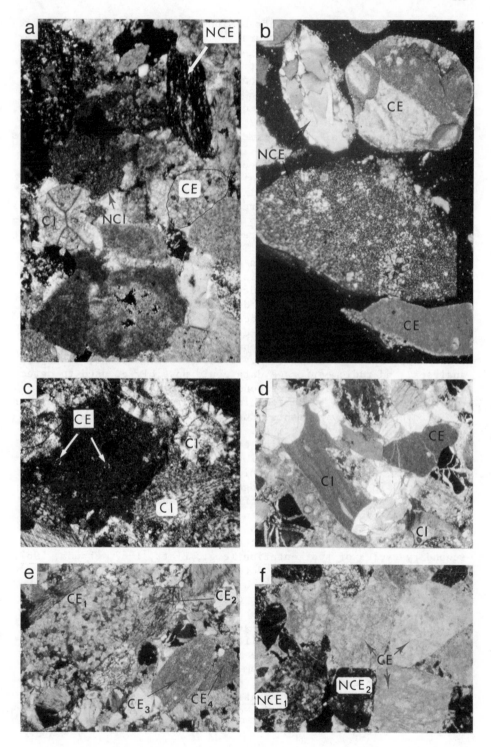

mixtures of terrigenous and intrabasinal carbonate particles. Where
diagenetic effects do not obscure the original grain features, carbonate particles may be recognized either as extraclasts or intraclasts by using several diagnostic criteria. Some general criteria
for discriminating between carbonate intraclasts and extraclasts
have been proposed by Zuffa (1980) and a guide is presented in Table 1 (also see Fig. 2). In all cases where it proves difficult to
distinguish them with confidence, carbonate particles should be
assigned to a neutral class (limeclasts). In these cases limeclasts
of intrabasinal or extrabasinal indeterminate origin have no place
in classification diagrams, but can be tentatively included within
the intrabasinal and/or extrabasinal component thus allowing speculation upon their possible implications either for sedimentary
basin and/or source area interpretation.

It is sometimes possible to determine the age of micritic particles of unknown origin by extracting small amounts of micritic
materials from the "limeclast". This can be done by using an uncovered slide on the stage of an optical microscope and examining
the nannofauna. The limeclasts is likely to be terrigenous when it
is considerably older than the deposit, and intrabasinal when it
is more or less coeval with the deposit. This technique is illustrated in an example from the Eocene Paludi Formation (Zuffa and
De Rosa, 1978, Longobucco Unit, eastern Calabria) which contains
limeclasts with coccoliths of Upper Cretaceous age thus indicating
a reworking cycle of about 20 m.y.. Cathodoluminescence, microprobe
analyses, and radiometric age data could possibly be used to detect
differing populations of terrigenous and intrabasinal sand particles. Experiments on these techniques are in progress by the author.

Geotectonic environments, such as mobile orogenic belts, can
create a continuous "recycling", reworking", or "cannibalism" of
sedimentary material by partial emergence and erosion of sedimentary sequences, or by redeposition of material from subaqueous tectonic highs. Good examples of this type of geotectonic setting are
to be found in the Apenninic orogenic deposits (Ricci Lucchi, this
volume). In such cases it is difficult to use the criteria proposed to account both for the origin of the particles and their age.
Similarly partial emergence exposing sedimentary deposits, which
is very common in carbonate platform sequences, can give rise to
secondary cycles of sedimentation (e.g., Blatt, 1967; Chanda, 1967).
Such evidence is difficult to detect but is nevertheless of secondary importance. However if sedimentary suites are buried, lithified and incorporated within an emergent orogenic belt it is usually possible to reconstruct the paleosource area and paleobasin
as the absolute age of the terrigenous and intrabasinal components
are likely to be sufficiently different to enable a useful distinction to be made. It therefore seems reasonable to consider as
"paleo" that detritus derived from formations originating in a tectonic setting pre-dating the one under investigation.

The assumption that extrabasinal and intrabasinal are respectively equivalent to paleo and neo does not always hold for marine

Table 1. Criteria for distinguishing between extrabasinal (terrigenous) and intrabasinal carbonate sand particles from shallow and deep-sea hybrid arenites (where diagenetic alteration of arenite is absent or slight). The following criteria can only be considered absolute in a few cases (i.e., 1, 2, and 3). When it is difficult to make a distinction with confidence, particles are assigned to a neutral class (LIMECLASTS).

MAIN TYPES OF CARBONATE SAND PARTICLES		LIMECLASTS	
		INTRABASINAL (NEO)	EXTRABASINAL (PALEO)
COMPOSITION	MINERALOGY	Calcite (Mg-calcite, aragonite, and rarely aphanitic dolomite for modern sediments)	Calcite, dolomite and ankerite
	FAUNA	Not older than the host sediment. Bioclasts may be abundant	Fossils may be present in carbonate rock fragments which are older than the host formation (1) - The occurrence of reworked single bioclasts in the host formation would be unusual
TEXTURE	GRAIN SIZE	May be different, typically coarser than terrigenous particles	Generally same as other terrigenous siliciclasts
	SPHERICITY AND ROUNDNESS	Shallow water: winnowing absent a) Peloids (McKee and Gutschik, 1969) and some fossil molds tend to have high values of sphericity b) Soft intraclasts occur with irregular and/or sinuous contours - also they exhibit an elongated and/or platy shape. Skeletal debris is generally angular Shallow water: winnowing present All allochemical components are well-rounded and tend to have high values of sphericity	Generally rounded and with high values of sphericity
	FABRIC	Common fabric of allochemical components (see Folk, 1974) Absence of particles with internal veins Absence of recrystallized particles	Particles with characteristics (i.e., color, oxidized contours) which can not be associated with the common characteristics of intrabasinal allochems Particles with internal veins are present Recrystallized particles (e.g., dolostones, metacarbonates) may be present (2)
INDIRECT EVIDENCE		IF BIOCLASTS OF THE SAME AGE AS THE DEPOSIT ARE PRESENT, THEN OTHER INTRABASINAL CARBONATE PARTICLES ARE LIKELY TO BE PRESENT	IF DOLOMITE AND/OR ANGULAR CHERT ROCK FRAGMENTS ARE PRESENT, THEN OTHER EXTRABASINAL CARBONATE PARTICLES ARE LIKELY TO BE PRESENT (3)
INFORMATION FROM FACIES STUDIES		TOTAL EVIDENCE OF THE DEPOSITIONAL ENVIRONMENT	TOTAL EVIDENCE OF THE DEPOSITIONAL ENVIRONMENT
GEOLOGY		GEOTECTONIC FRAMEWORK	GEOTECTONIC FRAMEWORK

arenites. Caliche grains produced in a fluvial environment and transported to the sea are extrabasinal because they derive from a subaerial source area but can be almost coeval (and therefore neo) to the age of the marine sedimentary sequence. In the same way, recycling of material from submarine tectonic highs in mobile orogenic belts may produce intrabasinal particles which can be significantly older (i.e. paleo) than those in the final depositional site.

RIP-UP CLASTS

Rip-up clasts are penecontemporaneous muddy grains eroded by high-energy current flows from depositional "parking areas" located between the source area and the final depositional site. For instance, they can be produced in fluvial environments when currents cut stabilized bars or overbank deposits or in submarine environments when turbidity currents erode channel-levee deposits. These rip-up clasts can have any grain size. If, because of compaction and partial lithification within a "parking area", sand particles are generated by a ripping-up of fine-particle aggregates problems may arise in provenance interpretation. In fact, shaley rock fragments generated by this process can appear to be part of the terrigenous framework when they should be recognized as intrabasinal clasts, i.e. they give information on the conditions of the depositional basin rather than on the source area.

There is generally no problem in recognizing intrabasinal shaley rip-up clasts because their mineralogical, chemical and geometric characteristic are different from those of extrabasinal sedimentary lithic grains. The mineralogical and chemical characteristics of rip-up clasts are similar to the extrabasinal pelitic matrix and generally distinct from those of argillite and shale terrigenous particles even when they form part of the sand framework. The textural features of rip-up clasts should include (i) a low degree of lithification, (ii) oversized elongated grains, and (iii) a less organized internal structure than that of the terrigenous lithified shale and argillites (Fig. 2, f). Finally, rip-up clasts tend, because of their low density, to concentrate as laminae in thin section and as distinct layers in turbidite beds (e. g., clay-chip aligments).

PROPOSED DISTINCTION BETWEEN PALEOVOLCANIC AND NEOVOLCANIC PARTICLES

Recognition of volcanic sand particles in arenites is perhaps the most intriguing task in optical modal analyses of arenites. By taking into account the origin, age and depositional mechanism of these particles (Fig. 1), four main framework groups are distinguished in Table 2. Each group has different implications

Table 2. Some criteria for distinguishing types of volcanic grains in deep-sea arenites (where diagenetic alteration of arenites is absent or slight).

		EXTRABASINAL (subaerial)			EXTRABASINAL (subaerial)	INTRABASINAL (subaqueous)
		1. PALEOVOLCANIC (V₁)	2. NEOVOLCANIC (V₂) (V_{2a})	(V_{2b})	3. NEOVOLCANIC (V₃)	4. NEOVOLCANIC (V₄)
MAIN TYPES OF VOLCANIC GRAINS		derived from erosion of old volcanic suites	derived from active volcanism located in the source area		derived from active volcanoes not located in the main source area	derived from active volcanoes located within the depositional basin
EMPLACEMENT PROCESSES		erosion+trasportation	erosion+transportation	direct ejection into the basin (air fall, surge and pyroclastic flow)	wind transportation and/or direct ejection	direct ejection and/or resedimentation from tectonic highs
TEXTURE — GRAIN SIZE		same as other terrigenous clasts	same as other terrigenous clasts if the predepositional history is similar	different from other terrigenous clasts (including paleovolcanic particles) more sorted than V_{2a} and V_1	different from other terrigenous clasts (including paleovolcanic particles) more sorted than V_{2b}	grains constitute a distinct grain size population (coarser or finer) not sorted
TEXTURE — GRAIN SHAPE (roundness + sphericity)		same as other terrigenous clasts	same as other terrigenous clasts if the predepositional history is similar	generally texturally immature with single crystal euhedral and shard forms (vitroclastic texture) common	generally texturally immature and shard forms (vitroclastic texture) common	texturally immature (particles are produced by explosive fragmentation and magma quenching in subaqueous environment)
COMPOSITION — MINERALOGY AND PETROLOGY		any composition	composition is generally different if compared with column 1; microlithic (lathlike), microgranular-felsitic, vitric and crystals (Dickinson, this volume)	composition is generally different if compared with column 1; mainly vitric to vitrolithic+ crystals	composition is generally different if compared with column 1 and 2; mainly vitric to vitrolithic	composition is generally different if compared with column 1, 2 and 3; mainly vitric and crystals
COMPOSITION — ALTERATION		moderate to strong, depending on the predepositional history and type of volcanism	slight (presence of mafic crystals)	absent (presence of mafic crystal)	slight or absent (presence of mafic crystals)	strong, with peculiar pattern due to subaqueous quenching
FIELD EVIDENCE				distinct pyroclastic layers	distinct pyroclastic layers	distinct well-graded volcaniclastic layers (distinct dispersal pattern)
MAIN GEOLOGIC FRAMEWORK		subduction complexes back-arc thrusted belts suture belts	magmatic arcs	magmatic arcs	magmatic arcs	oceanic domains

Figure 3.   Examples of Quaternary volcanic sand grains (2-0,0625
             mm) from the Nankai Trough (southwest Japan) and from
             the Upper Cretaceous Chugach Terrane (Alaska).
             For classification and symbols refer to Table 2.

             (a) Neovolcanic plagioclase grain with very delicate
             contours and brown-fresh glassy rims (V_{2b}). Polarized
             light (right), crossed nicols (left). Nankai Trough,
             Japan.

             (b) Zoned euhedral neovolcanic plagioclase grain (V_{2b_a}).
             The grain (2 mm long) is part of the framework of a
             fine-grained volcanic litharenite and it has been inter-
             preted as deposited by direct ejection into the basin
             from a volcanic arc that was active during deposition
             of the Upper Cretaceous Chugach turbidites, Alaska (Zuf-
             fa et al., 1980). Crossed nicols.

             (c) Neovolcanic quartz grain showing glassy rims (red
             arrows)and brown unaltered glass (black arrow) filling
             up the embayments (V_{2a} or V_{2b}). Polarized light. Nankai
             Trough.

             (d) Neovolcanic lithic grain with mafic fresh phenocrys-
             ts surrounded by brown unaltered glassy groundmass (V_2b).
             Polarized light. Nankai Trough.

             (e) Neovolcanic glass shards constituting distinct pyro-
             clastic layers of the Nankai Trough Quaternary sequence
             (V_{2b} or V_3). Polarized light. Size of grains is about
             0,125 mm.

             (f) Paleovolcanic altered lithic grain (V_1). Polarized
             light. Nankai Trough.

             (g) Altered paleovolcanic lithic grains (V_1) and neo-
             volcanic lithic grains (V_2) showing euhedral fresh
             plagioclase phenocrysts surrounded by unaltered brown
             glassy groundmass. The neovolcanic grains are thought
             to be derived from the active Bonin Arc, the paleovol—
             canic from erosion of the Shimanto Belt, Japan (De Rosa
             et al., in prep.)

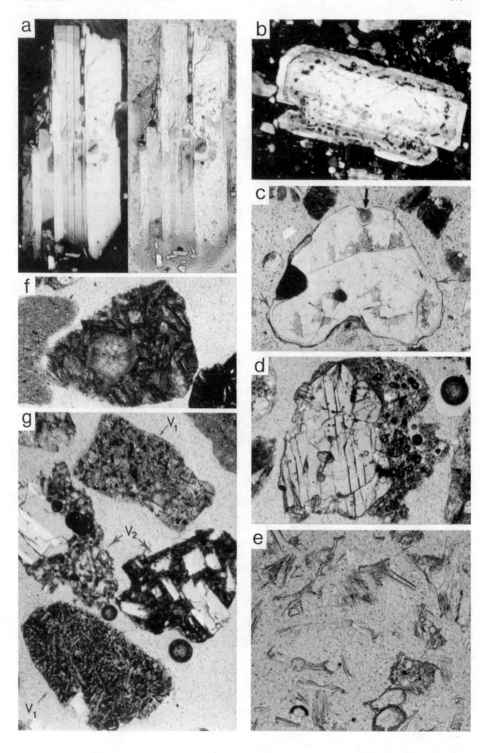

for determining source area and depositional basin configurations.
The geological setting together with compositional and textural
data concerning the volcanic particles in arenites unless strongly modified by diagenesis can be used to distinguish different
grain populations. Figure 3 illustrates examples of Quaternary volcanic sand grains from the Nankai Trough (southwest Japan) and from
the Upper Cretaceous Chugach Terrane (Alaska).

In spite of the fact that practical applications of the proposed criteria can be somewhat subjective, time consuming and only
provide semiquantitative data, it is essential that an attempt be
made to ascertain whether populations of volcanic grains of differing origin are mixed in the total sand framework. Even should
only qualitative information be obtained, a distinction between
neo-and paleovolcanic particles can shed a new light on provenance
interpretation.

METHODOLOGICAL APPROACH TO MODAL ANALYSIS WHICH MINIMIZES THE DE-
PENDENCE OF ARENITE COMPOSITION ON GRAIN SIZE

A first order methodological problem is that the more popular classifications of arenites require different classes of rock
fragments in the L (or R) pole of the common QFL (or QFR) diagrams.
Modal data from fifteen samples collected from the base to the top
of a turbiditic bed (Fig. 4) of the Albidona Formation (Upper

Figure 4.   Turbidite beds of the Albidona Formation (Upper Oligo-
            cene-Lower Burdigalian, southern Apennine, Italy).
            Arrows indicate a sampled bed which shows amalgamation
            to the upper right. Dots indicate the quotas of the fif-
            teen analyzed samples reported in Figure 5.

Oligocene-Lower Burdigalian, Italy) are plotted in Figure 5 according to the Folk (1974) and Gazzi-Dickinson (1966-1970) methods and by the method still used by many authors which includes all rock fragments in the L Pole.

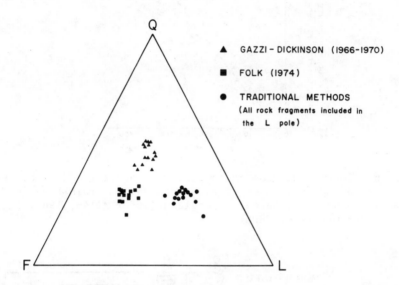

Figure 5. Composition of fifteen samples from a turbidite bed (Albidona Formation, Italy; Mancuso, 1981) plotted according to the Folk (1974), Gazzi-Dickinson (1966-1970) and traditional classifications.

Unfortunately several QFL or QFR diagrams in the literature are still unaccompanied by sufficient information concerning the criteria used in grouping compositional classes, especially rock fragments; information on grain size is also commonly lacking. Figure 6 shows that the content of coarse-grained rock fragments, composed of single crystals of more than 0.0625 mm in size, is strongly dependent on the grain size of arenites (Huckenholtz, 1963; Gazzi, 1966; Blatt, 1967; Fuchtbauer, 1967; Gazzi et al., 1973; Odom et al., 1976; Zuffa, 1969, 1980). The use of arenite classifications including coarse-grained rock fragments in the L pole leads to the paradox that the composition of samples collected at different levels of a single graded turbidite bed plot in quite different areas of the standard QFL diagram (Pettijohn et al., 1972)! Consequently it is only possible to use such diagrams to compare the composition of arenites when they have been obtained from samples of the same grain size (Basu, 1976; Mack, 1978).

A special counting technique (Gazzi, 1966; Dickinson, 1970;

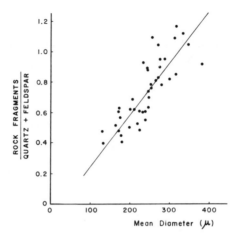

Figure 6.   Scatter diagram of composition versus grain size. De-
            pendence of rock fragment content on grain size in Cilee
            Sandstone, Devonian, England (after Allen, 1962, Fig. 9,
            courtesy of SEPM).

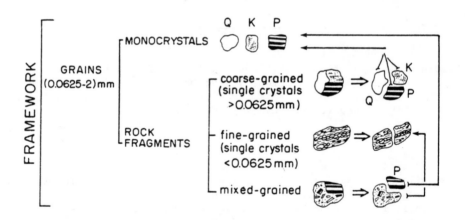

Figure 7.   Flow chart of counting technique to minimize the depend-
            ence of rock composition on grain size (see text for ex-
            planation). From Zuffa, 1980, courtesy of SEPM.

Zuffa, 1980; Ingersoll et al., 1984) minimizes the drawback due to
grain-size variation, and allows greater accuracy in classifying
arenites and determining their provenance. In this method the basic
criterion for counting the detrital framework (Fig. 7) consists of
separating coarse-grained lithic fragments (single crystals larger
than 0.0625 mm in size) from fine-grained lithic fragments (single
crystals less than 0.0625 mm in size). Because of disintegration

during sediment dispersal, the former tend to break into individual crystals larger than the grain-matrix size limit (0.0625 mm),
whereas the latter tend to break into still smaller fine-grained
lithic fragments. Therefore rather than counting the coarse-grained
fragments as such they are apportioned according to the mineral beneath the cross-hair.

In Figure 8 the composition of nine samples from three distinct graded turbidite beds analyzed by Gazzi (1966) are initially
plotted on the standard QFL diagram by including on the L pole only
fine-grained rock fragments (Gazzi-Dickinson method) and then by
including all rock fragments in the L pole. When the Gazzi-Dickinson
counting method is used, the points representing the data for the
different samples are far less scattered in the diagram. The Gazzi-
Dickinson point-counting method has been recently tested by Ingersoll et al., (1984). This study shows very clearly how representative points of five grain-size fractions of several arenites derived from the same coarse-grained source rock, are better cluster-
ed by the Gazzi-Dickinson method than by using traditional methods
(Fig. 9). The former does not however compensate for differences
due to the distribution of single minerals (like feldspars) among
the various grain-size fractions of a sandstone (Fig. 10). Where
significant variations occur, multiple triangular diagrams, each
denoting a specific mean size range are needed (Odom et al., 1976).

OPTICAL PETROGRAPHYC CLASSES FOR MODAL ANALYSES OF ARENITES

The criteria for distinguishing extrabasinal and intrabasinal
particles and minimizing the apparent dependence of arenite composition on grain size are illustrated in Table 3. Crystals of sand
size are recorded according to whether they are single crystals or
belong to a particular type of rock fragment. Thus, the classes
quartz, k-feldspar and plagioclase are each split into a number of
subclasses indicating whether the mineral is a single crystal or
constitutes part of a specific type of rock fragment as indicated
on the left side of Table 3 (modified from Zuffa, 1980).

CLASSIFICATION OF ARENITES

The criteria discussed in this paper enable four main sand
framework groups to be distinguished 1) noncarbonate extrabasinal,
2) carbonate extrabasinal, 3) noncarbonate intrabasinal, 4) carbonate intrabasinal (Table 3; NCE, CE, NCI, CI, respectively). Their
three dimensional configuration in the form of a tetrahedron es-
tablishes a first-level classification of the main types of arenites (Fig. 11, from Zuffa, 1980). Representative compositional points
within the tetrahedron can, in general, be considered hybrid arenites, but it is proposed here that only the volume of the interior
octahedron represents "hybrid arenites". The four triangular faces
represent three component compositions of the more common arenites,
and a rough subdivision of the tetrahedron into two parts is con-

Table 3. Optical petrographic classes for modal analysis of arenites. (x) Temporal criterion refers to the age of the various constituents of modal classes with respect to the time of formation of the deposit. NCE: noncarbonate extrabasinal grains; CE: carbonate extrabasinal grains; NCI: noncarbonate intrabasinal grains; CI: carbonate intrabasinal grains; V: neovolcanic grains; Lc: limeclasts; CCm: carbonate cement; NCCm: noncarbonate cement; NCMt: noncarbonate matrix; CMt: carbonate matrix; VSp: void spaces (modified from Zuffa, 1980).

CRITERIA		TEXTURAL		GENETIC	CHEMICAL
MAIN CONSTITUENTS	I	FRAMEWORK COMPONENTS (0.0625-2) (mm)		EXTRABASINAL	NONCARBONATE
					CARBONATE
				INTRABASINAL	NONCARBONATE
					CARBONATE
				EXTRABASINAL and/or INTRABASINAL	NONCARBONATE
				EXTRABASINAL and/or INTRABASINAL	CARBONATE
	II	INTERSTITIAL COMPONENTS	CEMENT		CARBONATE
					NONCARBONATE
			MATRIX 0,0625 (mm)	mainly EXTRABASINAL	MAINLY NONCARBONATE
				mainly INTRABASINAL	MAINLY CARBONATE
	III	VOID SPACES		PRIMARY and/or SECONDARY	

COMPOSITIONAL and/or TEXTURAL main optical petrographic classes	identification code	TEMPORAL (X)
Q | quartz		
F | k-feldspar plagioclase		
L | acidic volcanic rocks intermediate volcanic rocks basic volcanic rocks serpentinite fine-grained phyllite shale siltstone chertt etc. micas and chlorites other minerals	NCE	paleo
limestone dolostone	CE	
glauconite gypsum iron-oxides phosphate etc.	NCI	
intraclasts oolites fossils peloids	CI	neo
quartz k-feldspar plagioclase femic minerals felsitic volcanic rocks microlithic volcanic rocks ophitic volcanic rocks vitric volcanic rocks	V	
limeclasts	Lc	neo and/or paleo
calcite dolomite spar microspar ankerite and/or and/or etc. pseudospar	CCm	early and/or late
quartz phyllosilicate iron-oxides etc.	NCCm	
protomatrix orthomatrix	NCMt	paleo
micrite	CMt	neo
many textural classes are possible (see for ex. Fuchtbauer, 1974)	VSp	early and/or late

Subclass descriptions:

Q_s single crystals
Q_p in plutonic rock fragments (r.f.)
Q_M in high-grade metamorphic r.f.
Q_m in low-grade metamorphic r.f.
Q_{Va} in acidic volcanic r.f.
Q_{Vi} in intermediate r.f.

K_s single crystals
K_p in plutonic r.f.
K_M in high-grade metamorphic r.f.
K_m in low-grade metamorphic r.f.
K_{Va} in acidic volcanic r.f.
K_{Vi} in intermediate volcanic r.f.

P_s single crystals
P_p in plutonic r.f.
P_M in high-grade metamorphic r.f.
P_m in low-grade metamorphic r.f.
P_{Va} in acidic volcanic r.f.
P_{Vi} in intermediate volcanic r.f.
P_{Vb} in basic volcanic r.f.

This column indicates how the classes quartz, k-feldspar and plagioclase can be split into a number of subclasses indicating whether the mineral is a single crystal or constitutes part of a specific type of rock fragment. The same criterion can be utilized with the classes micas and chlorites, other minerals and the first four classes of group V when present as a principal framework component.

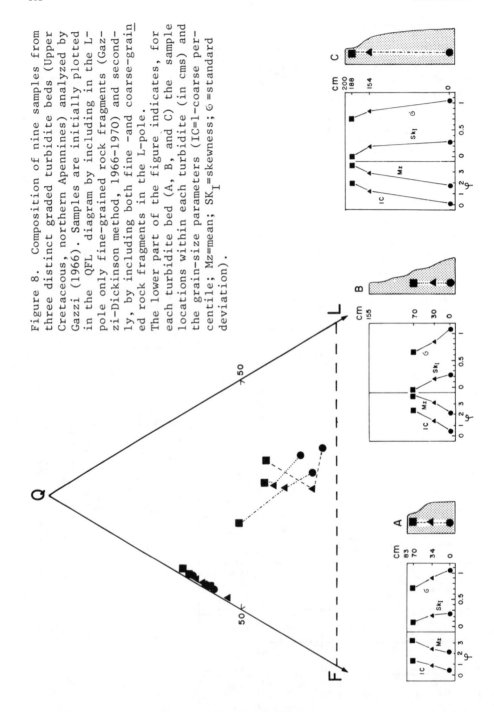

Figure 8. Composition of nine samples from three distinct graded turbidite beds (Upper Cretaceous, northern Apennines) analyzed by Gazzi (1966). Samples are initially plotted in the QFL diagram by including in the L-pole only fine-grained rock fragments (Gazzi-Dickinson method, 1966-1970) and secondly, by including both fine -and coarse-grained rock fragments in the L-pole. The lower part of the figure indicates, for each turbidite bed (A, B, and C) the sample locations within each turbidite (in cms) and the grain-size parameters (IC=1-coarse percentile; Mz=mean; SK$_I$=skewness; σ=standard deviation).

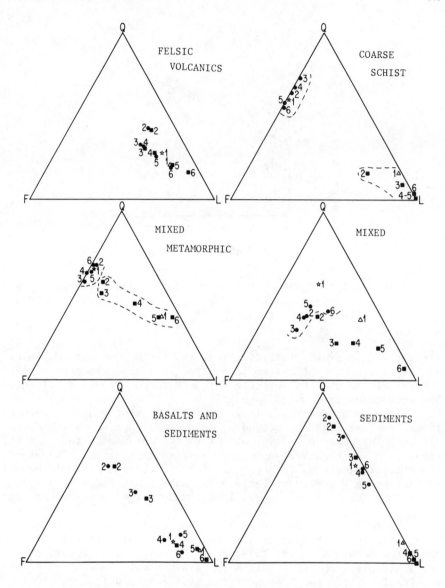

Figure 9.   QFL diagrams for sands of different provenance (indicated on the right side of the triangles). In each triangle are plotted: (1) the unsorted sample using the
Gazzi-Dickinson method (star) and the traditional method (triangle), (2) the sand fractions, 4-3φ, 3-2φ,
2-1φ, 1-0φ, and 0-(-1)φ of the same sample using the
Gazzi-Dickinson method (circles) and the traditional
method (squares): after Ingersoll et al., 1984, courtesy of SEPM.

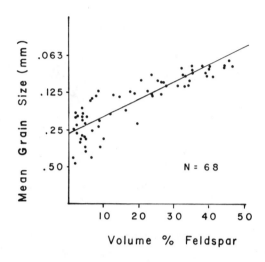

Figure 10.   Relationship of volume percent feldspar to mean size
             in some Cambrian sandstones (after Odom et al., 1976,
             Fig. 3); courtesy of SEPM.

sistent with a field classification. The terms shown at the poles
of the four-component system in Figure 11 (e.g., carbonate extra-
renite) can be used to indicate both the composition and origin of
the arenaceous framework. Once this basic first-level classification has been established, it is possible to progress to a second-
level of classification depending on whether representative compositional information is available for the rock under study. As a
result, rock categories such as arkose, quartz-arenite, litharenite, etc., indicate a second-level classification. It should be noted that group V and Lc of Table 3 can only be included in either
the NCE or NCI pole of Figure 11 when their origin and age has been
recognized.

      Figure 12 shows how these different classification levels can
be applied to the Paleogene arenites of the Ager Valley in the Pyrenees (De Rosa and Zuffa, 1979). A first-level classification enables a separation of the different grain populations which evolve
vertically within the sequence from tidal-bar platform deposits
(lower right corner) into fluvially-dominated littoral deposits
(I in Fig. 12). The second-level classification indicates the composition of the noncarbonate extrabasinal components (litharenite)
and of the carbonate extrabasinal grains (II in Fig. 12); finally
a tertiary-level classification specifies the nature of fine-grained extrabasinal rock fragments (III in Fig. 12).

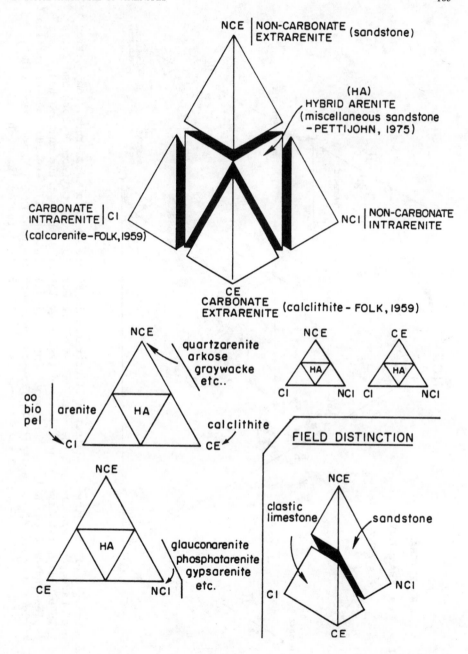

Figure 11. Main types of arenites as defined by optical analysis
of the arenaceous modes. From Zuffa, 1980, courtesy
of SEPM.

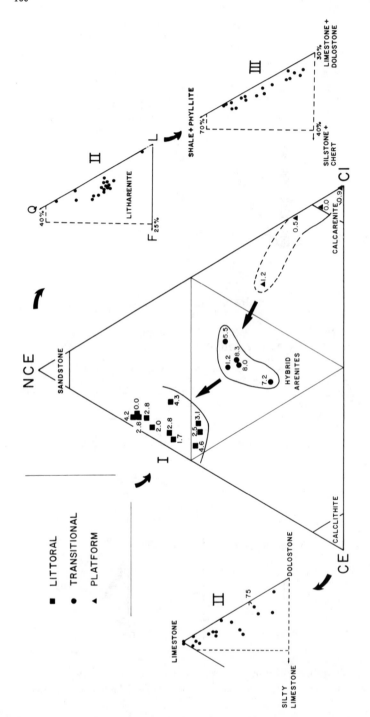

Figure 12. Example of applying the criteria illustrated in Figure 11 to a case study. Composition of Paleogene arenites of the Ager Valley (Pyrenees, Spain, data from De Rosa and Zuffa 1979). I: first-level classification (see Fig. 11 for symbols); the numbers beside samples indicate the NCI (non carbonate intrabasinal) content. II: second-level classification; Q, quartz; F, feldspar; L, fine-grained rock fragments. III: tertiary-level classification (fine-grained rock fragments types). See text for explanation.

AKNOWLEDGMENTS

I thank R. De Rosa for her thoughtful comments on the criteria for distinguishing volcanic and carbonate particles, D. Rio who conceived the idea of detecting carbonate extraclasts by nannofauna examination, R. Compagnoni, E. McBride, D. Fontana, T. Nilsen, and R. Valloni who provided criticism on Tables 1 and 2. A. Basu, A. Matter, S. Burley, W. Dickinson and T. Teale all contributed with comments and kindly improved the last version of the manuscript. The Italian Ministry of Education contributed toward publication costs of this article.

REFERENCES

ALLEN, J. R. L., 1962, Petrology, origin and deposition of the highest Lower Old Red Sandstone of Shropshire, England: Jour. Sed. Petrology, v. 32, p. 657-697.

BASU, A., 1976, Petrology of Holocene fluvial sand derived from plutonic source rocks: implications to paleoclimate interpretation: Jour. Sed. Petrology, v. 46, p. 694-709.

BLATT, H., 1967, Provenance determinations and recycling of sediments: Jour. Sed. Petrology, v. 37, p. 1031-1044.

CHANDA, S. K., 1967, Petrogenesis of the calcareous constituents of the Lameta Group around Jabalpur, M. P., India: Jour. Sed. Petrology, v. 23, p. 425-437.

DE ROSA, R., and ZUFFA, G. G., 1979, Le areniti ibride della Valle di Ager (Eocene, Pirenei centromeridionali): Mineralog. et Petrog. Acta, v. 23, p. 1-12.

DICKINSON, W. R., 1970, Interpreting detrital modes of graywacke and arkose: Jour. Sed. Petrology, v. 40, p. 695-707.

FOLK, R. L., 1959, Practical petrographic classification of limestones: Am. Assoc. Petroleum Geologists Bull., v. 43, p. 1-38.

FOLK, R. L., 1974, Petrology of sedimentary rocks: Austin, Texas, Hemphill's Bookstore, 182 p.

FÜCHTBAUER, H., 1967, Die Sandsteine in der Molasse nördlich der Alpen: Geol. Rundschau, v. 56, p. 266-300.

FÜCHTBAUER, U., 1974, Sediments and sedimentary rocks 1., Stuttgard, Schweizerbart'sche Verlagsbuchhandlung, 464 p.

GANDOLFI, G., and PAGANELLI, L., 1975, Il litorale toscano fra Piombino e la foce dell'Ombrone (area campione alto Tirreno) composizione provenienza e dispersione delle sabbie: Soc. Geol. Italiana Boll., v. 94, p. 1811-1832.

GANDOLFI, G., PAGANELLI, L., and ZUFFA, G. G., 1983, Petrology and dispersal pattern in the Marnoso-Arenacea Formation (Miocene, Northern Apennines), Jour. Sed. Petrology, v. 53, p. 493-507.

GAZZI, P., 1966, Le arenarie del flysch sopracretaceo dell'Appennino modenese; correlazioni con il flysch di Monghidoro: Mineralog. et Petrog. Acta, v. 16, p. 69-97.

GAZZI, P.,and ZUFFA, G. G., 1970, Le arenarie paleogeniche dell'Appennino emiliano: Mineralog. et Petrog. Acta, v. 16, p. 97-137.

GAZZI, P., ZUFFA, G. G., GANDOLFI, G., and PAGANELLI, L., 1973, Provenienza e dispersione litoranea delle sabbie delle spiagge adriatiche fra le foci dell'Isonzo e del Foglia: inquadramento regionale: Soc. Geol. Italiana Mem., v. 12, p. 1-37.

GRABAU, A. W., 1904, On the classification of sedimentary rocks: Am. Geologist, v. 33, p. 228-247.

HUCKENHOLTZ, H. G., 1963, Mineral composition and texture in graywackes from the Harz Mountains (Germany) and in arkoses from the Auvergne (France): Jour. Sed. Petrology, v. 33, p. 914-918.

INGERSOLL, R. V., BULLARD, T. F., FORD, R. L., GRIMM, J. P., PICKLE, J. D., and SARES, S. W., 1984, The effect of grain size on detrital modes: a test of the Gazzi-Dickinson point-counting method: Jour. Sed. Petrology, v. 54, p. 103-116.

MACK, G. H., 1978, The survivability of labile light mineral grains in fluvial, aeolian and littoral marine environments: the Permian Cutler and Cedar Mesa Formations, Moab, Utah: Sedimentology, v. 25, p. 587-604.

MACK, G. H., 1984, Exceptions to the relationship between plate tectonics and sandstone composition: Jour. Sed. Petrology, v. 54, p. 212-220.

McKEE, E. D., and GUTSCHICK, R. C., 1969, History of Redwall limestone of northern Arizona: Geol. Soc. America Mem., v. 114, 726 p.

MANCUSO, L., 1981, Influenza della metodologia nella determinazione per via ottica della composizione di torbiditi arenacee: Università della Calabria, unpublished thesis, 77 p.

ODOM, I. E., DOE, T. W.,and DOTT, R. H., Jr., 1976, Nature of feldspar-grain size relations in some quartz-rich sandstones: Jour. Sed. Petrology, v. 46, p. 862-870.

PETTIJOHN, F. J., 1949, Sedimentary rocks: New York, Harper and Brothers, 526 p.

PETTIJOHN, F. J., 1975, Sedimentary rocks (3rd ed.) New York, Harper and Row, 718 p.

PETTIJOHN, F. J., POTTER, P. E., and SIEVER, R., 1972, Sand and sandstone: New York, Springer-Verlag, 618 p.

SESTINI, G., 1970, Flysch facies and turbidite sedimentology: Sed. Geology, v. 4, p. 559-597.

WILSON, J. L., 1975, Carbonate facies in geologic history: New York, Springer-Verlag, 471 p.

ZUFFA, G. G., 1969, Arenarie e calcari arenacei miocenici di Vetto-Carpineti (Formazione di Bismantova, Appennino settentrionale): Mineralog. et Petrog. Acta, v. 15, p. 191-219.

ZUFFA, G. G., and DE ROSA, R., 1978, Petrologia delle successioni torbiditiche eoceniche della Sila nord-orientale (Calabria): Soc. Geol. Italiana Mem., v. 18, p. 31-55.

ZUFFA, G. G., 1980, Hybrid arenites: their composition and classification: Jour. Sed. Petrology, V. 50, p. 21-29.

ZUFFA, G. G., NILSEN, T. H., and WINKLER, G. R., 1980, Rock-fragment petrography of the Upper Cretaceous Chugach Terrane, Shoutern Alaska: U.S.G.S. Open-File Report 80-173, p. 1-28.

CATHODOLUMINESCENCE MICROSCOPY AS A TOOL FOR PROVENANCE
STUDIES OF SANDSTONES

Albert MATTER and Karl RAMSEYER[1]

Geologisches Institut,   1) Department of
Universität Bern,   Geological Sciences,
Baltzerstrasse 1,   University of California,
3012 Bern, Switzerland   Santa Barbara, California 93106

ABSTRACT

The origin of monocrystalline quartz, the dominant detrital
component of most sandstones, can still not be determined quantitatively with the petrographic microscope and published classifications. With the aid of a technically improved cathodoluminescence microscope that enables the study of low luminescent minerals, six classes of monocrystalline former high-quartz are distinguished and used as a guide to provenance. Quartz crystals
from plutonic rocks and phenocrysts from volcanites show a wide
variation of luminescence colours from blue through mauve to violet. Volcanic quartz phenocrysts usually show a zonation or irregular distribution of luminescence colours which generally enables
distinction from plutonic quartz. Red luminescing quartz is also
of volcanic origin and crystallizes at lower temperatures than
phenocrysts. Quartz crystals which have been plastically deformed,
as evidenced by strong undulatory extinction, luminesce bluish-
black whereas brown quartz is derived from regional metamorphic
rocks. Metamorphic quartz that recrystallised at high temperatures (hornfels, granulites) reverts to a blue luminescence colour comparable to plutonic quartz. Luminescence petrography also
allows discrimination between various types of feldspars and rock
fragments and to estimate quantitatively their abundance. Furthermore, cathodoluminescence enables detrital grains to be distinguished from syntaxial overgrowths in well cemented sandstones so
that the original grain size and roundness parameters can be determined.

G. G. Zuffa (ed.), Provenance of Arenites, 191–211.

1. INTRODUCTION

Luminescence appears to have been first described by Cascierolo, a 17th century alchemist who noticed that the rock of Bologna (barite) emitted light without external thermal stimulation (Zinkernagel 1978). However, it was not until 1879 when Crookes published his observations on solid materials emitting light after electron bombardment that cathodoluminescence was first described.

For over more than half a century cathodoluminescence remained a field of research for physicists and chemists. They were primarily interested in the physical principles of the origin of luminescence induced by cathode radiation and the application of their findings in the production of synthetic crystals, especially phosphors. The research on the nature of cathode radiation also led to the discovery of the electrons by Thompson and the X-rays by Röntgen.

The application of electron-exited luminescence as a petrological tool was suggested by Smith and Stenstrom (1965) who used an electron microprobe to study cathodoluminescence of quartz, feldspar and carbonates. However, it was only after Long and Agrell (1965) and Sippel (1965) combined the design of a simple luminoscope device with the petrographic microscope that cathodoluminescence microscopy became a valuable technique in sedimentary petrography.

After Sippel (1968) convincingly demonstrated that cathodoluminescence reveals many features which are invisible with the petrographic microscope, this new technique has rapidly gained popularity among sedimentary petrographers. Cathodoluminescence has been widely applied to the study of diagenesis in carbonate rocks (e.g. Meyers 1974, Nickel 1978, Richter and Zinkernagel 1981). More recently technically improved CL microscopes have been built which allow the study of low luminescent minerals, such as quartz, in sandstone petrography.

Luminescence petrography enables detrital quartz grains to be readily distinguished from overgrowth cement (Sippel 1968). Additionally, grains of monocrystalline quartz of diverse origins, rock fragments of various kinds and different types of feldspar can all be characterised on the basis of their luminescence colour and used in provenance studies (Zinkernagel 1978, Ramseyer 1983).

The amount of information obtained through quantitative analysis of thin-sections from sandstones and siltstones by cathodoluminescence is much greater than that gained from conventional petrographical analysis and therefore enables a better

discrimination of provenance, transport paths and also of the
diagenesis of the rocks concerned. However, whilst CL microscopy
has been widely applied in diagenetic studies, few authors (see
for example Thomas 1974, Richter and Zinkernagel 1975) have used
it as a tool for sediment provenance. According to Pettijohn et
al. (1973) one of the most difficult problems remaining in sedimentary petrography is the identification of specific provenance
areas particularly where source area weathering severely modifies the initial detrital composition and where problems of recycling are encountered. Similarly, provenance can be extremely
important in studying sediment transport in areas where only
core material is available and palaeocurrent measurements cannot
be undertaken. In these cases cathodoluminescence is a powerful
technique and it seems appropriate to evaluate the potential of
CL microscopy in solving problems of sediment provenance.

2. TECHNIQUES

Luminescence can be examined with the aid of an electron
microprobe (Smith and Stenstrom 1965), special cathodoluminescence microscopes (Sippel 1965, Zinkernagel 1978) or with the
SEM equipped by a CL detector (Coy-Yll 1969, Grant and White
1978). However, for petrographic investigations special CL microscopes are best suited because they allow the observation of a
relatively large area of the same sample in luminescent, plane-
polarised and cross-polarised light. The commercially available
Nuclide LUMINOSCOPE which is equipped with a cold-cathode produces unsatisfactory results with sandstones. Quartz, the dominant
mineral of most sandstones, has very low luminescence intensities which with the Nuclide LUMINOSCOPE produces a limited range
of very dull colours which are effectively impossible to record.
By re-arranging the position and the alignment of electron source,
objective and thin-section and using a hot-cathode, Zinkernagel
(1978) designed a new CL microscope which allows high-quality
colour pictures to be taken of luminescence features of sandstones. Based on Zinkernagel's technical principle (see Zinkernagel 1978 for details), Ramseyer (1982, 1983) designed a technically improved CL instrument that allows very faint luminescence colours having a relatively short decay time to also be
recorded.

Regular uncovered petrological thin sections are most practical to study cathodoluminescence because they allow a direct
comparison of the luminescence features with the observations
made in plane-polarised and cross-polarised light. A high-tempe-
rature epoxy resin is used for impregnating porous samples and
for slide fixing to reduce outgassing to enable a better polish
to be obtained and to aid in recognition of pores. The impregna-
tion is carried out with a blue stained epoxy resin using stan-

dard techniques (vacuum and pressure treatment). A good polish
is important because it enhances the quality of the luminescence
pictures and, even more so, those taken in plane-polarised and
cross-polarised light which fascilitates the interpretation of
the luminescence observations. The thin sections are cleaned
with alcohol in an ultrasonic bath and given a light coating of
aluminium which has several advantages over gold (e.g. lower
costs, higher reflectivity) that is used by Zinkernagel (1978).

The thin sections are bombarded with electrons using an
acceleration voltage of -30 kV at an operating vacuum of $< 10^{-5}$
torr. The luminescence features are photographed with an Ekta-
chrome 400 reversal colour film using exposure times for quartz
that vary from 1 1/2 to 3 1/2 minutes depending on the magnification of the objective used. Petrographical studies are carried
out from colour slides projected onto an 18x18 cm matt screen.
Interpretation of luminescence mineralogical and textural observations is complemented by comparison with plane-polarised and
cross-polarised transparencies of the same field of view. Quantification by modal point-count analysis can be easily undertaken using transparent grids of different grid size depending on
sandstone grain size.

3. NATURE OF LUMINESCENCE

3.1. Physical aspects

A theoretical treatment of the physical principles of the
origin of luminescence effects is beyond the scope of this review
as it would neither contribute to a better understanding nor to
a better interpretation of luminescence observations. In addition, the origin of luminescence is related to rather complicated physical mechanisms that have only in recent years been explained on the basis of modern concepts such as the ligand-field
theory. Luminescence results from the emission of photons in the
ultraviolet, the visible and the infrared range of the electromagnetic spectrum after excitation by charged particles (electrons,
protons, α- particles) or by high energy photons (ultraviolet
and γ-radiation). According to the mode of excitation different
kinds of luminescence are distinguished:

Excitation by UV radiation   → Photoluminescence
    "   "   chemical processes → Bio- and chemolumine-
                                          scence
    "   "   electrons   → Cathodoluminescence

Cathodoluminescence is generated by electrons that have
been accelerated to between about -8 and -50 kV. The sample current density varies according to the type of luminoscope used

from 0.1 to 25 000 uA/mm2. Using a crystal phosphor Adirowitsch (1953) gave a perceptual explanation of luminescence on the basis of the energy band model. According to this model, shown on Figure 1, the process of luminescence may be subdivided into 3 steps:

- the absorption of energy (arrows 1, 2) causing the transition of electrons from the valence band to the conduction band

- the short- or longer-lived storage of electrons (arrows 4, 5)

- the emission of photons ($\widehat{=}$ luminescence) produced by recombination of electrons from the conduction band with ionized activator elements (arrow 3).

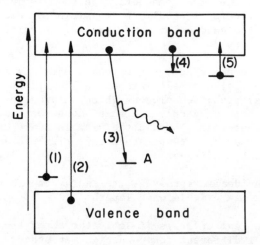

Fig. 1.   Schematic model of energy bands in phosphors, from Adirowitsch (1953).
A: Ionised activator term
(1) and (2) Exitation of electrons
(3) Recombination of electrons with ionised activator causes emission of characteristic light (cathodoluminescence)
(4) Electrons retained in electron traps
(5) Thermal influence brings electrons back into conduction band (thermoluminescence).

The absorption of energy can only take place in discrete steps according to the quantum-theory i.e. a minimal energy level which corresponds to the energy difference between valence and conduction band must be exceeded in order to excite the crystal. The short length of time ($\sim 10^{-8}$ sec) during which the crystal

can remain in the excited condition depends on several factors
such as the recombination probability and electron storage capacity. The crystal returns to its original (lower) energy state
by emitting photons. The energy of these photons is equal to the
energy difference between the valence band and the activator
term (Fig. 1). These activator terms are considered to be either
trace elements or defect centres in the lattice. As this value
is a constant for each crystal species and activator term, narrow emission "bands" are recorded. Cathodoluminescence is thus
caused by the transition of electrons from the valence band (2)
or activator terms (1) to the conduction band followed by recombination (3) as shown on Figure 1. Thermoluminescence in contrast is triggered by the transition of electrons stored at high
energy levels into the conduction band and subsequent recombination (3) according to the general scheme. The preceding discussion illustrates that in most crystals luminescence in the visible range of the electromagnetic spectrum is due to the presence
of activator terms. The direct transition of electrons from the
valence to the conduction band generally results in higher energy UV radiation.

Luminescence studies on carbonates mainly revealed that in
addition to activator elements so called killer elements (quen-
ching terms) are often present which, expressed in a simplified
way, may absorb precisely the same amount of energy that is emitted by luminescence. Thus in the visible range of the spectrum,
luminescence at room temperature may originate as a result of
two different conditions:

(1) in ideal crystals if the energy gap is very narrow
($\Delta E \cong$ visible wave length)

(2) in real crystals if activator terms are present. Several different ionic species may behave as activator
term (see Table 1).

The following causes of cathodoluminescence in different
real crystals species are known

- Presence of trace elements and unfilled lattice positions	Quartz (Sprunt 1981) Feldspar (Mariano, Ito and Ring 1973) Carbonates (Sommer 1972a,b)
- Lattice defects (twinning, lattice transformations, shock damage)	Quartz (Ramseyer and Mullis, in prep.) Feldspar (Sippel and Spencer 1970)
- Relative lattice order - Crystal modifications - Mechanical deformations	Quartz (Zinkernagel 1978) Quartz (Ramseyer 1983) Quartz (Dietrich, pers. comm. 1983)

3.2. Luminescence properties of major framework-building minerals in sandstones

The most important detrital minerals in sandstones are quartz, feldspars and carbonates. The luminescence behaviour of these minerals and the origins of the luminescence is briefly discussed below.

Carbonates. Carbonates often occur as detrital grains in sandstones although even more frequently they form pore filling and grain replacive cements. Cathodoluminescence of carbonates is in general related to the presence of impurities. Bivalent manganese, considered to be the most important activator, stimulates luminescence if present in a concentration of at least 80 ppm (Pierson 1981). The role of iron as a quencher of luminescence has been noted by various authors (e.g. Meyers 1974, Richter and Zinkernagel 1975, Pierson 1981, Amieux 1982). Richter and Zinkernagel (1975, 1981) suspected ferrous iron as the main inhibitor of luminescence whereas Amieux (1982) postulates that ferric iron only is effective (Table 1). Amieux (1982) also suspects that the blue luminescence colour of some calcites is stimulated by the lattice rather than activator elements.

Calcite and aragonite generally show yellow, orange, red and green luminescence colours (Sommer 1972 a, b, Schrank and Friedman 1975, Richter and Zinkernagel 1981, Amieux 1982). Amieux (1982) suggests that two different series of luminescence colours do exist within which the variation of the colours and the intensities are controlled either by

(a) the Mg^{2+} content (Schrank and Friedman 1975) or the oxidation state of manganese (Osiko and Maksimova 1960), or

(b) valence state and/or amount of iron.

Variations in the amount of the Mg^{2+} activator or the valence state of manganese result in yellow to orange-red changes in luminescence colours whereas variation in the valence state and/or amount of iron causes a colour shift from yellow through brown to non-luminescent as shown on Figure 2a.

Dolomite luminesces either yellow or red depending whether Mn^{2+} is present in Ca^{2+} or Mg^{2+} sites (Sommer 1972a). More rarely white, violet and pale blue luminescence have also been observed in dolomite.

The lower threshold of recognizable luminescence is 20-40 ppm Mn^{2+} (Richter and Zinkernagel 1981). These authors and Amieux (1982) also noted a change of luminescence colour from

orange-red over red-brown to brown-black with increasing concentration of iron (Fig. 2b). If the amount of iron exceeds 1.5 weight
percent (= 3.1 % $FeCO_3$) dolomite becomes extinct regardless of
the amount of Mn^{2+} present (Pierson 1981). According to Richter
and Zinkernagel (1981) the extinction is at about 15 % $FeCO_3$.
These greatly differing values may be due to the more sensitive
instrument of Zinkernagel which enables to record low intensities
of luminescence. Cathodoluminescence of dolomite may also be
caused by rare earth elements (Mukherjee 1948) or by Zn^{2+} and
possibly Pb^{2+} (Amieux 1982).

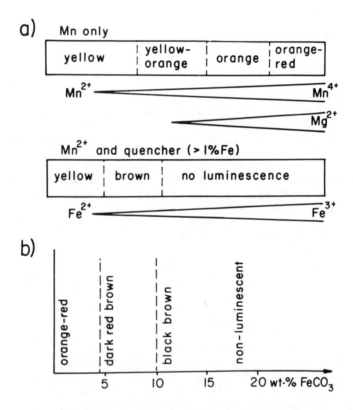

Fig. 2.   (a) Relationship of cathodoluminescence colours
               of calcite and valence state of manganese
               and iron. From Amieux (1982) modified.
               Influence of magnesium from Schrank and
               Friedman (1975).

           (b) Relationship between amount of Fe^{2+} shown
               as weight % $FeCO_3$ and luminescence colours
               in the dolomite-ankerite series. Modified
               from Richter and Zinkernagel (1981).

Magnesite luminesces red according to Sommer (1972a).
Lamiraux (cited in Amieux 1982) also reported pink and blue luminescence without giving further details on its cause (Table 1).

The iron-rich carbonates ankerite and siderite are non-luminescent (Table 1, Fig. 2b). Lamiraux (cited in Amieux 1982) also
recorded orange luminescence colours of siderite.

Feldspars. The causes of the large variety of colours of
luminescence which have been observed in feldspars are unknown
except for those of Ca-plagioclase and synthetic albite (Table 1).
The luminescence colour of Ca-plagioclase and synthetic albite
have been related by Mariano, Ito and Ring (1973) to trace concentrations of titanium, iron, copper, manganese and europium.
Other authors (Smith and Stenstrom 1965, Sippel 1968, Herzog et
al. 1970, Gorz et al. 1970, Kastner 1971) only reported the colour of luminescence. No data exist on the origin of the luminescence colours of potassium feldspars.

Quartz. In contrast to feldspars, a considerable amount of
information has been published on the causes of luminescence
colours of quartz (e.g. Hanusiak and White 1975, Grant and White
1978, Zinkernagel 1978, Sprunt 1981, Ramseyer and Mullis, in
prep.). Quartz shows a gradation of luminescence colours from
blue over different shades of violet to red in addition to brown
and non-luminescent.

Zinkernagel (1978), who was the first to investigate the
relationships between luminescence characteristics of quartz
from different source rocks and modes of formation, concluded
that the degree of lattice order controls the luminescence colour. He distinguished three types of luminescent quartz:

Non-luminescent α- quartz: ordered lattice
Brown quartz   : $\downarrow$
Violet quartz   : disordered lattice

Hanusiak and White (1975) noticed that the intensity of the
(blue) luminescence of α - quartz is much higher below -80°C
than at room temperature. From a comparison of the luminescence
behaviour of Al_2O_3 and MgO with that of α-quartz they concluded
that the blue luminescence colour of quartz is not caused by an
activator element but by the SiO_4^{4-} tetrahedron itself. According
to both our own and also to published data, shown on Table 1,
blue luminescence may be related to different emission maxima.
At least three different emission maxima at wavelengths of 430,
446 and 455 nm can be identified and indicate therefore that the
"blue" luminescent quartz actually comprises different blue colours. These results suggest that blue luminescence is caused by
different factors. With the exception of Sprunt (1981), it is

Mineral	Activator	Quencher	Wavelength (nm)	Cathodoluminescence colour	Reference
Calcite	Mn^{2+}		590	yellow (synthetic crystals)	SOMMER (1972a)
	Mn^{2+} > 20-40 ppm > 80 ppm				RICHTER and ZINKERNAGEL (1981) LONG and AGRELL (1965)
		Fe^{2+}			RICHTER and ZINKERNAGEL (1975)
		Fe^{3+}			AMIEUX (1982)
Low Mg-Calcite High Mg-Calcite				orange-yellow red	SCHRANK and FRIEDMAN (1975)
Aragonite	Mn^{2+}		540	green-yellow (synthethic crystals)	SOMMER (1972b)
Vaterite	Mn^{2+}		595	yellow (synthetic crystals)	SOMMER (1972a)
Dolomite	Mn^{2+}		675	red Mn^{2+} at Ca^{2+} sites	
			597	yellow-orange Mn^{2+} at Mg^{2+} sites	SOMMER (1972a)
		$Fe > 1\%$		non-luminescent	PIERSON (1981)
	Mn^{2+} >30 ppm				RICHTER and ZINKERNAGEL (1981)
	RE, Mn			white, yellow, orange, red violet	MUKHERJEE (1948)
	Zn^{2+} (Pb^{2+})			pale blue	AMIEUX (1982)
Magnesite	Mn^{2+}		677	red	SOMMER (1972a)
	?			blue	LAMIRAUX (1977)
Siderite	?			orange	LAMIRAUX (1977)
Strontianite	Mn^{2+}		590	yellow (synthetic crystals)	SOMMER (1972a)
	Eu^{2+}		410	blue	MARIANO and RING (1975)
Witherite	Mn^{2+}		677	red (synthetic crystals)	SOMMER (1972a)
	?			green	DUDLEY (1976)
Albite	Ti^{4+} Fe^{2+} Fe^{3+} Cu^{2+} Mn^{2+}		460 550 700 420 570	blue green red blue yellowish-green } synthetic crystals	MARIANO, ITO and RING (1973)
Plagioclase	Fe^{3+} >1000 ppm			red	
	Ti^{4+} 500-1000 ppm			blue	
	Mn^{2+} or Fe^{2+}		550	yellow-green	
Quartz	Lattice order		430 620-630	non-lum.| ordered blue brown | ↓ red violet | disordered	ZINKERNAGEL (1978)
	$Fe > Ti$ $Ti > Fe$			red blue	SPRUNT (1981)
	SiO_4^{4-} tetrahedron		455	blue	HANUSIAK and WHITE (1975)
	Al		620		GRANT and WHITE (1978)
	Na H interst.		289 374	ultraviolet ultraviolet	MITCHELL and DENURE (1973)
	Defect centre		430	blue	SIPPEL (1971)
	Si^{3+}		446	blue	MITCHELL and DENURE (1973)
	Mn, Al		650	red	CLAFFY and GINTHER (1959)
	Lattice defects		650-690	broad red peak in α-quartz	RAMSEYER and MULLIS (in prep.)
	h^+, ?Li^+ interst.		510 430-450	green short lived blue	
	O^{2-} vacancy		260	ultraviolet	JONES and EMBREE (1976)

Table 1   Cathodoluminescence colours with wavelengths (where measured) of various carbonate minerals, feldspars and quartz together with suspected activators and quenchers of luminescence. After AMIEUX (1982), modified.

generally assumed that the blue colour is caused by different
kinds of lattice defects. Sprunt (op.cit.) in contrast suggested
the blue luminescence colour of quartz is caused by high amounts
of titanium and low iron contents whilst the red colour resulted
from high iron and low titanium concentrations (Table 1). However, the red luminescence colour of quartz appears to be related to the presence of Al in trace amounts according to Grant
and White (1978).

Most recent studies by Ramseyer and Mullis (in prep.) on
alpine quartz crystals from open veins revealed that the origin
of luminescence colours of α-quartz is dependant on several
factors. These authors showed that the brown luminescence colour, which was produced artificially by electron bombardment,
correlated with the abundance of lattice defects recognized from
the number of surface etch pits. The change to brown colours was
also recognized spectroscopically by an increase of the red peak
at 650 nm (Table 1). Extinction zones in the crystals or zones
of low intensity of luminescence invariably contained higher
concentrations of aluminium ($>$ 1500 ppm Al). Moreover, Ramseyer
and Mullis (op. cit.) recorded the spectra of the short-lived
bluish and green luminescence colours which had been described
by Zinkernagel (1978). These colours clearly show a relationship
with the crystal structure of quartz and are probably stimulated
by positively charged trace elements (H^+ or Li^+) stored in structural channels parallel to the c-axis of the crystal.

In conclusion, the luminescence colours of quartz may be
related to a number of factors, including the degree of lattice
order (i.e. temperature), the stress rate, the Ti/Fe ratio, the
Al concentration as well as the occurrence of trace amounts of
positively charged ions with small ionic radii.

4. THE CATHODOLUMINESCENCE OF QUARTZ, FELDSPARS AND ROCK FRAGMENTS AS A GUIDE TO PROVENANCE

Quartz occurs in sandstones as monocrystalline and polycrystalline or composite grains that may be derived from plutonic, volcanic and metamorphic rocks or recycled from sediments.
Since Sorby (1880) first attempted to identify in thin sections
features of quartz from different source rocks, numerous attempts
were made to use quartz as a guide to provenance. Krynine (1946)
proposed an elaborate classification which is based on grain
shape, extinction and the character of the inclusions. Despite
many more recent investigations (e.g. Blatt 1967, Basu et al.
1975) it is still impossible to identify with confidence the
origin of specific grains of the common monocrystalline quartz
with the aid of the petrographic microscope. This precludes a
quantitative analysis of monocrystalline quartz grains of diffe-

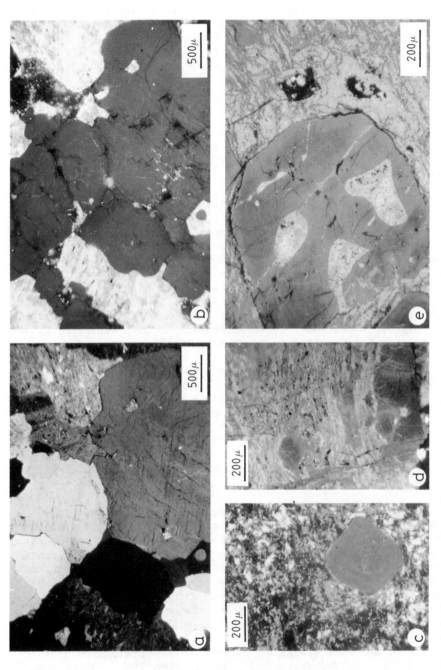

Plate 1 (a,b) Crossed-polars luminescence pair of Rapakivi Granite with blue plutonic quartz, white K-feldspar, yellow calcite in plagioclase (brown). Volcanic zoned or embayed phenocrysts (violet) in feldspathic matrix (c) and in ignimbrites (d,e) with red and red-brown matrix quartz.

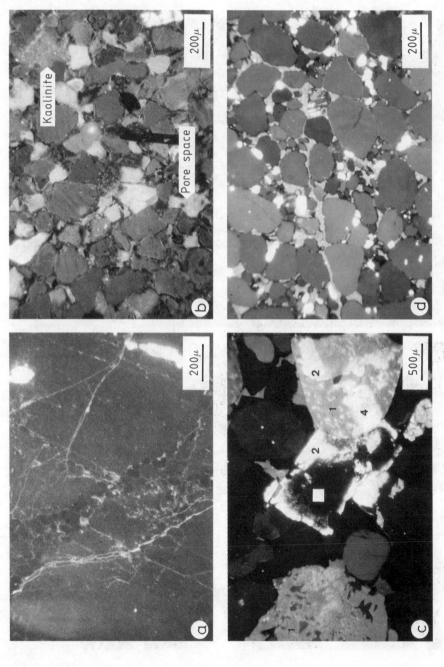

Plate 2 (a) Dull blue luminescing metamorphic quartz with brown subgrains; (b) Metamorphic quartz grains (brn) in Triassic sandstone; (c) Luminescence colours of feldspars:(1) K-feldspar (red, lt. brn),(2) high-albite (white, pink, viol),(3) low-albite (dk. brn),(4) oligoclase (yel); (d) Bimodally sorted sandstone with K-feldspar (white) concentration in fine mode.

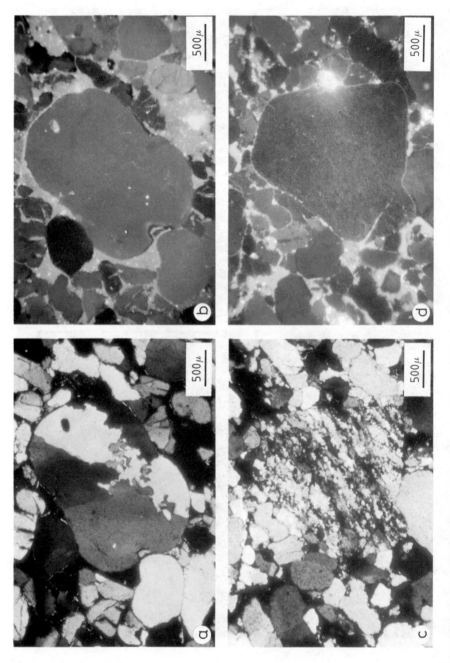

Plate 3 Crossed-polars luminescence pairs of polycrystalline quartz grains derived from (a,b) plutonic and (c,d) metamorphic source (cf. with Plates 1, 2).

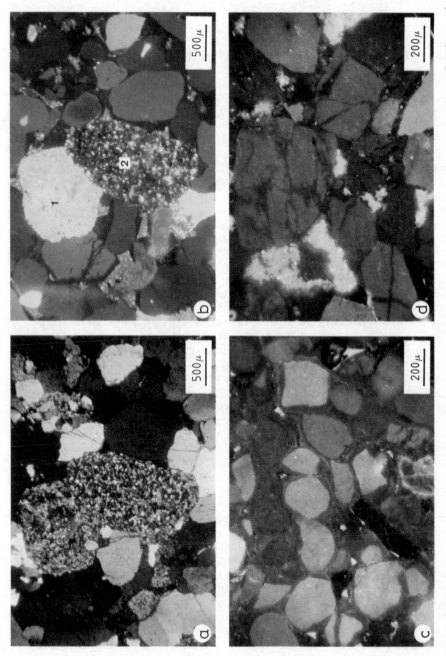

Plate 4 (a,b) Crossed-polars luminescence pair for rock fragment identification (1 felsic volcanic clast, 2 siltstone); (c) Originally well-rounded detrital quartz grains in multiphase quartz cemented sandstone; (d) Authigenic kaolinite (bright blue) replacing former K-feldspar grains (white luminescing relics).

rent origin in sandstones, although volcanic quartz can frequently be identified.

Zinkernagel (1978) using his new CL microscope distinguished
three classes of monocrystalline quartz which he assigned to specific groups of source rocks. Brown quartz is derived from metamorphic rocks, non-luminescent authigenic quartz from sediments
and the large class of violet luminescing quartz is from volcanic (blue-violet) plutonic (red-violet) or contact metamorphic
rocks.

We prefer a more detailed subdivision of luminescing high
temperature quartz (originally mostly the β-variety) into six
types, namely light blue to blue, bluish black, blue-violet (mauve), violet, red and brown. The provenance of these luminescence
types was identified by examination of the CL of quartz in specific types of crystalline rocks.

Plutonic quartz as well as quartz phenocrysts in volcanic
rocks show a wide variation of luminescence colours from blue
through mauve to violet (Plate 1a-e). This similarity between
plutonic and volcanic quartz is not surprising because phenocrysts in volcanic rocks crystallise early under hypabyssal or
plutonic conditions. In addition, quartz phenocrysts generally
show zoned or inhomogenous distribution of luminescence colour
which allows to distinguish plutonic from volcanic monocrystalline quartz. Matrix quartz in volcanic rocks luminesces red
which is believed to be related to its lower crystallization
temperature and/or rapid rate of crystallization (Plate 1d-e).

Bluish-black luminescence is observed in plutonic quartz
crystals which have been plastically deformed as shown by their
strong undulatory extinction. The change from the originally
blue to this dull colour is the effect of a reduced intensity of
luminescence. It appears that this type is most abundant in geo-
logically old sandstones such as for example the Torridonian
sandstone (Plate 2a). Polygonisation of quartz grains, producing
subgrains, progressively destroys undulous extinction (Voll 1969)
and also the original blue luminescence colour. The subgrains
attain a dull brownish-black colour (Plate 2a). Polygonisation
and the formation of subgrains is accompanied by an increase in
temperature and often recrystallisation according to Voll (1969).
The loss of the luminescence may therefore reflect cleaning of
the quartz lattice from trace elements and from lattice defects
(Ramseyer 1983). Brown quartz such as shown on Plate 2b is characteristic of regional metamorphic rocks as discussed by
Zinkernagel (1978). However, observations on hornfels and granulites indicate that quartz which recrystallised at elevated temperatures under the conditions of high-quartz crystallisation,
reverts to blue luminescence. When occurring as detrital mono-

crystalline grains this metamorphic quartz is often indistinguishable from plutonic quartz.

In the same way as for monocrystalline quartz CL microscopy
can be used to subdivide feldspars which according to electron
microprobe analysis have identical major element chemical composition (Ramseyer 1983). For example, we have so far distinguished
four luminescence types of potassium feldspar and six types of
high-albite in addition to various types of plagioclase (Plate
2c, d). Zoning, fracturing, twinning and intergrowth habits are
also visible on CL photographs. Such information aids in determining single or multiple sources and furthermore allows to identify the primary source rocks using the triangle of Trevena and
Nash (1979, 1981).

Rock particles are the most informative grains because they
carry their own signature of provenance. They can usually be
determined with greater confidence than the different types of
monocrystalline quartz using the petrographic microscope. Nevertheless, CL microscopy is of significant diagnostic value in
further refining the identification of rock particles with respect to their origin, particularly the fine-grained varieties,
including polycrystalline quartz (Plate 3). Grains of meta-plu-
tonic, -volcanic and -sedimentary origin as well as fine-grained
quartzites and acid felsic volcanic rocks which are difficult to
determine otherwise are also easily recognized (Plate 4a-b).
However, it should be pointed out that neither CL analysis nor
conventional petrographic analysis of thin sections with the
petrographic microscope by itself provides all the information
on provenance. It is the coupling of techniques, i.e. the quantitative modal analysis of CL images and simultaneous comparison
of the observed features with those visible in thin section under crossed nicols, that allows discrimination of the different
types of monocrystalline quartz, feldspars and rock fragments
and to estimate quantitatively their abundance.

5. STUDY OF DIAGENETIC MINERALS AND FABRICS AS AN AID TO RECONSTRUCTING ORIGINAL TEXTURE

With luminescence petrography techniques it is possible to
distinguish detrital grains from optically continuous quartz cement (Sippel 1968) so that size, roundness and shape of the grains
are clearly seen as shown on Plate 4c. The amount of quartz as
well as of other cementing minerals and porosity can be estimated
accurately using CL techniques. Moreover, the amount of pressure
solution is easily determined (Plate 3d). The amount of porosity
loss from pressure solution can then be estimated using the
pressure solution-porosity plots of Beard and Weyl (1973). In
cases where grain fracturing and healing of the fractures or

rotation of grains have occurred it can be seen with luminescence (Plate 4d). The age relationships of the fractures can be determined, particularly whether fracturing occurred prior or after deposition. Luminescence photographs also allowed to identify detrital grains which had been replaced by calcite which with other petrographic techniques would have been interpreted as pore-filling cement. Furthermore, dissolved grains and grains replaced by clay minerals were identified as former potassium feldspar grains from the presence of tiny relics only visible under luminescence (Plate 4d). These examples show that CL microscopy may also help to reconstructing primary mineralogy.

6. CONCLUSIONS

The application of sophisticated techniques such as CL microscopy and electron microprobe analysis, in addition to conventional petrographical investigations, provides a more detailed modal analysis. Furthermore, the determination of original grain size and roundness in diagenetically modified sandstones permits a more complete analysis of provenance. However, even such an advanced petrological analysis may not give a straight-forward solution to the difficult provenance problem because quartz grains may be of recycled origin. Cross-checking the petrographical information derived from the detailed study of different constituents and using stratigraphical and sedimentological data will contribute to the analysis of provenance. Future studies on the still incompletely understood origins of the different luminescence colours of the major detrital minerals may shed light on their conditions of crystallization and history of deformation.

ACKNOWLEDGEMENTS

We wish to thank S.D. Burley for discussions on the text and for improving the English, Ms. R. Gubler for typing the manuscript and Mr. A. Werthemann for photographic assistance. Support from the Swiss National Science Foundation (Grant No. 2.913-0.83) is gratefully acknowledged.

REFERENCES

ADIROWITSCH, E.I. 1953. Einige Fragen zur Theorie der Lumineszenz der Kristalle, Akademie Verlag Berlin, 298 pp.
AMIEUX, P. 1982. La cathodoluminescence: méthode d'étude sédimentologique des carbonates. Bull. Centres Rech. Explor. - Prod. Elf-Aquitaine 6, 737-483.

BASU, A., YOUNG, S.W., SUTTNER, L.J., CALVIN, W.C. and MACK, G.H. 1975. Re-evaluation of the use of undulatory extinction and polycrystallinity in detrital quartz for provenance interpretation. J. sediment. Petrol. 45, 873-882.

BEARD, D.C. and WEYL, P.K. 1973. Influence of texture on porosity and permeability of unconsolidated sand. Bull. amer. Assoc. Petroleum Geol. 76, 349-369.

BLATT, H. 1967. Original characters of clastic quartz. J. sediment. Petrol. 37, 401-424.

CLAFFY, E.W. and GINTHER, R.J. 1959. Red-luminescing quartz. J. opt. Soc. Amer. 49, 412-413.

COY-YLL, R. 1969. Quelques aspects de la cathodoluminescence des minéraux. Chem. Geol. 5, 243-254.

CROOKES, W. 1878. Contributions to molecular physics in high vacua. Phil. Trans. 170, 641-662.

DUDLEY, R.J. 1976. The use of cathodoluminescence in the identification of soil minerals. J. Soil Sci. 27, 487-494.

GORZ, H., BHALLA, R.J.R.S.B. and WHITE, E.W. 1970. Detailed cathodoluminescence characterization of common silicates. In: WEBER, J.W. and WHITE, E. (eds.), Space applications of solid state luminescent phenomena, 1-12.

GRANT, P.R. and WHITE, S.H. 1978. Cathodoluminescence and microstructure of quartz overgrowths on quartz. Scanning Electron Microscopy, 1978, 789-794. SEM Inc. AMF O'Hare, Chicago.

HANUSIAK, W.M. and WHITE, E.W. 1975. SEM cathodoluminescence for characterization of damaged and undamaged alpha quartz in respirable dusts. In: JOHARI, O. and CORVIN, I. (eds.). Scanning Electron Microscopy 1975, 125-132, III Research Institute, Chicago.

HERZOG, L.F., MARSHALL, D.J. and BABIONE, R.F. 1970. The Luminoscope - a new instrument for studying the electron stimulated luminescence for terrestrial, extraterrestrial and synthetic materials under the microscope. Pennsylvania State Univ., Spec. Publ. 70-101, 79-98.

JONES, C.E. and EMBREE, D. 1976. Correlations of the 4.77-4.28-eV luminescence band in silicon dioxide with the oxigen vacancy. J. Applied Physics 47, 5365-5371.

KASTNER, M. 1971. Authigenic feldspars in carbonate rocks. Amer. Mineralogist 56, 1403-1442.

KRYNINE, P.D. 1946. Microscopic morphology of quartz types. Proc. 2nd Pan-Am. Cong. Mining Engn. and Geology 3, 2nd Comm., 35-49.

LONG, J.V.P. and AGRELL, S.O. 1965. The cathodoluminescence of minerals in thin section. Mineral. Mag. 34, 318-326.

MARIANO, A.N., ITO, J. and RING, P.J. 1973. Cathodoluminescence of plagioclase feldspars. Geol. Soc. Am., Abstr. Programs, 5, 726.

MARIANO, A.N. and RING, P.J. 1975. Europium-activated cathodoluminescence in minerals. Geochim. cosmochim. Acta 39, 649-660.

MEYERS, W.J. 1974. Carbonate cement stratigraphy of the lake
   valley formation (Mississippian), Sacramento Mountains, New
   Mexico. J. sediment. Petrol. 44, 837–861.
MITCHEL, J.P. and DENURE, D.G. 1973. A study of SiO_2 layers on Si
   using cathodoluminescence spectra. Solid–State Electronics 16,
   825–839.
MUKHERJEE, B. 1948. Cathodoluminescence spectra of indian calcites, limestones, dolomites and aragonites. Ind. J. Phys. 22,
   305–310.
NICKEL, E. 1978. The present status of cathode luminescence as
   a tool in sedimentology. Minerals Sci. Engng. 10, 73–100.
OSIKO, V.V. and MAKSIMOVA, G.V. 1960. Valence of the mangenese
   activator in crystal phosphors. Optics Spectroscopy N.Y. 9,
   248.
PETTIJOHN, F.J., POTTER, P.E. and SIEVER, R. 1973. Sand and
   sandstone. Springer–Verlag, New York, Heidelberg, Berlin,
   618 pp.
PIERSON, B.J. 1981. The control of cathodoluminescence in dolomite by iron and manganese. Sedimentology 28, 601–610.
RAMSEYER, K. 1982. Cathodoluminescence Microscopy, a powerful
   tool in sandstone petrology. 11th Int. Congress on Sedimentology, Hamilton, Abstracts, 120.
RAMSEYER, K. 1983: Bau eines Kathodenlumineszenz-Mikroskopes
   und Diagenese-Untersuchungen an permischen Sedimenten aus
   Oman. Unpubl. Ph.D. thesis Univ. Berne, vol. 1 (text 152 pp.),
   vol. 2 (Figures, Tables).
RAMSEYER, K. and MULLIS, J. (in prep). Relationships between
   growth conditions and cathodoluminescence colours in natural
   alpha quartz.
RICHTER, D.K. and ZINKERNAGEL, U. 1975. Petrographie des
   "Permoskyth" der Jaggl-Plawen-Einheit (Süd-Tirol) und Diskus-
   sion der Detritusherkunft mit Hilfe von Kathoden-Lumineszenz-
   Untersuchungen. Geol. Rdsch. 64, 783–807.
RICHTER, D.K. and ZINKERNAGEL, U. 1981. Anwendung der Kathoden-
   lumineszenz in der Karbonat-Petrographie. Geol. Rdsch. 70,
   1276–1302.
SORBY, H.C. 1880. On the structure and origin of non-calcareous
   stratified rocks. Geol. Soc. London Proc. 36, 46–92.
SCHRANK, S.A. and FRIEDMAN, G.M. 1975. Lithified layers from
   subbottom carbonate sediments of the Red Sea: Deep Sea Drilling Project, Leg 23B. 9e Congr. int. Séd., Nice, Thème 7,
   198–206.
SIPPEL, R.F. 1965. Simple device for luminescence petrography.
   Rev. scient. Instrum. 36, 1556–1558.
SIPPEL, R.F. 1968. Sandstone petrology evidence from luminescence petrography. J. sediment. Petrol. 38, 530–554.
SIPPEL, R.F. 1971. Luminescence petrography of the Apollo 12
   rocks and comparative features in terrestrial rocks and meteo-
   rites. Proc. 2nd Lunar Sci. Conf. 1, 247–263.

SIPPEL, R.F. and SPENCER, A.B. 1970. Cathodoluminescence properties of lunar rocks. Science 167, 677–679.

SMITH, J.V. and STENSTROM, R.C. 1965. Electron-excited luminescence as a petrologic tool. J. Geol. 73, 627–635.

SOMMER, S.E. 1972a. Cathodoluminescence of carbonates: 1 – Characterization of cathodoluminescence from carbonate solid solutions. Chem. Geol. 9, 257–273.

SOMMER, S.E. 1972b. Cathodoluminescence of carbonates: 2 – Geological applications. Chem. Geol. 9, 275–284.

SPRUNT, E.S. 1981. Causes of quartz cathodoluminescence colours. Scanning Electron Microscopy 1981. SEM Inc. AMF O'Hare, Chicago.

THOMAS, J.B. 1974. Cathodoluminescence as applied to sandstone petrology: sediment-source rocks relationships. Nuclide Corp. Pubs. No. 1487-1174, 1–10.

TREVENA, A.S. and NASH, W.P. 1979. Chemistry and provenance of detrital feldspar. Geology 7, 475–478.

TREVENA, A.S. and NASH, W.P. 1981. An electron microprobe study of detrital feldspar. J. sediment. Petrol. 51, 137–150.

VOLL, G. 1969. Klastische Mineralien aus den Sedimentserien der Schottischen Highlands und ihr Schicksal bei aufsteigender Regional- und Kontaktmetamorphose. Technical University Berlin, 206 pp.

ZINKERNAGEL, U. 1978. Cathodoluminescence of quartz and its application to sandstone petrology. Contrib. Sedimentology 8, 69 pp.

ENVIRONMENTAL INTERPRETATION OF QUARTZ GRAIN SURFACE TEXTURES

David Krinsley
Patrick Trusty

Department of Geology, Arizona State University
Tempe, AZ 85287

ABSTRACT

The surface texture or roughness of quartz sand grains has
been examined with scanning electron microscopy in the secondary
electron mode.   Information obtained is used to define sedimentary environments and, to a lesser extent, post-depositional or
diagenetic conditions.   Sands from modern depositional environments were initially examined, and the features characteristic
of specific environments were classified.   These features were
then reproduced in laboratory experiments to produce textural
confirmation of modern environments and to provide new insights
into origin mechanisms.
The technique was then used to define ancient sedimentary
environments, generally in conjunction with other techniques.
A number of applications in the field of sedimentology and
other sciences are indicated.   These include stratigraphy,
structure, planetary geology, forensic sciences and soil science.

INTRODUCTION

This paper is concerned with sand grain surface textural
analysis as a method for identifying ancient sedimentary
environments and certain types of diagenesis.   Individual sand
grain surface roughness or texture is examined and classified
according to modern depositional environments and then used to
determine types of environments in the fossil record.   A
description of the technique, its value to geology and other
fields, and the problems inherent in the method are discussed.
The scanning electron microscope (SEM) is used in the secondary

G. G. Zuffa (ed.), Provenance of Arenites, 213–229.

electron mode to image surface topography, and in the cathodoluminescence mode to image distortions in the crystal lattice
and/or small amounts of impurity ions in combination with
surface texture. The former mode has been developed to a much
greater extent than the latter and will be discussed here almost
exclusively; it frequently permits the separation and identification of ancient depositional and transport environments
representing wind, water, glaciation and weathering, in addition
to diagenesis (Krinsley and Doornkamp, 1973). Certain subenvironments have also been recognized. The technique has been
applied to sand grains as old as the late Precambrian; generally
the method is applicable as long as the sediment is at least
partially unconsolidated.

Properties of contemporary environments of sedimentation
can be divided into three groups; the first include properties
that exist only in contemporary sediments and not in the
lithology, while the second group contains properties which
can be analyzed in both modern sediments and lithology, but
which change diagenetically with time. The third group encom-
passes properties which are the same in both modern sediments
and lithology. Surface textures fall mostly into the third
group and occasionally into the second. This third type of
measure is extremely important in environmental reconstruction,
as the correspondence between modern and ancient parameters is
quite close and permits environmental correlations to be made,
assuming parameter origin is understood with a high degree of
certainty. This is one of the major reasons why the origin of
sand grain textures should be studied in great detail; knowledge
of the physical and chemical manner in which features originate
can be directly related back to the lithology.

HISTORICAL BACKGROUND

Probably the first comments in the literature on quartz
sand surface textures appeared in Henry Clifton Sorby's address
to the Geological Society of London in 1880. He discussed
rounding of sand grains in various environments and described
certain chemical and mechanical textures observed with the light
microscope. Pettijohn in his first edition of Sedimentary Rocks
(1949), was concerned with the "surface character" of sedimentary grains; he believed that interpretation was less than
satisfactory and that more experimental and observational data
were needed before a satisfactory understanding of the subject
could be achieved. All of the work up to that time - - and it
was very little - - had been done with the light microscope.
It was difficult to see much textural detail when examining
individual grains, as resolution was poor ($\sim 2000 \overset{\circ}{A}$). Depth of
field also left a great deal to be desired, so much so in fact
that it was impossible to view a grain of 100 micron diameter

at reasonably high magnification with every part of the sand grain in focus. Additionally, the very tiny markings on the surfaces of the grain could not be observed with reasonable clarity. This difficulty was overcome to some extent when geologists began to use the transmission electron microscope (TEM) to study surface features on sand grains. Sand grain surfaces at the high magnifications and resolutions permitted by the TEM were examined for the first time in 1962 (Biederman, 1962; Porter, 1962; Krinsley and Takahashi, 1962). Groups of specific environmental surface textures were established to identify eolian, beach, and glacial environments; observational and simulation experiments were used to accomplish this end. However, there were a number of problems due to the fundamental character of the TEM. It was necessary to construct replicas of individual sand grains, a tiresome and time-consuming process. The original surfaces were not examined, and as a result, certain features were misidentified or ignored. The replica technique did not duplicate features which contained a great deal of relief; certain of these which were not observed using TEM replicas were subsequently imaged and shown to be definitive when examined with the scanning electron microscope (SEM). The latter device was not developed until the late 1960's, and it was used for the first time to study sand grains in 1969 (Krinsley and Margolis, 1969). All subsequent work of any note was done with SEM.

Diagenesis was also a problem when examining the textures on ancient grains, but it was often possible to study diagenesis itself and, as a bonus, to determine sedimentary environments through which the grains had passed. A tremendous amount of detail was observed on every grain, so much so that researchers were overwhelmed with data. Additionally, several grains from a deposit which appeared to be uniform often did not show exactly the same distribution of textures. This led workers to suspect that there was so little textural uniformity that it might be extremely difficult to characterize the deposit as a whole. How many grains per deposit should be studied? How large an area of each grain should be examined, and at what magnifications? Did the size of the features studied make a difference in the final analysis? Many of these questions and others are unanswered, even today.

Initially, quartz sand grains from a number of different environments such as eolian, beach, glacial, and river were selected, examined, studied and classified for characteristic surface topography or texture with TEM and subsequently SEM. Experiments were then performed in an attempt to duplicate the textures observed; it was hoped that the experience gained by producing the textures artificially might provide insight into the natural mechanisms involved (Krinsley and Takahashi, 1962; Kaldi et al., 1978; Linde and Mycielska-Dowagiallo, 1980). These experiments were successful in a gross way, but detailed statistical analyses were not performed.

The technique was then used to identify ancient environments, and checked against other, independent data from known
environments. Again results were usually satisfactory in
unconsolidated or semiconsolidated samples if diagenesis had not
been too intense. Ideally, SEM should be used in conjunction
with other environmental analysis methods, but if no other
method is practicable it may be used as a last recourse.

One difficulty is that occasional intense weathering in the
modern environment following mechanical action (K. Pye, personal
communication) and/or long-term post-depositional solution and
precipitation may remove all traces of mechanical action,
eliminating the possibility of environmental interpretation.
However, if large numbers of sand grains are studied, it is
often possible to find locations where chemical action has not
masked the original mechanical features. This is the case
because chemical action is so variable that, given enough
searching, several episodes of both chemical and mechanical
action can often be observed. Frequently the relative age
relations of different textures can be determined by studying
their contacts.

It is also sometimes possible, in the case of weathered or
otherwise unsuitable surfaces, to etch down to the original
surface (assuming that it has not been removed) or to use
cathodoluminescence (CL). This latter technique gives subsurface information to depths of at least a micron or so
(10,000 $\overset{o}{A}$), as compared to secondary electron depths of several
hundred angstroms. It may thus be possible to study surfaces
covered with precipitated silica. The kind of information
obtainable with CL, however, is somewhat different, so that the
correspondence between the two methods may be difficult to
determine.

Another problem is the distinction between modern surface
chemical action and diagenetic alteration. This is difficult,
as criteria have not been established which will distinguish
between the two during early diagenesis. A number of differing
solution and precipitation features have been noted in the early
stages (Waugh, 1965, 1970), but they cannot be separated into
pre- and post-depositional groupings. However, a later stage,
in which quartz terminations are formed, only seems to occur in
diagenesis. Presumably individual quartz grains would grow
together as the terminations interlocked and a dense quartzite
would result. This subject warrants further work.

TEXTURAL FEATURES

The types of features detected on quartz grain surfaces can
be classified as mechanical and chemical features, or combinations of the two. The mechanisms producing these features are

conchoidal fracture, cleavage (contrary to popular opinion, quartz does have cleavage or slip), and solution-precipitation. The abundances, intensities and spatial arrangements of the above three features can be used successfully to delimit a number of weathering, depositional and diagenetic environments.

DEPOSITIONAL ENVIRONMENTS

Weathering and Youthful River Textures. Quartz grain textures from granites and their weathering residues were studied in addition to grains from youthful rivers which presumably contained modified original granitic textures (Coch and Krinsley, 1976). Detailed examination was made of granite grus (a weathering residue) and quartz from a single youthful stream draining the granite, both from the Black Hills of South Dakota. Grus grains could generally be identified by microblock textures; a number of other minor features were present (Figure 1). As the granite quartz grains were carried downstream, the microblocks were progressively destroyed and their place taken by upturned plates (Margolis and Krinsley, 1971) (Figure 2). The plates are larger and more irregular on river grains than on grains of eolian origin. The relative distance of transport could be determined by the ratio between microblocks and upturned plates.

Grain frosting as observed with the naked eye and the light microscope is a function of this ratio. Microblocks, being larger and smoother, permit light to pass through without much scattering; thus the grains appear to be clear. Upturned plates are small and their spacing approaches the wave length of visible light; thus the tiny ridges diffract or scatter visible light and a frosted appearance results.

Subaqueous Textures. Subaqueous features include rounded and smoothed grain edges (Figure 3); these tend to be found on littoral, shelf and turbidite sands. Grains between about 1000 to 1500 μm contain the best examples of these features; abrasion has less effect on smaller grains because of momentum considerations. Grains of less than 200 μm contain few subaqueous features of any kind.

Characteristic mechanical V-shaped patterns and grooves are found on rounded and smoothed grain edges; the former generally have a density greater than 3 V's/μm^2 (Figure 4). The V's appear to be a series of notches cut in upturned cleavage plates (parallel, thin, flat ridges, oriented at some angle to the grain surface) which are ubiquitous over many subaqueous abraded surfaces (Krinsley and Doornkamp, 1973). It is probable that the large V's represent single mechanical events with the smaller ones the result of irregular grinding of loose debris created by the original impact or perhaps ultrasonic energy generated during collision. Fracture propagation of energy

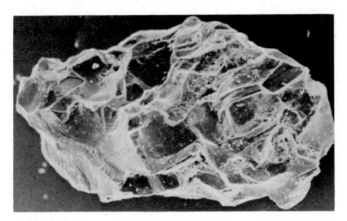

Fig. 1. Angular quartz grain from the black hills granite of South
Dakota, USA; grain has been carried several miles downstream
from the source. Note flat, unabraded microblocks   and
microblocks which have apparently suffered abrasion
Distance across bottom of micrograph, ∿600 microns.

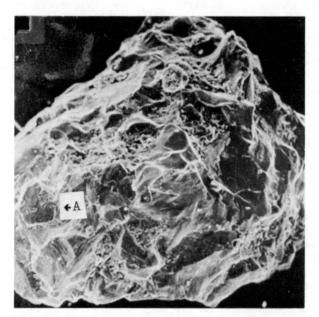

Fig. 2. Same location and grain types as in Plate 1. Note the clear
microblocks   and those with abrasion on their faces.
This grain appears to have suffered more abrasion than the
grain pictured in Plate 1, as more microblocks appear to
have been abraded. A is a biotite grain. Distance across
bottom of micrograph ∿900 microns.

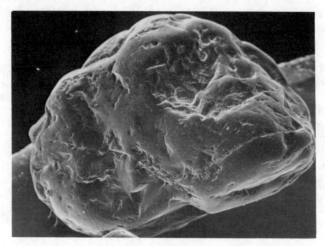

Fig. 3. Rounded quartz grain from high energy beach zone near Sandy
         Hook, N.J., USA. Grain has numerous rounded protrusions on
         edges, a characteristic of high energy waves. Characteris-
         tic V (indented) shaped patterns are found on these edges.
         Distance across bottom of micrograph 440 microns.

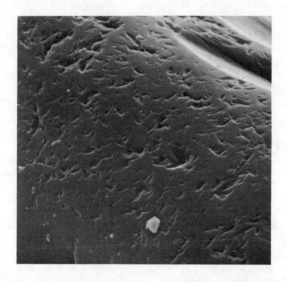

Fig. 4. Surface of quartz sand grain of Upper Miocene age, Argentine
         Shelf, South Atlantic Ocean. Note characteristic indented
         V-shaped patterns and straight depressions which in places
         represent one arm of a series of V's. This pattern is
         characteristic of subaqueous action and in this case
         probably represents movement in the littoral or beach zone.
         Distance across the bottom of the micrograph 20 microns.

across crystallographic plates probably produced the upturned cleavage plate topography. At high magnifications, rounded surfaces no longer appear smooth; cleavage plates and V-shaped patterns are generally ubiquitous on all abraded areas.

Straight or slightly curved grooves are observed scattered over subaqueous grain surfaces at the same magnifications which best show the mechanical V patterns; the grooves are never as numerous as the V's and frequently include satellite V's, the depressions forming one side of a given V (Figure 4). The grooves are depressed slightly below the surface and may be fairly straight, curve once or twice, or may even extend in ellipses with satellite V's along their lengths. They are not present on smaller grains (less than about 200 μm in diameter) perhaps because there is a decrease in imparted abrasion energy.

Although etch pits or V's formed by chemical etching may occur in any environment, they are most common in subaqueous situations and may be either syndepositional or diagenetic; the former is much more common. They tend to have very regular sides as compared to the somewhat irregular mechanical V's. Etch pits are all oriented almost exactly the same and are located on either rhombohedral or prismatic quartz faces.

Depressions of varying size which contain evidence of solution and precipitation of silica occur irregularly scattered across subaqueous grain surfaces. Mechanical V's are not present within the depressions suggesting that the latter are protected from abrasion because of their location below the general grain level. Occasionally depressions are observed which contain conchoidal breakage patterns (probably representing a previous breakage cycle) indicating that they have not been acted upon chemically. However, these features contain precipitated, amorphous silica in more than 90% of the cases observed.

As subaqueous quartz sand grains of smaller and smaller size are examined, chemical solution and precipitation replace abrasion markings. Grains less than 100 to 200 μm in diameter almost never contain evidence of abrasion. Chemical action includes etching with the formation of etch pits as described above, formation of quartz crystal terminations on the surfaces of upturned plates and perhaps most numerous, irregular layers of precipitated silica covering most or all of a given grain with irregular solution markings scattered about the grain surface. The layers described may be very thin so that it is possible to see the underlying masked topography. The extent to which precipitation occurs is a function of a number of variables including time spent in the environment, surface characteristics (chemical and physical) of a given grain, the chemical environment of the surrounding water envelope, temperature and perhaps several others. After a grain surface has been abraded, it is chemically reactive due to the presence of broken bonds and thus solution and/or precipitation may occur rapidly.

Eolian Textures. There are five types of textural features that are characteristic of eolian quartz grains from modern hot deserts. The most obvious and common feature is grain rounding, which ranges from perfectly spherical to somewhat angular, but always with some spherical edges (Figure 5).

"Upturned plates" commonly cover the surfaces of most grains greater than 300 to 400 μm in diameter (Krinsley and Doornkamp, 1973). These plates appear as more or less parallel ridges ranging in length from about 0.5 to 10 μm and are the result of breakage along cleavage planes in the quartz lattice (Figure 6). Sand grains of this relatively large size generally travel as saltating or creeping bed load experiencing a succession of high-velocity collisions. At the time of collision, the kinetic energy of each particle is at least partly converted to elastic energy in the grain. When typical eolian velocities are compated with grain velocities during aqueous transport (Bagnold, 1941), it becomes evident that the kinetic energy of a wind-moved particle (varying with the square of its velocity) must often be several hundred times greater than that of a particle moved by water. The results of these high-energy elastic collisions appear to be "abrasion fatigue" (Pascoe, 1961), and the upturned plates are thought to be resulting cleavage scarps. These plates are frequently modified in desert environments by solution and precipitation (Ricci Lucchi and Casa, 1970; Margolis and Krinsley, 1971).

Equidimensional or elongate depressions, 20 to 250 μm in maximum dimension in size on smaller grains, are caused by the development of conchoidal fractures on the grain surface (Figure 5). They may result from direct, as opposed to glancing impact between saltating or creeping grains.

Smooth surfaces occur on smaller grains (90 to 300 μm diameter); they are caused by precipitation (Waugh, 1970) and solution of silica and are unaffected to any great extent by abrasion (Krinsley and Doornkamp, 1973). Grains of this size are more normally carried in suspension rather than by creep or saltation (Bagnold, 1941) and are, therefore, unlikely to collide. When collisions occur, development of features caused by abrasion fatigue are less common.

Arcuate, circular or polygonal fractures are most commonly found on smaller (90–150 μm diameter) grains (Ricci Lucchi and Casa, 1970; Krinsley and Doornkamp, 1973). These features may be the result of physical or chemical weathering, possibly including the crystallization of salts. Although weathering action might be expected to occur at times of rest on grains of all sizes, the textural evidence for it would be removed by abrasion of the larger saltating or creeping grains.

Eolian sand grains from coastal and periglacial environments have been examined (Margolis and Krinsley, 1971), and significant variations in the occurrences of the above four features as compared to grains from desert environments have been noted. These

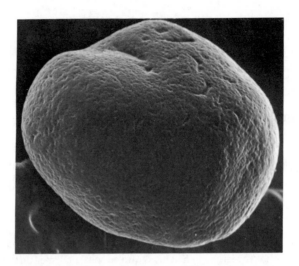

Fig. 5. Well-rounded quartz sand grain from the Lybian Saharan
        Desert. Note almost complete lack of corners, and fairly
        regular, fine texture not present on source and subaqueous
        grains. Also note large elongated depressions and
        indentation at top of grain. Distance across bottom of
        micrograph 225 microns.

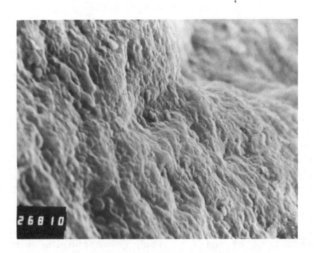

Fig. 6. Detailed surface photograph of a well-rounded quartz sand
        grain from the Lybian Saharan Desert. Note approximately
        parallel, "upturned plates", which appear to be ubiquitous
        on hot desert grains, and which also are found on both
        coastal and cold dune sands. The plates in this photograph
        have been subdued with a thin layer of precipitated silica.
        Distance across bottom of micrograph 31 microns.

generalizations are based on examination of 20 samples from each environment; about 25 grains were studied in each sample.

Coastal eolian sand grains contain upturned plates, but on only on small patches of grain surface. In contrast, hot desert grains show plates over most of their surfaces. Equidimensional or elongate depressions occur very occasionally on coastal eolian grains but are almost always present on hot desert sand grains larger than 500 μm in diameter. Smooth surfaces and arcuate, circular or polygonal cracks on smaller grains are found frequently on hot desert sand but again are very rarely found on coastal eolian sand.

Periglacial eolian sand also contains upturned plates, but only in small patches on grain surfaces. Equidimensional or elongate depressions occur with greater frequency on periglacial sand than on coastal eolian grains, but their frequency is less than on hot desert sand grains. Smooth surfaces on smaller grains and arcuate, circular or polygonal cracks are seldom found on periglacial eolian sand.

Glacial Textures. Whalley and Krinsley (1974) have described the types of textures on quartz grains from glacial environments. It appears that many of the typical fracture patterns described previously in the literature such as arc steps, parallel and subparallel steps and arc-shaped steps or grooves (Krinsley and Donahue, 1968; Krinsley and Margolis, 1969) probably originated either in the parent rock or when quartz grains were removed from that rock by weathering (Figure 7). In the case of "wet-base" glaciers, grinding in an environment where a thin layer of water surrounds the grains probably produces microblock and upturned plate textures. "Dry-base" glaciers may produce the various conchoidal patterns, as arc-shaped steps, parallel and subparallel steps and arc steps indicated above. No surface textures described so far reliably characterize any particular glacial subenvironments which are extremely variable. On the other hand, this great variability (in terms of type and size of feature) is generally character- istic of glacial deposits. In particular, the variation in the amount of precipitation on grain surfaces and the degree to which small debris become attached or cemented is fairly definitive. Thus, although glacial subenvironments cannot as yet be characterized by surface textures, the glacial environ- ment can.

Diagenetic Textures. Diagenetic textures on quartz grains have been studied in detail by Waugh (1970) in connection with the general study of diagenesis in continental red beds via SEM. A number of other authors have also described diagenetic alteration of quartz grains, which generally involve various stages of overgrowth formation and pressure solution (Waugh, 1965, 1970; Austin, 1974; Riezebos, 1974; Whalley and Chartres, 1976). The study of quartz diagenesis and its various forms cannot readily be separated from the examination of all of the various phases within a sandstone and the relations between them.

Fig. 7. Quartz sand grain from modern fiord in Norway. Note large,
         irregular and conchoidal fractures. Also note angularity of
         grains; numerous sharp corners are in evidence. However,
         the grain has been slightly smoothed with a very thin layer
         of precipitated silica. Distance across bottom of micrograph
         325 microns.

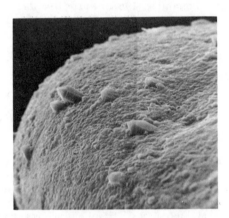

Fig. 8. Quartz sand grain from Navajo Sandstone of Jurassic age,
         Arizona, USA. An overgrowth is beginning to develop on the
         surface of this grain. Note all the quartz crystal
         terminations pointing in the C axis direction; these
         terminations grow on the edges of "upturned plates." The
         grain apparently was originally abraded in an eolian
         environment and the terminations were added after abrasion
         ended. The final stage in the development of this process
         is a grain that appears to be a single quartz crystal
         similar to the small terminations present on the grain
         surface. Distance across bottom of micrograph 90 microns.

Quartz overgrowths start as numerous incipient crystals on detrital grains (Pittman, 1972). They then develop into overgrowths with well-defined crystal faces, depending upon conditions of time, temperature and physiochemical conditions (Figure 8). There is little evidence that these sequences have provided much information concerned with the various diagenetic processes. In addition, pressure solution, which has been studied in much less detail with the SEM than has overgrowth formation, has also been of little value in diagenetic studies, which attempt to determine how temperatures, pressures and chemical reactions in rocks varied with time. Probably the examination of these features should be combined with an in-depth study of the chemical and physical conditions involved in diagenetic reactions. Backscattered electron techniques in scanning electron microscopy, which permit the study of very fine particles in thin section with the aquisition of Z (atomic number) contrast, should do a great deal to advance the study of fine-grain rocks (with quartz) and fine-grain cements in quartz containing rocks.

APPLICATIONS

The technique is useful in a number of ways. These include:
1. The selection of several environmental episodes in a sample of sand grains and the determination of their respective relative ages (Krinsley, 1981).
2. Studies of how diagenesis occurs at the microlevel, e.g., cementation and pressure solution (Waugh, 1965; 1970).
3. Comparison of source area grains as compared to samples from sedimentary basins. Studies of multiple sources. Statistical analysis (Bull, 1978).
4. The tracing of particular environments up the sedimentary column by comparing specific textures in a series of progressively younger samples (Krinsley, 1981).
5. Classification of fault gouge in terms of structure and origin (Kanori, 1980; Kanori et al. 1980).
6. Determination of environment on land at specific times in earth history from information in ocean cores (Krinsley and McCoy, 1977).
7. Study of ancient wind velocities (Krinsley and Wellendorf, 1980).
8. Studies of eolian and other environmental abrasion mechanisms in the laboratory (for eolian action see, for example, Krinsley et al. 1979).
9. Studies of the origin of fine silt and clay particles in the geological column (Krinsley and McCoy, 1978).

10.   Cathodoluminescence in quartz grains indicating previous
      detailed fracturing history (e.g. Krinsley and Hyde, 1971;
      Krinsley and Tovey, 1978; Tovey and Krinsley, 1981).
11.   Simulation studies of artificial martian eolian sands
      (Krinsley et al. 1979).
12.   Forensic geology in relation to sand grains (e.g. Murray,
      1979).
13. Rb^{87}/Sr^{86} dating of quartz grains and their relation to
      surface texture (Powers et al. 1979).
14.   Microcorrelation from one geographic area to another with
      respect to individual sedimentary layers. This has not as
      yet been done.
15.   Origin of brittle materials, crack propagation, plastic
      deformation of quartz in the material sciences.
16.   Relation of quartz sands to various soil horizons and types
      of soils.

DISCUSSION AND CONCLUSIONS

        The technique has been in operation for a number of years,
but has never been used extensively for sedimentary environmental analysis. There are a number of reasons for this state of
affairs. The technique is expensive and time consuming; a
scanning electron microscope must be made available for a
considerable period of time. In addition, some of the features
described may represent several environments, although a combination of features is usually very specific. More work should
be done in this area. A good deal more collecting in and
characterization of modern environments needs to be accomplished.
In general, most workers simply use the methods that have been
provided in Krinsley and Doorkamp (1973) and do not attempt to
gather additional data and apply it to new situations. This
publication is out of date and nothing is available to replace
it, although a new volume should be published within a year or
so (Marshall, Sheridan and Whalley, in preparation). Finally,
very little work has been done on quantification of surface
textures; since a large number of presumably characteristic
textures exist, and at least 20 to 30 grains must be examined to
characterize a single sample, quantitative work would obviously
be the way to proceed.
        It would, therefore, appear that the four points above must
be examined in some detail before the technique can universally
be valuable to sedimentologists.

REFERENCES

Austin, G.S. 1974. Multiple overgrowths on detrital quartz sand
    grains in the Shakopee Formation (Lower Ordovician) of
    Minnesota. J. Sed. Petr. 44, 358-362.
Bagnold, R.A. 1941. The physics of blown sand and desert dunes.
    Methuen and Co., Ltd., London.
Biederman, E.W. 1962. Destruction of shoreline environments in
    New Jersey. J. Sed. Petr. 32, 181-200.
Bull, P.A. 1978. A quantitative approach to scanning electron
    microscope analysis of cave sediments. SEM in Study of
    Sediments. E., Whalley, W. Geo Abstracts, Norwich, 201-226.
Coch, C., and Krinsley, D.H. 1976. Surface textures of quartz
    sand grains from Coolidge Creek, South Dakota (abstract).
    Program, Northeast-Southeast Annual Meeting, Geological
    Society of America, Arlington, Virginia, 152-153.
Kaldi, J., Krinsley, D., and Lawson, D. 1978. Experimentally
    produced eolian surface textures on quartz sand grains from
    various environments. SEM in the Study of Sediments. Ed.,
    Whalley, W. Geo Abstracts, Norwich, 261-274.
Kanori, Y. 1980. Surface textures of quartz grains from fault
    gouges (Pt. II). Formational age and fracturing mode.
    Repts. of the Central Electricity Board, Res. Rept. 381007,
    1-35.
Kanori, Y., Kawaguchi, I., and Mizutani, S. 1980. Surface
    textures of grains from fault gouges (Pt. III). Experi-
    mentally soluted surfaces of sand grains. Repts. of the
    Central Electricity Board, Res. Rept. 381008, 1-21.
Krinsley, D., and Doornkamp, J.C. 1973. Atlas of Sand Grain
    Surface Textures. Cambridge Univ. Press, Cambridge, 91 p.
Krinsley, D., Greeley, R., and Pollack, J. 1979. Abrasion of
    wind blown particles on Mars - erosion of quartz and
    basaltic sand under simulated Martian conditions. Icarus
    39, 364-384.
Krinsley, D., and Hyde, P. 1971. Cathodoluminescence studies of
    sediments. Proceedings 4th Annual SEM Symposium, IIT
    Research Institute, Chicago, 409-416.
Krinsley, D.H., and Donahue, J. 1968. Environmental inter-
    pretation of sand grain surface textures by electron
    microscopy. Geol. Soc. of Am. Bull., 79, 743-748.
Krinsley, D., and Margolis, S. 1969. Scanning electron
    microscopy: a new method for studying sand grain surface
    textures. Trans. NY Acad. Sci. 31, 457-477.
Krinsley, D., and McCoy, F.W. 1977. Significance and origin of
    surface textures on broken sand grains in deep-sea sedi-
    ments. Sedimentology 24, 857-862.
Krinsley, D., and McCoy, F.W. 1978. Eolian quartz sand and silt.
    SEM in the Study of Sediments. Ed., Whalley, W. Geo
    Abstracts, Norwich, 249-260.

Krinsley, D., and Takahashi, T. 1962. The surface textures of
    sand grains: an application of electron microscopy.
    Science 135, 923-925.
Krinsley, D., and Tovey, N.K. 1978. Cathodoluminescence in
    quartz sand grains. SEM/1978/Vol. I, Los Angeles, CA,
    887-894.
Krinsley, D., and Wellendorf, W. 1980. Wind velocities
    determined from surface textures of sand grains. Nature
    283, 372-373.
Krinsley, D. 1981. Grain surface textural analysis. In Grain
    Surface Studies of Vacherie Dome Samples. Ed., Kolb, C. and
    Clarke, E. File Report QR 02.4.5 Institute for Environ-
    mental Studies, Baton Rouge, LA, 12-17; B1-B26.
Linde, K., and Mycielska-Dowgiallo, E. 1980. Some experimentally
    produced microtextures on grain surfaces of quartz sand.
    Geografisk. Ann. 62A, 171-184.
Margolis, S.V., and Krinsley, D.H. 1971. Submicroscopic
    frosting on eolian and subaqueous sand grains. Geological
    of America Bulletin 82, 3395-3406.
Murray, R.G. 1979: Forensic Geology.
Pettijohn, F.J. 1949. Sedimentary Rocks. New York, Harper and
    Brothers, 718 p.
Pittman, E. 1972. Diagenesis of quartz in sandstone as revealed
    by scanning electron microscopy. J. Sed. Petr. 42, 507-519.
Porter, J.J. 1962. Electron microscopy of sand surface textures.
    J. Sed. Petr. 32, 124-135.
Powers, L., Brueckner, H., and Krinsley, D. 1979. Rb-Sr ages
    from weathered and stream transported quartz grains from
    the Harvey Peak Granite, Black Hills, South Dakota.
    Geochim. et cosmochim. Acta 43, 137-146.
Ricci Lucchi, F., and Casa, G. 1970. Surface texture of desert
    quartz grains: A new attempt to explain the origin of
    desert frosting. Gior. di Geologia, Series 2, 36, 761-776.
Riezebos, P.A. 1974. Scanning electron microscopal observations
    on weakly cemented Miocene sands. Geologie en Mijnbouw
    53, 109-122.
Sorby, H.C. 1880. On the structure and origin of non-calcareous
    stratified rocks. Qt. J. Geol. Soc. London XXXVI, 46-92.
Tovey, N.K., and Krinsley, D. 1980. A cathodoluminescent
    study of quartz sand grains. J. Microscopy 120, 279-289.
Waugh, B. 1965. Preliminary electron microscope study of the
    development of authigenic silica in the Penrith Sandstone.
    Proc. Yorkshire Geol. Soc. 35, 59-69.
Waugh, B. 1970. Formation of quartz overgrowths in the Penrith
    Sandstone (Lower Permian) of northwest England as revealed
    by scanning electron microscopy. Sedimentology 14,
    309-320.
Whalley, W.B., and Krinsley, D.H. 1974. A scanning electron
    microscope study of surface textures of quartz grains
    from glacial environments. Sedimentology 21, 87-105.

Whalley, B., and Chartres, C.J. 1976. Preliminary observations
    on the origin and sedimentological nature of Sarsen
    stones. Geologie en Mijnbouw 55, 68-72.

READING PROVENANCE FROM DETRITAL QUARTZ

Abhijit Basu

Dept. of Geology, Indiana University
Bloomington, Indiana 47405 U.S.A.

ABSTRACT

Quartz is the most abundant and durable mineral in all siliciclastic sediments representing almost all common parent rocks. In multicycle sediments some quartz may survive from several generations away and represent some remote parent rocks. Therefore, quartz has the greatest potential of all detrital minerals for reading provenance of arenites.

To this day, Krynine's genetic classification of detrital quartz as modified by Folk remains as the focal point for many petrologic investigations. Continuous lattice dislocation produces undulosity in quartz crystals, whereas discontinuous high density lattice dislocation gives rise to polycrystallinity. Because lattice dislocation in quartz is much more common in metamorphic rocks than in igneous rocks, strongly undulose and finely polycrystalline detrital quartz grains in sediments tend to indicate dominance of metamorphic source rocks. Controlled studies show that abundance of such detrital grains in the medium sand-size range, if plotted in the diamond diagram proposed by the Indiana group, can successfully discriminate the dominance of different source rocks. Additionally, the internal fabric of detrital poly-crystalline quartz grains may also indicate the relative importance of stress causing lattice dislocation and of thermal annealing in parent rocks.

231

G. G. Zuffa (ed.), Provenance of Arenites, 231–247.
© 1985 by D. Reidel Publishing Company.

Although the shape of detrital quartz grains is probably not diagnostic of specific parent rocks, availability of new well tested morphometric methods such as Fourier grain shape analysis or fractal geometry may have some promise for provenance interpretation. Trace and minor element abundances in individual detrital quartz grains hold the promise of assigning each grain to a specific rock type, but we have to wait for the sophisticated technology to be routinely available to sedimentary petrologists before such endeavours are made.

INTRODUCTION

Quartz is by far the most abundant mineral in siliciclastic sediments and rocks. Even in sediments which are comprised of a high abundance of "rock fragments", quartz commonly constitutes a major proportion of the rock fragments. This is not to say that quartz is the dominant mineral in every siliciclastic sediment or rock; but quartz is usually much more abundant than feldspars, biotite, muscovite, kaolinite, illite, and any other mineral. This holds true for all detrital and, by and large, also for diagenetic components of an arenite. No doubt that this phenomenon is due to the highly resistate nature of quartz compared to other minerals in the sedimentary mileu. Durability and abundance of quartz also assure that nearly all parent rocks containing quartz are represented by detrital quartz in their daughter sediments. If common parent rock associations (e.g. an island arc suite) are considered instead of a single parent rock (e.g. andesite), the probability of a better representation of parent rock associations by quartz increases many times. It follows that multicycled sediments may carry a signature of some remote parent rocks. Therefore, study of detrital quartz, at least theoretically, can provide an insight to the ultimate sources of arenites. In this paper we shall examine a few ways with which such a goal can be accomplished; we shall also consider the modification of original characteristics of quartz during the processes of sedimentation and lithification; and, comment briefly on the preservation potential of these characteristics during the post-lithification stage of an arenite.

HISTORICAL BACKGROUND

Ever since Henry Clifton Sorby started studying a mountain with a microscope in 1849, Europe has been the cradle of microscopic petrology. Systematic use of microscopic properties of detrital minerals for interpreting provenance, however, was initiated by Krynine (1937; 1940; 1950) in the United States. Krynine's teaching has vastly influenced and shaped all provenance studies to follow. Krynine relied mostly on what he perceived as the original optical characteristics of detrital quartz, such as extinction types, nature of subgrain contact in poly-crystalline particles, grain and crystal shapes, and inclusions. Bokman (1952), in a short paper, critically examined some of Krynine's criteria and proposed, in addition, that the statistical dis-tribution of length/width ratios of detrital quartz can be used to discriminate between igneous and meta-morphic source rocks. Even today Krynine's genetic classification of quartz as modified by Folk (1980) remains as the starting point for many provenance studies. The most important point to note in Folk's work (1980) is the recognition of a continuum of different properties of quartz with respect to its ultimate origin.

In the last two decades or so, there have been two major breakthroughs in interpreting detrital sediments to infer provenance. First, Conolly (1965) and Blatt (1967a,b) noted the importance of grain size dependance of undulosity and polycrystallinity of detrital quartz. This meant that comparisons between arenites would be valid only if observations were limited to a specified range of grain size. Second, Dickinson (1972, 1974) noted that there are preferred associations of rocks in different plate tectonic regimes. This meant that it makes more geological sense to interpret a parent rock associa-tion than to try to isolate a single parent rock for every grain in a siliciclastic sediment. However, for our immediate specific purpose of reading prove-nance from detrital quartz, Blatt's aforementioned work is of greater importance because Dickinson's contribution is general in nature.

OPTICAL PROPERTIES

I assume that the reader is familiar with the

classical ideas of Krynine and Folk -- ideas which
are still mostly valid. Therefore, I choose to
discuss the more important refinements of such
thought. First, we need some common vocabulary.
Briefly, a polycrystalline quartz grain is a detrital
particle with two or more subgrains; the subgrains
are defined by different extinction positions; a
sedimentary rock fragment consisting of quartz alone
is recognized as such only if bedding or other lithification evidence is preserved; chert is recognized
separately on the basis of extremely small crystal
size (usually < 1 um). A monocrystalline quartz
grain is a detrital particle in which no subgrain can
be discerned although different parts of the grain
may have different extinction positions, and the
extinction shadows sweep gradually across the grain
with rotation of the microscope stage. Undulosity is
roughly defined as the amount of rotation of a microscope stage that is necessary to obtain complete
extinction in all parts of a monocrystalline quartz
grain. However, one must realize that undulosity
and, to a large extent, polycrystallinity are expressions of a common phenomenon viz. lattice dislocation. Dislocation in a single quartz crystal may be
such that the c-axis is shifted between lattices at
the submicroscopic scale and the microscopic expression could be undulosity. On the other hand, dislocation density may vary and across certain boundary
surfaces within a quartz crystal the c-axis may be
shifted appreciably. In such cases, the microscopic
expression would be to exhibit subgrains in otherwise
"single crystals". In addition, there may be
instances where separate crystals or crystallites
grew together and what now appears as a polycrystalline quartz grain could be more truly called a monomineralic rock fragment. In other words, there are
genetically two kinds of polycrystalline quartz
grains which are not, and perhaps cannot be, dif-
ferentiated in a population of siliciclastic sediments. Another important point to realize is that
the extinction of any domain in a quartz grain
depends on the location of the c-axis with respect to
the axis of the microscope, the directions of the
polarizer and the analyzer, and very importantly, the
orientation of the plane of the thin section with
respect to the optic axis of the crystalline domain
under investigation. If a quartz crystal has several
domains with different orientations of c-axes, one
could imagine a hypothetical optic axial plane
between the two most divergent domains. If this

hypothetical optic axial plane is parallel to the
plane of the thin section, undulosity as defined
earlier, would measure the actual divergence of the
c-axes. Such an orientation of the optic axial plane
can be easily obtained with an universal stage, but
is impossible to obtain with the conventional flat
stage. If the hypothetical optic axial plane is
normal to the plane of the thin section, undulosity
would be zero. In all intermediate positions of the
hypothetical optic axial plane undulosity may be any
random value (for a fuller explanation see Basu et
al., 1975; and, Blatt and Christie, 1963).

It is precisely this inevitibility of randomness
in measured undulosity values of quartz on a flat
stage that made Blatt and Christie (1963) rightly
criticize Krynine's methods. They showed that sand
sized quartz grains in parent plutonic igneous, and
metamorphic rocks are not different in their undulosity values, although extrusive rocks contain more
non-undulose quartz. This may seem surprising to a
conventional petrographer, but the restriction of
grain size between 0.062 to 2.000 mm in parent rocks
excludes a large number of grains from all coarse
grained metamorphic and plutonic igneous rocks; in
low grade schists where complete recrystallization of
the protoliths has not taken place, this grain size
restriction may also provide a preferential sample of
relict grains; and, in volcanic rocks the sample size
would be inevitably small. Therefore, as Blatt and
Christie (1963) concluded, the impression one gets
from a microscopic examination of all quartz in
plutonic, volcanic, and schistose rocks is not valid
for the subset of sand sized quartz grains in these
parent rocks. In addition, they showed that the
abundance of sand-size undulose quartz in sedimentary
rocks is roughly proportionate to the maturity of the
sediments. If all this was not enough, they also
showed that undulosity in quartz in sedimentary rocks
increases with the age of the rock, presumably in
response to post-depositonal deformation. Because
sand sized quartz grains in parent igneous and metamorphic rocks do not adequately represent the
original clastic material derived from the parent
rocks, Blatt (1967a) gathered similar data from sand
sized quartz in clastic grains derived exclusively
from different parent rocks. His new data
corroborated his earlier conclusions. In addition,
between these two studies it was also evident that
polycrystallinity of detrital quartz grains were much

more dependent on grain size than on their source
rocks.   The Krynine-Folk criteria of quartz undulosity and polycrystallinity for provenance determination were buried by Blatt.

   Philosophically it is hard to accept that the
most abundant component of siliciclastic sediments
would be so impotent as not to provide any information on genesis.   A re-examination of the issue has
shed some light (Basu et al., 1975).   Take the
geometrical consideration of strain, c-axis bending
and the consequent undulosity in quartz.   Although it
is true that on a flat stage measured undulosity may
be random, there could be some statistical control on
the probability of some cluster.   Otherwise, even in
parent rocks no one should have spotted any empirical
difference which led Krynine to his original genetic
classification of quartz.   Let us consider unstrained
quartz grains in which the c-axis is not bent at all;
in any random section undulosity is always zero.
Next, consider strained quartz.   Depending on the
degree of divergence of the c-axis in any grain of
strained quartz, the probability of obtaining
randomly different undulosity other than the true
divergence angle of c-axes will increase.   However,
in nature quartz can only have a finite strain before
brittle fracture.   It has been found that the actual
divergence of c-axis in quartz is usually $<20^{\circ}$ and
rarely exceeds 30°.   Under the circumstances it is
reasonable to expect that the so called random values
of undulosity would be restricted in range and would
not be distributed evenly between 0° and 90°.   As a
matter of fact, undulosity values of monocrystalline
quartz in grus reported by Blatt (1967a;his fig. 9)
hardly exceed 16°!   Moreover, the complete sand size
fraction (0.062-2.000 mm) is a large range and, like
polycrystallinity, measured values of undulosity are
also dependent on grain size.   Perhaps a narrower
size range can provide more meaningful comparison.

   Nevertheless, the work of Blatt (1967a) and
Blatt and Christie (1963) has clearly shown that a
single detrital grain of quartz cannot be assigned to
a parent rock on the basis of undulosity or polycrystallinity.   If these properties could be shown to
be somewhat dependent on parentage, only then there
may exist a possibility of infering the dominance of
some parent rocks as major suppliers of detritus to
an arenite.

Following the above line of reasoning, Basu et
al. (1975) made a study of undulosity and polycrystallinity of medium sand sized (0.25-0.50 mm)
detrital quartz in Holocene fluvial sands derived
from known source rocks. They measured undulosity
values of about 900 monocrystalline quartz and also
measured the true angle of c-axis divergence of these
grains on a universal stage. In addition, they
examined over 7600 polycrystalline quartz grains,
also in the medium sand size fraction, and courted
the number of subgrains in each of them. Their
results show that statistically a higher proportion
of both moderately to strongly undulose monocrystalline quartz grains (undulosity > 5°) and a higher
proportion of polycrystalline quartz grains with 4 or
more subgrains, in the medium sand size range, do
characterize a metamorphic source, especially lower
grade metamorphic sources. Plutonic rocks tend to
provide non-undulose and weakly undulose (undulosity
< 5°) monocrystalline quartz grains, and, poly-
crystalline quartz grains with only two or three
subgrains. Basu et al. (1975) backed up their study
by testing their criteria to infer the dominant
source rock of Upper Carboniferous, Triassic, and
Oligocene arenites, the source rocks of which had
been inferred independently.

Briefly, the method of Basu et al. (1975)
requires that a representative number of quartz
grains in an immature (= first cycle?) arenite be
counted using the classification below:

Monocrystalline quartz

  Undulosity < 5°

  Undulosity > 5°

Polycrystalline quartz

  2 or 3 subgrains

  > 4 subgrains

The data are plotted in a double triangle (fig. 1).
If > 25% of all polycrystalline quartz grains consist
of > 4 subgrains each, then the lower triangle is
used; otherwise the counts are plotted in the upper
triangle. On the basis of data from Holocene fluvial
sands (Basu et al., 1975), this double triangle is

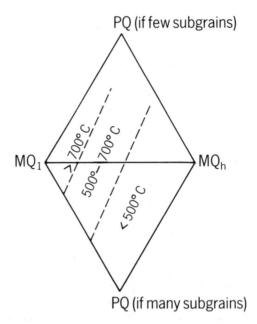

Fig. 1. A modified version of the double triangle proposed by Basu et al. (1975) for the determination of sources of detrital quartz grains using modal data of sand grains in the 0.25-0.50mm range. PQ = poly-crystalline quartz; MQ_1 = monocrystalline quartz of low undulosity; MQ_h = monocrystalline quartz of high undulosity. If more than a quarter of all poly-crystalline quartz grains contain four or more sub-grains, the lower triangle is used for plotting the modal data; otherwise the upper triangle is used. The broken lines represent approximate boundaries between detrital quartz populations derived from low grade metamorphic rocks, medium to high grade meta-morphic rocks, and from granites or granulites. These boundaries are suspected to correspond approxi-mately to the reaction:

    chlorite + muscovite + quartz =

    cordierite + biotite + andalusite + water

which occurs at about 500°C (Hirschberg and Winkler, 1968), and the stability of an hypersthene + sillimanite + quartz assemblage (Newton, 1972). Stated another way, the quartz populations in the three regions of the double triangle may be considered to have been derived from source rocks which equilibrated in equivalent of temperature ranges shown at 1 bar.

divisible into three regions indicating low rank
metamorphic, high rank metamorphic, and
plutonic/granulitic source rocks. The most important
point regarding the use of this diagram is that its
application is limited to the medium sand-size fraction of first cycle sands and sandstones. Further,
this method only provides a clue to the dominance of
the kind of source rock; parent rocks of sands
derived from multiple sources cannot be
distinguished. Additional problems may be caused by
post-depositional processes such as diagenesis, and
tectonism.

Polycrystalline quartz grains are also directly
useful in the interpretation of source rocks. Folk
(1980) has identified a continuum of undulosity and
polycrystallinity in detrital quartz grains from
different source rocks. Wilson (1973) has traced a
sequence of lattice deformation and annealing in
progressive metamorphism of the Mt. Isa quartzite,
Australia. Young (1976) synthesized the information
and the ideas, supported by new data, and showed that
indeed the internal fabric of each detrital polycrystalline quartz grain could be applied to infer-
ring the approximate source rock. Young's work
(1976) also shows that his continuum of detrital
quartz is compatible with dislocation theory as it
applies to quartz.

SHAPE

Folk (1980) has continued to reiterate Krynine's
faith in the shape of quartz grains, especially that
of the subgrains in polycrystalline quartz as an
indicator of source rock. Indeed Bokman (1952)
showed that the elongation of quartz grains in the
weathered products of granites and schists were different, the latter producing a larger abundance of
elongate monocrystalline quartz grains. However,
Blatt (1967a) showed that in a given size range the
elongation of detrital quartz grains was probably
more a function of maturity and age of the sandstone
(due to post-lithification changes) than a function
of source rock. Blatt (1967a) furthered his argu-
ments by claiming that the destruction of clastic
quartz grains during soil forming processes and
during sediment transportation was significant enough
to obscure the parent rock signature of size or shape
of detrital quartz grains (see also Cleary and

Conolly, 1971, 1972). It became clear that classical methods of measuring or estimating shape e.g. Zingg's classification (Pettijohn, 1957) or that developed by Sneed and Folk (1958) could not be used for provenance interpretation.

Ehrlich and Weinberg (1970) have introduced another method for estimating grain shape -- a rigorous quantitative approach termed Fourier grain shape analysis. Basically the method is a curve fitting procedure; fixed shapes of curves in the form of flower petals or clover leaves are fitted to the grain perimeter. For example, an elongate grain with an approximately elliptical cross section could be modelled by fitting two "petals" in the form of the numeral eight; a triangular cross section of a grain would need three "petals" and a squarish grain would need four petals; and so on. In other words, the more the number of "corners" in the shape of a grain, the higher is the required number of petals. For two dimensional projections of a grain or for the two dimensional form of a grain in a thin section, this method could be applied fairly easily with a computerized device. Briefly, the method calls for digitizing the outline of a grain perimeter in polar coordinates with the center of the grain as the origin. A Fourier expansion of the polar coordinates provides an estimate of the relative contribution of a "two petal flower", a "three petal flower", a "four petal flower", ... and "n-petal flower" to the overal shape of the grain. In the language of this "Fourier Grain Shape Analysis", the contributions show up as amplitudes at 2, 3, 4 ...n harmonics (Ehlrich and Weinberg, 1970). Thus each grain would have an amplitude-harmonic spectra of its own. For a population of grains, the spectra could be statistically sorted out in families. The technique is useful to sedimentologists on the assumption that grain shape, including the details of surface morphology, is dependent upon provenance and transportation processes.

Because different minerals tend to break and fracture in different ways in response to similar processes, it is important that measurements be made on only one mineral type for comparison between two populations. Because of the ubiquitous nature of quartz in siliciclastic sediments and arenites, quartz is by far the best mineral suited for Fourier grain shape analyses. In a series of papers, Ehrlich and his students have shown that Fourier grain shape analysis of the population of detrital quartz grains in sands and sandstones could be used successfully to determine if different sources supplied material to a sand body or if material from similar sources were transported by different mechanisms and agents to a depositional site (Ehrlich et al., 1980; Ehrlich and Chin, 1980; Full et al., 1984; Mazzullo et al., 1982; Mazzullo and Ehrlich, 1983; Wagoner and Younker, 1982; etc.). It is important to notice, however, that Ehrlich and his coworkers did not make any distinction between polycrystalline and monocrystalline quartz grains. It is likely that Fourier grain shape analysis, if restricted to only monocrystalline quartz grains, may prove to be even more useful.

The theoretical basis of the relation of the amplitude spectra of quartz grain shape to those in quartz grains in source material, or, to that of the transportation process is not understood. Therefore, there is room for much skepticism. Presently, Fourier grain shape analysis is only a more exact method of quantifying grain shape than other methods. If one goes to higher harmonics, even surface texture as observed under an SEM could also be quantified. Such quantification may be useful in provenance studies of unconsolidated material because lithification and diagenesis can obliterate much of inherited surface texture. However, note that this method can only provide a grouping of quartz grains into populations which might have been derived from different sources. The method does not identify source rocks in any way.

One major limitation in the practice of the Fourier method is that grains with reentrant angles in the periphery have to be discarded. This poses serious problems for grains with embayments, for example, as seen in some quartz phenocrysts in rhyolites. Crenulations on grain margins, such as chatter marks, also cannot be modelled. A complementary method of using fractal geometry may be used in such cases (Mandelbrot, 1982; Whalley and Orford, 1983; Orford and Whalley, 1983). Basically, the method calls for measuring undissceted segments of a perimeter as a fraction of a pre-chosen scale and then expresses the "fractal dimension", a variable dependent on a normalized form of the fractional step lengths necessary to describe the perimeter. This method of using fractal geometry for quantifying

particle morphometry should be particularly suitable
for the analysis of quartz grain surface texture at
about the level of micron-scale or even smaller
scales. Krinsley (this volume) shows the importance
of the feature despite possible overprinting by diagenesis; the new methodology for quantification is
also at hand. I think that provenance interpretation
of the morphometric properties of detrital quartz
will receive a significant boost in near future.

CHEMICAL PROPERTIES

Physical properties of quartz, either optical or
morphometric, are limited in their use as provenance
indicators. For immature or first cycle arenites one
might be able to identify the dominant source rock
clan, but one cannot be any more specific. Surviving
rock fragments shed more light on the specific parent
rock, but they may be rare. For the purpose of more
meaningful provenance interpretation, one may have to
turn to, for example, chemical compositions and
structural states of feldspars (Trevena and Nash,
1981; Basu, 1977; Suttner and Basu, 1977). Surely,
it is much more satisfying to know as much about
specific parent rocks as possible from daughter sediments.

Trace element content of quartz is determined
principally by the physico-chemical environment
during crystallization of a quartz grain (Dennen,
1967). Suttner and Leininger (1972) have shown that
the trace element distribution in detrital quartz
derived exclusively from a shallow level
granodioritic batholith is different from that found
in detrital quartz derived from the volcanic cover of
the same batholith. Certainly the dividend in terms
of specifying source rock from the trace element
content of detrital quartz would be very high if one
could isolate the abundance patterns of trace
elements in quartz from common source rocks. However, there are two shortcomings in applying the
above. First, the population of detrital quartz
grains in arenites is not the equilibrium or near-
equilibrium assemblage found in igneous or
metamorphic rocks. Therefore, the trace element
content of all the quartz grains in an arenite will
only be a mixture from all the different sources; the
ensuing problem, however, is no different from those
inherent in studying a general population and is

identical to that in applying the double triangle as discussed earlier. Second, one has to be extremely careful about separating only detrital quartz grains from an arenite and avoid all diagenetic quartz. Prior to analysis it is also necessary to leach all the grains so as to remove the outer rind of each grain and ensure that there is no diagenetic contamination. These shortcomings are not unsurmountable. One may have to be patient and careful in picking quartz grains for analysis; and, mixing calculations can be performed to estimate the relative contributions of common parent rocks once the end member compositions are established. Tempting as it may be, we sedimentary petrologists have not yet reached that state of analytical sophistication. Again, I think that this is an area of research which will see large advances in the near future.

Whereas the trace element content of a population of detrital quartz grains in an arenite may be difficult to interpret, it may not be very difficult to interpret the trace element content of individual quartz grains. If supersensitive microanalyzers, such as proton microprobes or laser microprobes were available to sedimentary petrologists, isotopic or elemental analysis of individual quartz grains and/or inclusions therein would be possible (e.g. applications by Woolum et al., 1983, and by Norris et al., 1981). Ideally such data have the promise of assigning each individual quartz grain to a specific parent rock.

It has been emphasized at the beginning that nearly all parent rocks are represented in the quartz population of siliciclastic arenites. Therefore, (1) if it could be shown that trace element distribution in quartz is truly source rock sensitive, and (2) if trace element determination from quartz grains could be done routinely, then detrital quartz would be the most important and reliable guide to provenance interpretation. That day is clearly not here, neither is it imminent. But given the rapid pace of the advance of technology, I reckon that we shall be attempting such procedures within a decade.

ACKNOWLEDGEMENTS

I am grateful to G. G. Zuffa and W. C. James for reviewing an early draft of this paper. Part of the

work reviewed in this paper was performed under the
guidance of Lee J. Suttner. This paper was prepared
with the support of Indiana University Foundation and
through the dedication of the staff of the Department
of Geology.

REFERENCES

Basu, A., S. W. Young, L. J. Suttner, W. C. James,
    and G. H. Mack, 1975, Re-evaluation of the use
    of undulatory extinction and polycrystallinity
    in detrital quartz for provenance interpreta-
    tion: Jour. Sed. Petrology, v. 45, p. 873-882.

Basu, A., 1977, Provenance and Al/Si order-disorder
    of detrital alkali feldspars: Jour. Geol. Soc.
    India, v. 18, p. 477-492.

Blatt, H., 1967a, Original characteristics of clastic
    quartz grains: Jour. Sed. Petrology, v. 37, p.
    401-424.

Blatt, H., 1967b, Provenance determinations and
    recycling of sediments: Jour. Sed. Petrology,
    v. 37, p. 1031-1044.

Blatt, H. and J. M. Christie, 1963, Undulatory
    extinction in quartz of igneous and metamorphic
    rocks and its significance in provenance studies
    of sedimentary rocks: Jour. Sed. Petrology, v.
    33, p. 559-579.

Bokman, J., 1952, Clastic quartz particles as indices
    of provenance: Jour. Sed. Petrology, v. 22, p.
    17-24.

Cleary, W. J. and J. R. Conolly, 1971, Distribution
    and genesis of quartz in a piedmont-coastal
    plain environment: Geol. Soc. America Bull., v.
    82, p. 2755-2766.

Cleary, W. J. and J. R. Conolly, 1972, Embayed quartz
    grains in soils and their significance: Jour.
    Sed. Petrology, v. 42, p. 899-904.

Conolly, J. R., 1965, The occurrence of polycrystal-
    linity and undulatory extinction in quartz in
    sandstones: Jour. Sed. Petrology, v. 35, p.
    116-135.

Dennen, W. H., 1967, Trace elements in quartz as indicators of provenance: Geol. Soc. America Bull., v. 78, p. 125-130.

Dickinson, W. R., 1972, Evidence for plate-tectonic regimes in the rock record: Am. Jour. Sci., v. 272, p. 551-576.

Dickinson, W. R., 1974, Plate tectonics and sedimentation: in Dickinson, W. R., ed., Soc. Econ. Paleontologists and Mineralogists Special Pub., v. 22, p. 1-27.

Ehrlich, R., P. J. Brown, J. M. Yarus, and R. S. Przygocki, 1980, The origin of shape frequency distributions and relationship between size and shape: Jour. Sed. Petrology, v. 50, p. 475-484.

Ehrlich, R. and M. Chin, 1980, Fourier grain-shape analysis: a new tool for sourcing and tracking abyssal silts: Marine Geol., v. 38, p. 219-231.

Ehrlich, R. and B. Weinberg, 1970, An exact method for characterization of grain shape: Jour. Sed. Petrology, v. 40, p. 205-212.

Folk, R. L., 1980, Petrology of sedimentary rocks: Texas Hemphill's Book Store, Austin, 182 p.

Full, W. E., R. Ehrlich, and S. K. Kennedy, 1984, Optimal configuration and information content of sets of frequency distributions: Jour. Sed. Petrology, v. 54, p. 117-126.

Hirschberg, A. and H. G. F. Winkler, 1968, Stabilitatsbeziebungen zwischen chlorit, cordierit und alamandin bei der metamorphose: Contrib. Mineral. Petrology, v. 18, p. 17-42.

Krynine, P. D., 1937, Petrology and genesis of the Siwalik Series: Am. Jour. Sci., v. 34, p. 422-446.

Krynine, P. D., 1940, Petrology and genesis of the Third Bradford Sand: Penn. State Coll. Min. Inds. Expt. Sta. Bull., v. 29, p. 50-51.

Krynine, P. D., 1950, Petrology, stratigraphy and origin of the Triassic sedimentary rocks of

Connecticut: Connecticut Geol. Survey Bull., v. 73, 293 p.

Mandelbrot, B. B., 1982, The Fractal Geometry of Nature: Freeman, San Francisco, 460 p.

Mazzullo, J. M. and R. Ehrlich, 1983, Grain-shape variation in the St. Peter Sandstone: a record of eolian and fluvial sedimentation of an Early Paleozoic Cratonic Sheet Sand: Jour. Sed. Petrology, v. 53, p. 105-119.

Mazzullo, J., R. Ehrlich, and O. H. Pilkey, 1982, Local and distal origin of sands in the Hatteras Abyssal Plain: Marine Geol., v. 48, p. 75-88.

Newton, R. C., 1972, An experimental determination of the high-pressure stability limits of magnesian cordierite under wet and dry conditions: Jour. Geol., v. 80, p. 398-420.

Norris, S. J., P. W. Brown, and C. W. Pillinger, 1981, Laser pyrolysis for light element and stable isotope studies (abstract): Meteoritics, v. 16, p. 369.

Orford, J. D. and W. B. Whalley, 1983, The use of the fractal dimension to quantify the morphology of irregular-shaped particles: Sedimentology, v. 30, p. 655-668.

Pettijohn, F. J., 1957, Sedimentary Rocks: Oxford, Calcutta, 718 p.

Sneed, E. D. and R. L. Folk, 1958, Pebbles in the Lower Colorado River, Texas: a study in particle morphogenesis: Jour. Geol., v. 66, p. 114-150.

Suttner, L. J. and A. Basu, 1977, Structural state of detrital alkali feldspars: Sedimentology, v. 24, p. 63-74.

Suttner, L. J. and R. K. Leininger, 1972, Comparison of the trace element content of plutonic, volcanic, and metamorphic quartz from southwestern Montana: Geol. Soc. America Bull., v. 83, p. 1855-1862.

Trevena, A. S. and W. P. Nash, 1981, An electron

microprobe study of detrital feldspar: Jour. Sed. Petrology, v. 51, p. 137-150.

Wagoner, J. L. and J. L. Younker, 1982, Character-ization of alluvial sources in the Owens Valley of eastern California using Fourier shape analysis: Jour. Sed. Petrology, v. 52, p. 209-214.

Whalley, W. B. and J. D. Orford, 1983, Analysis of SEM images of sedimentary particle form by fractal dimension and Fourier analysis methods: Scanning Electron Microscope, v. II, p. 639-647.

Wilson, C. J. L., 1973, The prograde microfabric in a deformed quartzite sequence, Mt. Isa, Australia: Tectonophysics, v. 19, p. 39-81.

Woolum, D. S., D. S. Burnett, C. J. Maggiore, and T. M. Benjamin, 1983, REE fractionation in St. Severin phosphates: implications for Pu-REE coherence: Lunar Planet. Sci. XIV, Lunar Planet. Instt., Houston, p. 859-860.

Young, S. W., 1976, Petrographic textures of detrital polycrystalline quartz as an aid to interpreting crystalline source rocks: Jour. Sed. Petrology, v. 46, p. 595-603.

HEAVY MINERALS IN PROVENANCE STUDIES

Andrew C. Morton

Stratigraphy and Sedimentology Research Group
British Geological Survey
Keyworth, Notts. NG12 5GG, UK

ABSTRACT

   Heavy minerals are sensitive indicators of provenance.
Over 30 translucent detrital species are of common occurrence,
many of which have characteristic parageneses. However,
because they are a'so sensitive to the processes of weathering,
transportation, deposition and diagenesis, the heavy-mineral
suite of an aren' :e does not necessarily accurately reflect
the source-area mineralogy. ..n most cases, the effects of
weathering at so ce and transportation are minor, but
hydraulic controls at the time of deposition and subsequent
diagenesis (intrastratal solution) can cause major modifications. Hydraulic effects may be countered by the determination
of hydraulic ratios or by using only those minerals whose
proportions are not gr n ize controlled, as shown by scatter
plots. In tratal soluion is diagnosed by grain-surface
etching and y the occurrence of more diverse suites in low-
permeability units (eg concretions, shales) compared with
adjacent friable sands; different diagenetic settings are
characterised by different orders of mineral stability. The
most effective method to counteract problems of intrastratal
solution, and indeed of hydraulic controls, is to examine the
varieties of one mineral species (a 'varietal study') because
this minimises density and stability contrasts. Varietal
studies may be addressed either by the classical methods of
optical differentiation, or by more sophisticated techniques
such as cathodoluminescence, radiometric dating, or particularly
electron microprobe analysis.

G. G. Zuffa (ed.), Provenance of Arenites, 249–277.
© 1985 by D. Reidel Publishing Company.

1. INTRODUCTION

Heavy minerals, so called because their specific gravities
are greater than those of the major framework constituents, are
volumetrically minor components of arenites, usually forming
less than 1% of the rock. However, because of their diversity
(over 30 translucent detrital species of common occurrence) and
often characteristic parageneses, they have always occupied a
key role in the interpretation of sediment provenance. Heavy-
mineral analysis was at its most popular in the early part of
this century, but during the 1930's it became recognised that
factors other than provenance exert fundamental controls on
heavy-mineral distribution. This, coupled with the development
of newer sedimentological and correlation techniques, resulted
in a severe decline in popularity which has essentially continued
up to the present day. Nevertheless heavy minerals remain
extremely sensitive indicators of provenance, and providing
adequate account is taken of the limiting factors their study
can and will continue to provide crucial information for palaeogeo-
graphical reconstructions. This review, therefore, concentrates
on these limiting factors, which are present at both the analytical and interpretative stages, and outlines methods to combat
such problems.

2. ANALYSIS

Analysis takes place in five stages: sampling, preparation,
separation, counting and data treatment. At each stage in the
procedure it is possible to inadvertently introduce factors
which could alter our perception of the heavy-mineral suite.

2.1 Sampling

As discussed later, many heavy minerals are unstable, and
subject to dissolution through weathering processes. Analysis
of weathered sandstones could yield assemblages which are not
representative of the formation. It is therefore imperative
that analysis is carried out only on fresh samples.

2.2 Sample preparation

Prior to separation, sands need to be in a disaggregated
and cleaned state. As far as possible, sandstones should be
disaggregated by mortar and pestle, taking care not to grind the
grains. However, well-cemented material may require chemical
treatment, introducing a potential problem: acids can and do
modify heavy-mineral suites, apatite being particularly suscepti-
ble. Dilute acetic acid is least destructive and should be used
in preference to all other corrosives, but cements other than
calcite will require treatment with stronger agents (1).

Care should obviously be taken when comparing samples thus treated with those which have not. An alternative method is to crush the sandstone, especially if silica-cemented, but this fragments heavy-mineral grains and makes estimation of their proportions unreliable.

Adhering clays are most rapidly and thoroughly removed using an ultrasonic probe, but this can cause modifications to grain-surface textures. Alternative methods need to be followed if surface textures are to be studied.

Because heavy minerals occupy a large density range the composition of the heavy-mineral suites will vary according to sediment grain size. Variability is minimised, however, if a single, relatively narrow, size fraction is examined (1,2). The selection of a single narrow size fraction also eases problems of identification, by reducing the variability in optical properties. However, although this procedure has been adopted by many heavy-mineral workers, a standard size fraction has not been established. Consequently the literature contains a great deal of data which are not strictly comparable. As a general guideline, however, the 63-125 μm fraction is ideal for heavy-mineral work: the range is sufficiently narrow to minimise both the hydraulic effect and the variations in optical properties, and the fraction is likely to be present in most sandstones unless they be coarse and very well sorted.

2.3 Separation

Heavy-mineral separation in the laboratory is generally achieved using high density liquids, either by gravity-settling or by centrifuge. Panning is a useful field tool, but is too crude for quantitative work, and magnetic separation is not suitable for the recovery of a heavy-mineral suite because of the wide range in magnetic susceptibilities, although it is frequently used to separate a particular mineral group.

Tribromomethane (bromoform) and tetrabromoethane are the most widely used heavy liquids, because their specific gravities (2.89 and 2.97 respectively) allow complete separation from quartz and feldspar whilst ensuring the largest possible density range in the residue. Diiodomethane (methylene iodide) and thallous formate-malonate solution (Clerici's solution) have higher specific gravities (3.32 and 4.30 respectively) and are therefore not suitable for conventional heavy-mineral analysis. All heavy liquids used in the laboratory are toxic and must be handled with care, following recent guidelines (3). Both separation techniques (gravity-settling and centrifuge) have their individual merits, the centrifuge method being slightly more complex, requiring the use of partial freezing to recover the heavy minerals. The gravity-settling process is longer and loss of heavy liquids by evaporation is a greater problem. There is no significant difference in end product.

2.4 Analysis

Once separation has been achieved the heavy minerals are mounted on glass slides and examined under the petrographic microscope. Mineral proportions are estimated by making grain counts along randomly-chosen ribbon traverses. Several manuals are available to assist in the identification of heavy-mineral grains (4-7). For most purposes analysis is restricted to the non-opaque detrital suite because this group is most diagnostic of provenance, although opaques can be used to good effect (8). Micas are generally omitted from the count, for a number of reasons: their density range straddles that of bromoform and tetrabromoethane, their platy habit tends to cause them to float in the liquid, and they are hydraulically equivalent to quartz of a finer grain size (9) so that their behaviour during trans- port and deposition is quite different to that of the other heavy minerals, which are hydraulically equivalent to quartz grains larger than themselves. Diagenetic grains are also omitted, as they give no indication of provenance. Difficulty may be encountered in distinguishing detrital from diagenetic grains, particularly in the case of rutile and anatase, which frequently occur in both forms in sandstones.

Counting a total of 200-300 detrital non-opaque grains is usually considered to give reasonable estimations of the mineral proportions (10). A lower count than this will yield data with wide confidence limits, and counting more does not increase the precision sufficiently to compensate for time spent.

Identification of problematic grains may be achieved either by X-ray diffraction, or by electron microprobe.

2.5 Treatment of data

The raw data obtained from the optical analysis, expressed as number %, are frequently sufficiently characteristic to allow differentiation of sand units in both a regional and a stratigraphical sense, and so to enable the mapping of sedimen- tary petrological provinces. A 'sedimentary petrological province' is defined as a 'complex of distinctive homogeneous sediments which forms a natural unit in terms of age, origin and geographical distribution'. The criteria used may be as simple as mere presence/absence, or may depend on proportions of individual minerals, ratios of two or more minerals, or on the varieties of one specific mineral group.

Where heavy-mineral assemblages show complex variations, distribution patterns are probably best elucidated by vector analysis, as in the case of the Recent sediments on the Orinoco-Guyana shelf (11). Conventional mapping of heavy-mineral associations produces coast-perpendicular trends, suggesting river discharge is the dominant depositional agent, but vector analysis produces coast-parallel trends, indicating that longshore drift is dominant. Of the two forms of vector analysis, Q-mode is best suited to treatment of heavy-mineral data because it deals with frequencies expressed as percentages. R-mode can be used if mineral quantities are known in absolute terms, such as weight.

Once mineral distribution and petrological provinces are delineated, interpretation of sediment source areas and dispersal patterns can begin. Before conclusions can be drawn, however, all the factors that shape heavy-mineral assemblages must be taken into account. These not only include provenance, but also the nature and degree of weathering in the source area, abrasion during transport, hydraulic conditions during deposition, and diagenesis.

3. FACTORS INFLUENCING HEAVY MINERAL SUITES

3.1 Provenance

The parent rocks provide the initial control on the development of heavy-mineral suites in sandstones, and the aim of provenance studies is to isolate this factor by stripping off the other influences that modify the suite during the sedimentation and post-sedimentation processes. The advantage that heavy-mineral analysis has over light-mineral studies is the great diversity of the species which can occur in sandstones, and their often characteristic parageneses, detailed in several sedimentary petrology textbooks (5,6,7,10).

Provenance of Recent sediments can often by read directly from heavy-mineral data, because here, at least, the problems of intrastratal solution are eliminated. Studies of major river-systems such as the Rhine (12) or the Nile (13) have been particularly fruitful, especially with regard to the relative contributions of the various tributaries (Fig 1). Recent marine sediments are equally suitable; for example, eight petrological provinces can be mapped in Quaternary sediments of the Gulf of Mexico and western North Atlantic, and the latter area can be further divided into five sub-provinces (10).

Reading provenance in the geological record is generally

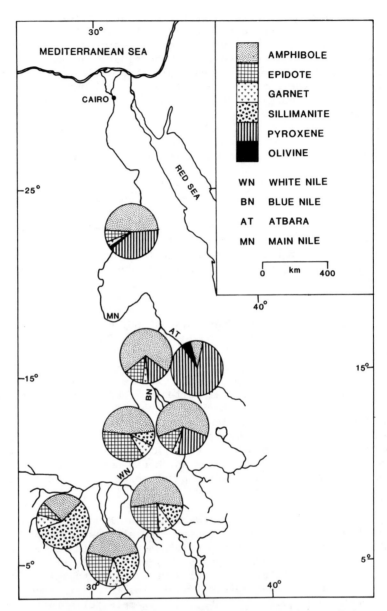

Fig. 1.   Heavy minerals of modern River Nile sediments.   Note
          the marked effects on composition caused by the
          confluence of the Blue Nile and the White Nile, and,
          further downstream, by the confluence of the Nile and
          the Atbara (13).

more difficult because of the influence of diagenesis.   However,
if the heavy minerals in a source area are sufficiently distinc-

tive and well-known, the influence of the source can be traced
directly into the adjacent sediments. The classic example here
is the Dartmoor granite of SW England. The heavy minerals of
this granite are known in some detail (14) making it possible
to unravel its erosional history by examining the surrounding
sediments (15).

The assessment of the degree of sediment recyling is a
problem which has never been satisfactorily addressed, but, up
to a point, the concept of the zircon-tourmaline-rutile (ZTR)
index (16) is a valid approach. Because these minerals have
high chemical and mechanical stability, they tend to become
concentrated during recycling, so that a high ZTR index may be
regarded as a measure of maturity. However, both a high degree
of source-area weathering and intense intrastratal solution also
increase the ZTR index. Consequently a high ZTR index can only
diagnose recycling if these other factors can be eliminated.

Another provenance problem that heavy minerals may resolve
is the judgement of the input from air-fall pyroclastics. Heavy-
mineral grains of air-fall origin are generally euhedral and
often form a distinctive mineral suite. Euhedral zircon, apatite
and sphene are the most common indicators of air-fall pyroclastic
activity (17) but amphiboles, biotite and pyroxenes may also be
diagnostic. A particularly characteristic suite, of euhedral
aegirine, apatite, arfvedsonite, magnesio-kataphorite, sphene
and zircon, occurs in the Palaeocene Thanet sandstones of SE
England (18), and is quite distinct from the rest of the assemblage, of rounded epidote, garnet and hornblende.

3.2 Intensity of source area weathering

There is no doubt that during the weathering and soil-
forming processes, the mineralogy of the bedrock may be greatly
altered (19-21). In theory, the degree to which this occurs in
a sediment source area essentially depends on the climate and
the rate of erosion, but there are few studies on the actual
effects of intense source-area weathering on the mineralogy of
the sediment taken into transport. However, Recent sediments
of Puerto Rico and the Orinoco basin, derived from areas subjected
to extremely intense weathering, nevertheless contain an abundance of minerals unstable in weathering profiles, such as
pyroxenes and amphiboles (22). Similarly, although the central
European massifs were subject to deep kaolinitic weathering
under warm humid conditions during the late Cretaceous, the
sands derived therefrom contain minor amounts of apatite, amphibole, epidote and pyroxene (23) whilst being characterised by a
high ZTR index.

The inference is that providing relief is such that the

rate of erosion is sufficiently rapid, some unstable minerals
will be transported into the basin even under the most extreme
degrees of weathering. In areas of low relief, where rates of
erosion are insufficiently great to prevent total breakdown
of unstable minerals, run off is likely to be very slow,
inhibiting transport of sand-size particles into the basin.
Therefore, in most cases the absence of a mineral from an
arenite is unlikely to be the result of intense source-area
weathering, and is only to be expected in sands derived from
low-lying tropical areas. Clearly, however, variations in
degree of source-area weathering will result in changes in
mineral proportions, and this must be borne in mind when
comparing heavy-mineral suites in sands of different ages
derived from the same source. This is exemplified by a comparison of Eocene and Cretaceous sands from the Great Valley of
California, both derived from the Sierra Nevada (24). Eocene
sands are relatively depleted in garnet, hornblende and sphene
indicating a greater degree of source-area weathering at this
time.

3.3 Effects of abrasion in transit

Once sand grains have been liberated from the source rock,
they are subject to mechanical abrasion during transit before
deposition in the basin. This, theoretically, can modify heavy-
mineral suites and was, at one time, assumed to be one of the
important controls on heavy-mineral suites (25). Several
experiments have been made on the susceptibility of heavy-mineral
grains to mechanical abrasion (26-30). The order of relative
susceptibilities as determined by Friese and Thiel are
remarkably similar (Table 1), with only augite showing markedly
different properties (but compare diopside in Friese's experiments with augite in Thiel's). The results of Dietz are somewhat at variance with the earlier work (Table 1), particularly
in the relative susceptibility of staurolite, tourmaline and
kyanite, but this reflects the different parameters calculated.
Thiel and Friese determined actual weight loss in the experiments,
whereas Dietz determined elongation ratios and roundness indices.

Natural examples, however, do not demonstrate appreciable
loss of heavy minerals in transit. Studies on the major river
systems such as the Mississippi (3), the Nile (13) and the
Rhine (12) reveal no decrease in mineral diversity with
distance from source. Even those minerals with highest susceptibilities to mechanical abrasion (andalusite, kyanite,
apatite) remain constituents of the assemblages. Nor is there
proof of destruction of minerals through abrasion in the beach
environment (32,33). It must be concluded that the effects of
abrasion in transit on heavy-mineral suites are minimal, and
will only be significant if sands are subjected to high energy
conditions over a prolonged period.

Table 1.   Experimentally-derived orders of mechanical stability
           of heavy mineral grains. Species most resistant to
           abrasion at foot of table. Minerals appearing in
           brackets in columns 1 and 2 are those which are not
           common to both orders of stability. Those marked by
           an asterisk (*) have variable stability, dependant on
           composition; their position in this table refers to
           their mean resistance.

Friese (27)	Thiel (28,29)	Dietz (30)
(Monazite)	(Enstatite)	Andalusite, Olivine
(Diopside)		Staurolite, Tourmaline
(Andalusite)		Hornblende, Garnet, Rutile
Kyanite	Kyanite	Zircon, Kyanite
	Augite	
Zircon		
Apatite	Apatite	
(Olivine*)	(Hypersthene)	
	Rutile	
	(Hornblende)	
	Zircon	
Epidote	Epidote	
Garnet*	Garnet	
(Topaz)	(Sphene)	
Augite, Staurolite	Staurolite	
Rutile		
(Spinel*)		
(Corundum)		
Tourmaline*	Tourmaline	

3.4   Effects of hydraulic conditions during deposition

        Studies of grain size distribution in modern sands (2,32,
34,35) show that the mean grain size of any given heavy-mineral
species tends to be smaller than the mean size of the sample.
Consequently, heavy-mineral grains are said to be hydraulically
equivalent to larger quartz grains. The hydraulic equivalent
size is defined as the difference in size between a given heavy-
mineral species and the size of a quartz sphere with the same
settling velocity in water (2). This parameter is most
conveniently determined by measuring the size difference between
the modal classes of the light minerals, determined by sieving,
and each heavy mineral (36), with the size distribution of the
heavy minerals calculated by converting number frequencies to
weight percent over a series of narrow size fractions.

Table 2.   Relationship between mineral density and hydraulic
           size in Rio Grande River sands (35) and New Jersey
           beach sands (37). Note good correlation (r = 0.91
           and 0.87 respectively).

Rio Grande River Sands			New Jersey Beach Sands		
	HYDRAULIC SIZE	DENSITY		HYDRAULIC SIZE	DENSITY
Magnetite	1.0	5.2	Zircon	0.9	4.7
Ilmenite	1.0	4.7	Opaques	0.6	4.1
Zircon	0.9	4.6	Garnet	0.6	3.8
Garnet	0.6	3.8	Staurolite	0.6	3.7
Baryte	0.5	4.5	Leucoxene	0.4	3.8
Sphene	0.5	3.5	Tourmaline	0.4	3.1
Hypersthene	0.4	3.4	Epidote	0.3	3.4
Diopside			Hypersthene	0.2	3.4
(Colourless)	0.4	3.3	Sillimanite	0.2	3.2
Apatite	0.4	3.2	Hornblende	0.1	3.2
Kyanite	0.3	3.6	Glauconite	0.1	2.9
Pyroxene	0.3	3.4	Diopside	0.0	3.3
Diopside (Brown)	0.3	3.4			
Hornblende	0.2	3.2			
Tourmaline	0.2	3.1			
r = 0.91			r = 0.87		

Density is the most important control on hydraulic size,
a positive correlation existing between these two parameters
(35,37) (Table 2). In many sands, however, hydraulic sizes
differ markedly from those predicted on theoretical grounds, with
three factors apparently involved. The first of these is grain
shape, the effects of which are particularly noticeable when
marked differences in morphology exist between minerals of
similar specific gravities. Taken to extremes, this can cause
reversals in the relationship between quartz and the denser
minerals; the platy habit of mica, for example, causes it to be
hydraulically equivalent to quartz particles with smaller grain
diameters (9). The grain size availability of mineral species
is another factor: some minerals, particularly zircon, tend to
occur as small crystals in the source rock. Consequently, their
mean size in a coarse sand can be less than that predicted on a
theoretical basis. The third factor, and the one which has
received most attention in recent years, is the entrainment
potential of mineral grains, anomalous hydraulic equivalence
relationships resulting from differences in entrainment potential

between grains which have similar settling velocities but
different density and size (38,39). The nature of the
depositing medium also affects hydraulic size, wind- and water-
lain sediments having different hydraulic equivalence relationships.

The significance of this behaviour in a provenance context
is that any change in hydraulic conditions during the deposition
of a sand unit (as reflected by variations in sediment grain
size, sorting etc), will cause fluctuations in heavy-mineral
proportions between samples. Therefore, changes in heavy-
mineral proportions do not necessarily diagnose changes in sediment source: before such conclusions are reached, the hydraulic
effects have to be stripped off. Several procedures have been
outlined to resolve this problem.

Grain size variations can be countered by the determination
of hydraulic ratios, defined as '100 times the weight of a given
heavy-mineral in a known size range divided by the weight of
light minerals of equivalent hydraulic size' (35). Providing
that the method is applied to fine or very fine sands, because
of the problems of availability of certain heavy minerals in the
coarser size grades, regional variation in hydraulic ratios of
a mineral can be ascribed to source area variability. Because
of the grain size availability factor, any study utilising the
hydraulic ratios technique requires calculation of hydraulic
size as a prerequisite. This method, although extremely time-
consuming, can be effective where hydraulic differentiation is
severe: for example, 4 heavy-mineral provinces in Recent beach
sands of New Jersey (37) can be defined on this basis (Fig. 2).

Effects of hydraulic differentiation can also be eliminated
by only using those minerals which show variations that are
independent of grain size, as shown by scatter plots of mean
grain size against mineral proportions. For example, proportions
of the six common heavy-minerals in the Pliocene Paso Robles
Fm. of California are independent of grain-size (40) and can
therefore be used to map petrological provinces. On the other
hand, garnet proportions in the $3\emptyset-4\emptyset$ fraction of Middle Jurassic
Etive Fm. sandstones from a well in the northern North Sea show
a strong correlation with grain-size (Fig. 3), with high garnet
contents in coarse sands and low contents in fine sand, except
for two aberrant samples with anomalously low garnet contents.
On this basis, the aberrant samples might be ascribed to a
different source, but this possibility is eliminated by plotting
garnet contents against the sorting coefficient: the two
aberrant samples are clearly the result of better sorting, and
all samples show a strong correlation between sorting and
garnet content. This example demonstrates the importance of
plotting other grain size parameters as well as mean size when
adopting the scatter plot method.

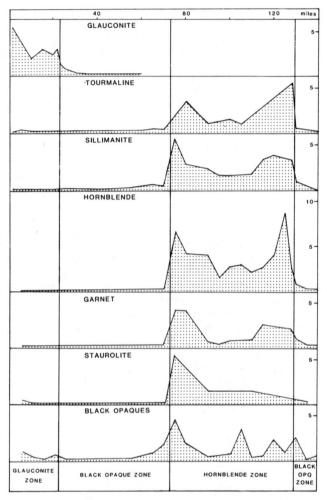

Fig. 2. Hydraulic ratios of specific heavy minerals from beach
         sands of New Jersey, defining four petrological
         provinces along the coastline (37).

     Gross variations over a study area can be determined by
averaging a large number of samples of variable grain size (10),
but this method is suspect if concomitant regional grain size
trends occur. The method could also mask local variations which
may be significant in provenance terms.

     Because it is unlikely that hydraulic differentiation could
be so severe as to entirely eliminate a particular heavy-mineral
species from a suite, those minerals showing regional patterns
of presence/absence can be utilised, providing of course that
this pattern is not the result of intrastratal solution.

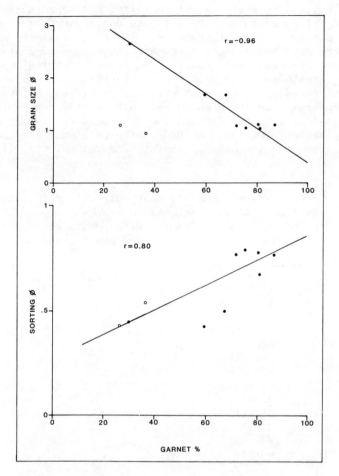

Fig. 3.   Correlation between garnet content and grain size
          parameters in Etive Fm. sands from a well in the North
          Sea, showing that the anomalously low garnet content
          of two samples in the plot of mean size vs. garnet
          proportions is the result of better sorting.

The most reliable technique, however, is to rely on the
varieties of one or more individual mineral species, thus
eliminating density effects. Varietal studies are also the
best techniques in studies of sand altered by intrastratal
solution, and are discussed at length later in the text.

3.5 Effects of intrastratal solution

Intrastratal solution is the single most important factor,
other than provenance, affecting heavy-mineral distribution in

arenites. Taken to extremes, it can alter a diverse suite
consisting of 20 or more species to one containing the stable
group zircon, rutile and tourmaline. The process was first
recognised when etched grain surfaces, indicative of post-depositional corrosion, were observed on several detrital species,
including amphibole, epidote, kyanite, pyroxene, sphene and
staurolite (41-43). Subsequently it was firmly established
that intrastratal solution could radically modify heavy-mineral
suites (44-46), with the formulation of a universally-applicable
stability series (Pettijohn's 'order of persistence') to describe
the behaviour of heavy mineral in intrastratal solution (Table
3).

Table 3.   Order of persistence of heavy-mineral grains after
           Pettijohn (46). Least stable minerals appear at the
           top of the table. Compare with orders of stability
           in epidiagenesis (Table 4) and anadiagenesis (Table 5).

Olivine
Sillimanite
Pyroxene
Sphene
Andalusite
Amphibole
Epidote
Kyanite
Staurolite
Apatite
Garnet
Zircon
Tourmaline
Rutile

     The potential for intrastratal solution is present in any
sandstone subject to porefluid movement, and it can, therefore,
occur at any stage in diagenesis. Of the diagenetic stages
defined by Fairbridge (47), heavy-mineral dissolution is known
to take place in epidiagenesis and anadiagenesis, each
characterised by different orders of stability (48), rather than
obeying the single order of persistence shown in Table 3.

3.5.1 Intrastratal solution : epidiagenesis

     Epidiagenetic changes are those brought about by descending
meteoric groundwaters (47). Any arenite, therefore, deposited
in a non-marine environment, or indeed in a marine environment
and subsequently exposed during phases of regression or uplift,
may suffer dissolution of heavy minerals by this process. The

composition of the meteoric groundwaters may vary between acidic
and alkaline, depending on climatic conditions. Cool temperate
or humid climates tend to give rise to acidic groundwaters,
whereas high pH conditions characterise areas with warm dry
climates (47). Heavy mineral dissolution due to the circulation
of low pH groundwater has been extensively documented in recent
years, and is particularly well-illustrated in the Miocene sands
of Denmark (49-52). For example, the Lavsjerg borehole of
Jutland, Denmark, shows a pattern of decreasing heavy-mineral
diversity upwards to immediately below a brown coal sequence,
correlating with an increase in etching severity on the minerals
affected (52). Amphibole is most strongly affected, being
depleted over a 35 m interval below the brown coal. Epidote
depletion occurs over 17 m, and garnet over about 3 m. Kyanite
and staurolite proportions are also lower immediately below the
coal, and show dissolution features in similar settings elsewhere in the Danish Miocene (50). The Palaeocene Thanet sands
of SE England were deposited in a shallow marine setting but
following a late Palaeocene regression they were exposed and
subjected to descending meteoric groundwaters (48) producing a

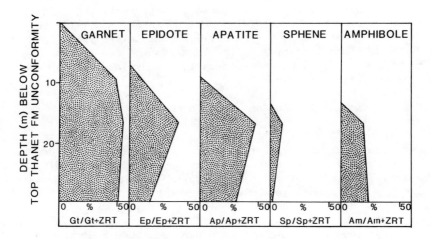

Fig. 4.   Heavy-mineral variability in the Thanet Fm. of the
          Stanford-le-Hope Borehole, Essex, UK. This pattern
          results from mineral dissolution caused by downward-
          percolating acidic groundwaters during a period of
          exposure prior to deposition of the overlying Woolwich
          and Reading Fm. (18). ZRT = Zircon + rutile + tourma-
          line.

similar dissolution pattern to that observed in the Lavsbjerg
borehole (Fig. 4). Similar patterns have been documented by
other authors (53,54,55) and have been produced experimentally,

Table 4.   Order of stability of heavy minerals subjected to
           flushing by low pH groundwaters in weathering
           (epidiagenesis) (48).   Least stable minerals appear
           at top of the table.

Olivine, Pyroxene
Amphibole
Sphene
Apatite
Epidote, Garnet
Chloritoid, Spinel
Staurolite
Kyanite
Andalusite, Sillimanite, Tourmaline
Rutile, Zircon

---

using solvents of pH 5.6 (56).   The similarity of these various
orders of stability allows the formulation of a generalised order
of stability for heavy minerals flushed by low pH fluids during
epidiagenesis (48), shown in Table 4.   This order of stability
is unlikely to be applicable in situations where groundwater
chemistry, particularly pH, is markedly different, but little
systematic work has been carried out on mineral stability in
such environments.   The instability of certain heavy minerals
under conditions conducive to the formation of red-beds has been
demonstrated (57,58) but no information regarding the response
of the entire suite is available.

3.5.2   Intrastratal solution : anadiagenesis

     Although several authors (59-63) have demonstrated that
heavy-mineral diversity decreases with increasing burial depth,
the possibility of variations in provenance has made it difficult
to isolate the effects of intrastratal solution.   However, late
Palaeocene sandstones in the central North Sea are known to be
of Scottish Highlands derivation throughout (64) and show
considerable variability in burial depth, being less than 200 m
below sea floor in the Moray Firth but over 3000 m at the basin
centre.   Therefore, they constitute an ideal natural package
for examination of the effects of intrastratal solution in
anadiagenesis (48).

     With increasing depth, heavy-mineral assemblages show a
marked decrease in diversity, correlating with an increase in
etching severity on grains of affected species (Fig. 5).   The
order in which the minerals disappear (hornblende at shallow
depth, followed by epidote, sphene, kyanite and staurolite)
gives a measure of their relative stability.   The decline in
garnet proportions with depth, and the increasing degree of

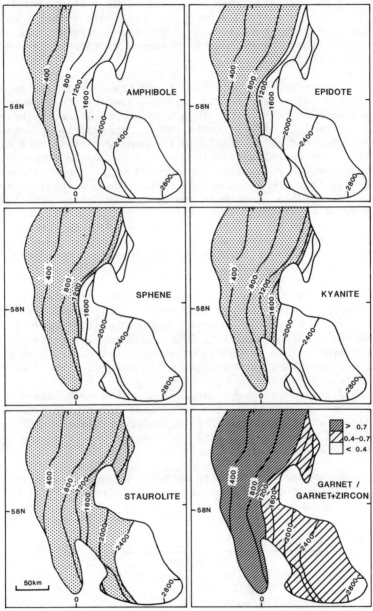

Fig. 5.   Distribution of heavy minerals in friable late Palaeocene sands of the central North Sea. The pattern of
decreasing mineral diversity with burial depth (shown
in metres) results from intrastratal solution, as
evidenced by increasing corrosion of minerals with
depth and the presence of more diverse assemblages in
adjacent low-porosity units (48).

corrosion, indicates this mineral is less stable than apatite, rutile, tourmaline and zircon, although more stable than staurolite. The burial-depth control on heavy-mineral dissolution indicates that the process is governed by temperature and pressure of the porefluids.

The widespread nature of intrastratal solution in the North Sea Palaeocene is also demonstrated by the presence of zones wherein heavy minerals are preserved at depths beyond their normal stability limits (48). Both calcite-cemented sandstones and silty mudstones frequently contain more diverse suites than encasing friable sands, in a manner comparable to that described by earlier workers (44,65). Zones with smectite and analcime cements generated by breakdown of basic vitric glass preserve epidote, amphibole and sphene at depths well below their limits of persistence in friable sands. Infiltration of oil into sandstones has also halted or inhibited further dissolution. Thinly-bedded or lenticular sand bodies tend to have more diverse assemblages than thick, massive units because of their poorer fluid communication.

The order of heavy-mineral stability deduced from North Sea sandstones is closely comparable to patterns observed elsewhere (59,62,66). However, experimental attempts to repro-duce the order of stability found in deep burial (56) were not successful. This is because the solvent used had a pH of 10.6, which does not compare well with the natural situation, because connate waters tend to be only mildly alkaline (47). However, limited experiments were carried out using pH 8 solvents and the order of stability thus determined closely accords with that observed in deep burial. It is, therefore, possible to formulate a general order of stability for heavy minerals during anadia-genesis (48) as shown in Table 5.

Table 5. Order of stability of heavy minerals subjected to deep burial (anadiagenesis) (48). Least stable minerals appear at top of the table. Interacting fluids are likely to be saline and mildly alkaline (47).

Olivine, Pyroxene
Andalusite, Sillimanite
Amphibole
Epidote
Sphene
Kyanite
Staurolite
Garnet
Apatite, Chloritoid, Spinel
Rutile, Tourmaline, Zircon

The two orders of stability, one characterising anadiagenesis, the other epidiagenesis (specifically, under low pH
conditions) are quite distinct. Of the minerals considered,
epidote, amphibole, pyroxene and olivine are moderately or
extremely unstable under both sets of conditions, whereas
staurolite, rutile, tourmaline and zircon are moderately or
extremely stable. However, chloritoid, garnet, sphene, spinel
and especially apatite are noticeably less stable in epidiagenesis
than anadiagenesis, with the reverse being true for andalusite,
kyanite and sillimanite.

The order of stability of heavy minerals is, therefore,
governed principally by the chemistry of the interstitial
fluids, particularly pH. This is dependent on the diagenetic
setting, but may also be affected by gross sandstone composition.
In any given situation, the limits of persistence of a particular
species are governed by dissolution rate, itself dependent on
fluid temperature (a function of burial depth and heat flow),
the amount of fluid movement through the sediment (controlled
by porosity and permeability, sandbody geometry and overburden
pressure) and the length of time to which the sediment has been
subject to these conditions.

Intrastratal solution is a pervasive process, and one
that must be carefully considered during provenance reconstructions. Clearly, delineation of source area and sediment
dispersal patterns for Palaeocene sands of the central North
Sea would have been totally erroneous if the effects of this
factor had not been appreciated. Zones of dissolution caused
by epidiagenetic flushing with low pH groundwaters are generally
of limited vertical extent, and can be diagnosed by zones of
apatite dissolution. Consequently, the effects of epidiagenesis
can usually be recognised and compensated for, but the effects
of anadiagenesis are less easily resolved. All arenites at
even relatively moderate burial depths will have been subject
to the process, unless the formation was 'sealed' from circula-
ting porefluids at an early stage. This has, for example,
occurred in parts of the Alpine Molasse (67), preserving such
rare and valuable index minerals as lawsonite and sodic amphibole.

If heavy minerals show regional patterns of presence/
absence which follow the order of stability in anadiagenesis,
intrastratal solution must be suspected; the presence of
etched grain surfaces is indicative of the process. Low-permea-
bility units (cemented horizons, horizons with pervasive clay
matrix, shales) or sands with poor fluid communications should
be examined to determine whether they contain richer assemblages
than the encasing friable sands. Even if the heavy-mineral suites
in friable sands are as diverse as those in enclosed concretions,

this does not prove the lack of intrastratal solution, because cementation is commonly a late-stage phenomenon.

The difficulty of establishing the true nature of a heavy-mineral suite at the time of deposition is frequently so great that alternative methods should be sought. One such method is to rely on the proportion or ratios of the minerals regarded as stable in the sands in question. If this technique is adopted, tests for hydraulic differentiation should first be made, because the minerals of the stable group in deep burial fall at the two extremes of the density range (tourmaline 3.1, apatite 3.2, rutile 4.2, zircon 4.6). Another approach is to map presence/absence of minor minerals known to be stable in deep burial (chloritoid, spinel): as already suggested, hydraulic controls are rarely so severe that they completely eliminate a mineral group from a deposit.

The third approach is to deal with the varieties of one particular mineral or mineral group. As already stated, varietal techniques also minimise the problems of hydraulic differentiation, and therefore are considered to provide the most reliable heavy-mineral data for provenance reconstruction.

4.   VARIETAL STUDIES

The study of the types of one mineral or mineral group in a heavy mineral suite is termed a varietal study. The value of a varietal study is that the effects of hydraulic differentiation and intrastratal solution are eliminated, or, at least minimised, because only grains of a certain density and stability are examined.

The most frequently employed method of varietal study is to distinguish grains optically on the basis of colour, form, habit or inclusions. This approach has been applied to many mineral species. Zircon types can be used to distinguish detritus from the Dartmoor granite (15). Tourmalines are particularly valuable in this respect, showing a great deal of variability in sediments (68). Ratios of blue-green to brown hornblende were of value in mapping petrological provinces in the shelf sands of the Gulf of Mexico (69). Similarly ratios of purple-colourless to orange-red garnets have been used to distinguish tills in eastern North America (70,71).

Although this is clearly a successful method for distinguishing sand bodies of differing provenance, it is a highly subjective approach (72). Another problem is that it is possible for the same colour variety to be caused by different combinations of end-members in solid solution.

The advent of rapid micro-analytical techniques such as the electron microprobe now means that it is possible to extend the scope of varietal studies by actually determining the compositional range of a mineral or mineral group in a sediment. Not only will this give an objective means of distinguishing sand bodies and mapping them regionally, but will also provide more detailed controls on the nature of the source material. Many heavy-mineral species can be treated in this way, notably pyroxenes, amphiboles, epidotes, garnets and tourmalines, all of which have potentially wide compositional ranges in sediments. The opaque suite is perhaps also best analysed by this technique.

Electron microprobe analysis in varietal studies is still in its infancy, and its full potential has not yet been realised. Studies to date have concentrated on the amphiboles: for example, the presence of glaucophane and crossite in shore and river sand of southern Turkey implies the presence of blueschists in the source area (73). In the Molasse of Savoy, France, Lower Chattian sediments contain Al-rich glaucophane and ferroglaucophane, whereas Burdigalian strata are characterised by crossite and Mg-rich glaucophane (67). This marked evolution of amphibole composition provides valuable information about the source area, which can be identified as the St. Bernard-Brianconnais Zone of the Alps.

Amphibole suites have also been useful in assessing the evolution of the passive continental margin of the SW Rockall Plateau (NE Atlantic) drilled during DSDP Leg 81 (74). Two amphibole suites, defined on the basis of variations in Si, $Mg/Mg + Fe$ and $(Na + K)_A$ can be recognised in the lower Palaeogene beds (Fig. 6). One suite, consisting of actinolite, actinolitic hornblende and magnesio-hornblende, is associated with epidote and piedmontite, and was derived from the metamorphic basement in southern Greenland. The other comprises edenites and pargasites, and was sourced from the southern Rockall Plateau. The cut-off of Greenland - derived detritus in the early Eocene marks the opening of a seaway between Rockall and Greenland following the onset of sea-floor spreading.

This method has also been applied to detrital pyroxenes from Palaeozoic rocks of eastern Australia (75), with several different populations being recognised on the basis of their TiO_2, MnO and Na_2O contents. These parameters are particularly useful from a provenance viewpoint because they are believed to distinguish different plate-tectonic settings (76).

However, the value of varietal studies of amphibole and pyroxenes is relatively limited, because of their instability in diagenesis. This also applies to a lesser extent to epidote. Tourmalines, being ultrastable, are theoretically the most

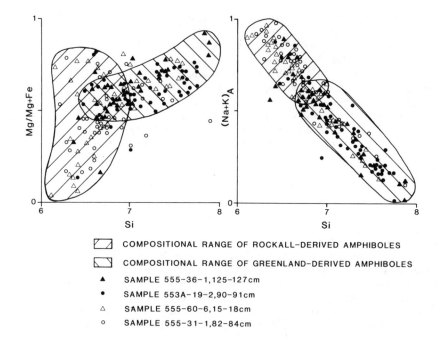

COMPOSITIONAL RANGE OF ROCKALL-DERIVED AMPHIBOLES

COMPOSITIONAL RANGE OF GREENLAND-DERIVED AMPHIBOLES

▲   SAMPLE 555-36-1,125-127cm

●   SAMPLE 553A-19-2,90-91cm

△   SAMPLE 555-60-6,15-18cm

○   SAMPLE 555-31-1,82-84cm

Fig. 6.   Composition of detrital amphiboles in lower Palaeogene
          beds of the SW Rockall Plateau, as determined by
          electron microprobe (74).

suitable, but have their own difficulties, because one of the
elements showing significant variations is boron, which is not
detectable with an energy-dispersive analytical system. Although
boron can be accurately determined with a wavelength-dispersive
system, this requires preparation of polished sections, which
can be avoided without sacrificing too much accuracy if the
energy-dispersive system is used. Analysis time is also
greater, and furthermore, many tourmaline grains are internally
heterogeneous: zoned types are common, as are grains with
abraded overgrowths.

     Therefore, the most valuable contribution likely to be
made in this field will probably be from investigations of
garnet, which is relatively stable, particularly in anadiagenesis.
Preliminary investigations into the Middle Jurassic Brent sands
of the northern North Sea demonstrate major stratigraphical
and regional variations in garnet chemistry (77). Garnets
group into three associations - Association 1 being almandine-
pyrope garnets with low grossular and spessartine contents,
Association 2 containing two groups, one pyrope-rich, the other
spessartine-rich, and Association 3, with moderate grossular
and pyrope contents, generally low in spessartine. The interplay

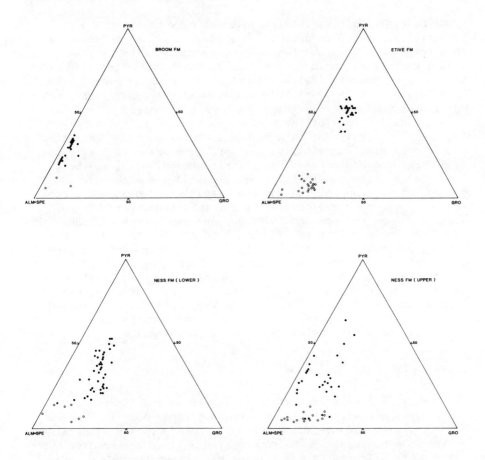

Fig. 7. Composition of detrital garnets in Middle Jurassic
         Brent Fm. sandstones of the Murchison oilfield, North
         Sea, as determined by electron microprobe.

of these Associations allow a four-fold breakdown of the
sequence in the Murchison oilfield (Fig. 7). The lowest unit
(Broom Formation) is exclusively composed of Association 1,
whereas the overlying Etive Formation is dominated by Association
2. Association 3 is dominant in the lower part of the overlying
Ness Formation, but Associations 1 and 2 reappear near the top
of the sequence. Microprobe studies of garnet suites are clearly
of great potential in the unravelling of the palaeogeography,
provenance and dispersal patterns of sandstone bodies. The
technique has also proved useful in locating kimberlites, by
tracing Mg-and Cr-rich garnets in superficial deposits (78).

Several other varietal techniques have been, or are
potentially, useful in provenance studies. It may be worthwhile
to attempt to use cathodoluminescence: apatites, for example,
show a variety of luminescence colours which may be related
to their paragenesis (79). Radiometric methods have been used
on a number of occasions; for example ^{40}Ar/^{39}Ar dating of
detrital tourmalines confirmed the Cornubian origin of tourmaline suites in parts of the Wealden succession of southern
England (80). However, this method deals with the whole suite,
rather than single grains, so that problems arise if the tourmalines are of mixed provenance. Fission track dating is therefore, an exciting possibility, because it deals with individual
grains rather than the whole suite. Apatite, sphene and zircon
are among the possible candidates for this treatment. Three
apatite groups are present in Palaeogene sandstones of DSDP
site 555 (Rockall Plateau), at about 56 Ma, 100 Ma and 200 Ma;
sphenes group at 62 Ma and 1380 Ma (81). Fission track dating
of zircon populations from Wealden sandstones of southern
England confirms that sediment was supplied from at least two
distinct source areas (82). In a similar vein, single-grain
U-Pb dating of detrital zircons is a very powerful tool (83,84)
but is limited in application to sediments from the early
Palaeozoic or older, owing to detection problems.

5. SUMMARY

Heavy minerals are extremely sensitive indicators of
source area and will remain of paramount importance in studies
of sediment provenance and dispersal. However, heavy-mineral
suites in sandstones are not controlled only by provenance, but
may also be affected by source-area weathering, processes of
transportation and deposition, and post-depositional alteration.
Source-area weathering and transportation rarely cause elimination of a species from a detrital heavy-mineral suite, but
hydraulic conditions during deposition and subsequent diagenesis
cause major modifications which must be assessed and compensated
for before reconstructing provenance and dispersal patterns.
Problems of hydraulic differentiation can be overcome by the use
of hydraulic ratios, by mapping only those minerals showing
regional presence/absence patterns, or by using scatter plots
to judge the degree of grain size control. The possibility of
diagenetic modifications can be discounted if the sediments
are Recent and unweathered, or if early cementation has sealed
the formation from further diagenesis. If these criteria are
not met, minerals that show regional patterns of appearance
which are comparable to determined orders of stability must not
be used in provenance reconstructions. In such circumstances,

provenance studies are best addressed by varietal techniques, by optical differentiation, cathodoluminescence, radiometric methods or electron microprobe analysis, on a single mineral species from the residue. This eliminates the effects both of density variations, thereby minimising the hydraulic control, and of stability variations, thereby removing the diagenetic control.

ACKNOWLEDGEMENTS

    The UK Department of Energy funded research into the factors controlling heavy-mineral distribution in North Sea reservoir sandstones. The author is grateful to Dr Robert Knox for critically reviewing this contribution, and to the Director of the British Geological Survey (NERC) for permission to publish.

REFERENCES

1. Carver, R.E. 1971. Procedures in sedimentary petrology (R.E. Carver, Ed.,), Wiley-Interscience, New York, pp. 427-452.

2. Rubey, W.W. 1933. J. Sed. Pet. 3, pp. 3-29.

3. Hauff, P.L. and Airey, J., Circ. US Geol. Surv. 827, pp. 1-24.

4. Galehouse, J.S. 1971. Procedures in sedimentary petrology (R.E. Carver, Ed.), Wiley-Interscience, New York, pp. 385-407.

5. Krumbein, W.C. and Pettijohn, F.J. 1938. Manual for sedimentary petrography, Appleton-Century, New York.

6. Milner, H.B. 1962. Sedimentary petrography, Vol. II, Allen & Unwin, London.

7. Tickell, F.G. 1965. Dev. Sedimentol. 4, pp. 1-220.

8. Riezebos, P.A. 1979. Sediment. Geol. 24, pp. 197-225.

9. Doyle, L.J., Carder, K.L. and Steward, R.G. 1983. J. Sed. Petrol. 53, pp. 643-648.

10. Hubert, J.F. 1971. Procedures in sedimentary petrology (R.E. Carver, Ed.), Wiley-Interscience, New York, pp. 453-478.

11. Imbrie, J. and Van Andel, T.H. 1964. Bull. Geol. Soc. Am.
     75, pp. 1131–1156.

12. Van Andel, T.H. 1950. Provenance, transport and deposition
     of Rhine sediments. PhD thesis, Univ. of Groningen,
     pp. 1–129.

13. Shukri, N.M. 1949. Quart. J. Geol. Soc. Lond. 105,
     pp. 511–529.

14. Brammall, A. 1928. Proc. Geol. Assoc. 39, pp. 27–48.

15. Groves, A.W. 1931. Quart. J. Geol. Soc. Lond. 87,
     pp. 62–96.

16. Hubert, J.F. 1962. J. Sed. Petrol. 32, pp, 440–450.

17. Weaver, C.E. 1963. J. Sed. Petrol. 33, pp. 343–349.

18. Morton, A.C. 1982. Proc. Geol. Assoc. 93, pp. 263–274.

19. Goldich, S.S. 1938. J. Geol. 46, pp. 17–58.

20. Dryden, L. and Dryden, C. 1946. J. Sed. Petrol. 16,
     pp. 91–96.

21. Jackson, M.L. and Sherman, G.D. 1953. Advan. Agron. 5,
     pp. 219–318.

22. Walker, T.R. 1974. Bull. Geol. Soc. Am. 85, pp. 633–638.

23. Skocek, V. and Valecka, J. 1983. Palaeogeog, Palaeoclima-
     tol, Palaeocol. 44, pp. 71–92.

24. Allen, V.T. 1948. J. Sed. Petrol. 18, pp. 38–42.

25. Mackie, W. 1923. Trans. Edinb. Geol. Soc. 11, pp. 138–164.

26. Cozzens, A.B. 1931. Wash. Univ. Studies, Sci. Technol,
     n.s. 5, pp. 71–80.

27. Friese, F.W. 1931. Min. Pet. Mitt. 41, pp. 1–7.

28. Thiel, G.A. 1940. J. Sed. Petrol. 10, pp. 103–124.

29. Thiel, G.A. 1945. Bull. Geol. Soc. Am. 56, pp. 1207.

30. Dietz, V. 1973. Contrib. Sedimentol. 1, pp. 69–102.

31. Russell, R.D. 1937. Bull. Geol. Soc. Am. 48, pp. 1307–1348.

32. Pettijohn, F.J. and Ridge, J.D. 1933. J. Sed. Petrol.
     3, pp. 92-94.

33. Van Andel, T.H. 1959. J. Sed. Petrol. 29, pp. 153-163.

34. Russell, R.D. 1936. J. Sed. Petrol. 6, pp. 125-142.

35. Rittenhouse, G. 1943. Bull. Geol. Soc. Am. 54, pp. 1725-
     1780.

36. Briggs, L.I. 1965. J. Sed. Petrol. 35, pp. 939-955.

37. McMaster, R.L. 1954. Bull. New Jersey Dept. Conserv.
     Econ. Dev., Geol. Series, 63, pp. 1-239.

38. Lowright, R., Williams, E.G. and Dachille, F. 1972. J.
     Sed. Petrol. 42, pp. 635-645.

39. Slingerland, R.L. 1977. J. Sed. Petrol. 47, pp. 753-770.

40. Galehouse, J.S. 1967. Bull. Geol. Soc. Am. 78, pp. 951-978.

41. Edelman, C.H. 1931. Fortschr. Min. Krist. Pet. 16,
     pp. 67-68.

42. Edelman, C.H. and Doeglas, D.J. 1931. Min. Pet. Mitt.
     42, pp. 482-490.

43. Edelman, C.H. and Doeglas, D.J. 1934. Min. Pet. Mitt.
     45, pp. 225-234.

44. Bramlette, M.N. 1941. J. Sed. Petrol. 11, pp. 32-36.

45. Smithson, F. 1941. Geol. Mag. 78, pp. 97-112.

46. Pettijohn, F.J. 1941. J. Geol. 49, pp. 610-625.

47. Fairbridge, R.W. 1983. Dev. Sedimentol. 25B, pp. 17-113.

48. Morton, A.C. 1984. Clay. Min. (in press).

49. Friis, H. 1974. Sediment. Geol. 12, pp. 199-213.

50. Friis, H. and Johannesen, F.B. 1974. Bull. Geol. Soc.
     Denm. 23, pp. 197-202.

51. Friis, H. 1976. Bull. Geol. Soc. Denm. 25, pp. 99-105.

52. Friis, H., Nielsen, O.B., Friis, E.M. and Balme, B.E.
     1980. Danm. Geol. Unders. Arbog 1979, pp. 51-67.

53. Weyl, R. and Werner, H. 1951. Proc. 3rd Int. Congr. Sedimentology, Groningen-Wageningen, pp. 293-303.

54. Grimm, W.D. 1973. Contrib. Sedimentol. 1, pp. 103-125.

55. Hester, N.C. 1974. J. Sed. Petrol. 44, pp. 363-373.

56. Nickel, E. 1973. Contrib. Sedimentol. 1, pp. 1-68.

57. Walker, T.R. 1967. Bull. Geol. Soc. Am. 78, pp. 353-368.

58. Walker, T.R., Waugh, B. and Crone, A.J. 1978. Bull. Geol. Soc. Am, 89, pp. 19-32.

59. Wieseneder, H. and Maurer, J. 1959. Eclog. Geol. Helv. 59, pp. 1155-1172.

60. Gazzi, P. 1965. J. Sed. Petrol. 35, pp. 109-115.

61. Yerkova, R.M. 1970. Sedimentology 15, pp. 53-68.

62. Scavnicar, B. 1979. Zbornik Radova, Sekcija za primjenu geologije, geofizike i geokemije, serija A, 6/2, pp. 351-382.

63. Morton, A.C. 1979. J. Sed. Petrol 49, pp. 281-286.

64. Morton, A.C. 1982. Bull. Am. Ass. Petrol. Geol. 66, pp. 1542-1559.

65. Blatt, H. and Sutherland, B. 1969. J. Sed. Petrol. 39, pp. 591-600.

66. Füchtbauer, H. 1974. Sediments and sedimentary rocks. E. Schweizerbart'sche, Stuttgart.

67. Mange-Rajetzky, M.A. and Oberhansli, R. 1982. Schweiz. Min. Pet. Mitt. 62, pp. 415-436.

68. Krynine, P.D. 1946. J. Geol. 54, pp. 65-87.

69. Van Andel, T.H. and Poole, D.M. 1960. J. Sed. Petrol. 30, pp. 91-122.

70. Dreimanis, A., Reavely, G.H., Cook, R.J.B., Knox, K.S. and Moretti, F.J. 1957. J. Sed. Petrol. 27, pp. 148-161.

71. Connally, G.G. 1964. Science 144, pp. 1452-1453.

72. McDonald, B.C. 1968. J. Sed. Petrol. 38, pp. 956-957.

73. Mange-Rajetzky, M.A. 1981. J. Geol. Soc. Lond. 138, pp. 83-92.

74. Morton, A.C. Init. Repts. DSDP 81 (in press).

75. Cawood, P.A. 1983. Bull. Geol. Soc. Am. 94, pp. 1199-1214.

76. Nisbet, E.G. and Pearce, J.A. 1977. Contrib. Min. Pet. 63, pp. 149-160.

77. Morton, A.C. Sedimentology (in press).

78. Hearn, jr. B.C. and McGee, E.S. 1983. Bull. US Geol. Surv. 1604, pp. 1-33.

79. Smith, J.V. and Stenstrom, R.C. 1965. J. Geol. 73, pp. 627-635.

80. Allen, P. 1972. J. Geol. Soc. Lond. 128, pp. 273-294.

81. Duddy, I.R., Gleadow, A.J.W. and Keene, J.B. Init. Repts. DSDP 81, (in press).

82. Hurford, A.J., Fitch, F.J. and Clarke, A. 1984. Geol. Mag. 121, pp. 269-277.

83. Gaudette, H.E., Vitrac-Michard, A. and Allegre, C.J. 1981. Earth Plan Sci. Lett. 54, pp. 248-260.

84. Scharer, U. and Allegre, C.J. 1982. Can. J. Earth Sci. 19, pp. 1910-1918.

SIGNIFICANCE OF GREEN PARTICLES (GLAUCONY, BERTHIERINE, CHLORITE) IN ARENITES

G.S. Odin

Université Curie - Dpt. Géol. Dyn. - Paris, France.

Abstract : The green particles encountered in marine sediments allow us to deduce some aspects of their depositional environment. Critical to this diagnosis are : (i) a precise identification of the green facies, (ii) a thorough understanding of the conditions leading to the genesis of the particular green facies encountered, and (iii) the distinction of in situ, perigenic and reworked grains. In present day seas, three facies (glaucony, marine "berthierine" and marine chlorite, all characterised by Fe-rich clays) are being generated. By a consideration of their various environments of deposition, we discuss a general model for the genesis of green particles and propose a scheme to categorise the different facies in terms of global sedimentary sequences.

When studying sediments, geologists frequently meet green facies usually in the form of green particles either dispersed in various kinds of rocks (clays, chalk, limestones or sands) or, more frequently, concentrated in horizons decimetric or metric in scale. The present contribution will focus on arenites. Because green particles are presently being formed in many Recent sediments, their study serves to formulate models to understand ancient formations.

The main point here will be to show that these green particles are diverse in their mineralogical nature and that their geological significance is also diverse. A general scheme will be proposed in order to distinguish easily the different natures and therefore the different meanings of these green particles.

279

G. G. Zuffa (ed.), Provenance of Arenites, 279–307.
© *1985 by D. Reidel Publishing Company.*

A) VARIETY OF GREEN PARTICLES

Within the last 15 years, a large variety of Cambrian to Recent so called "glauconitic sediments" have been investigated. The common point of all these sediments was obviously the green colour, although the term "glauconite" was not always properly used. The first step in the analysis of such green sediments is to concentrate a sufficient quantity of the pigment in order to facilitate observations and analyses. A very useful common property of all the green particles encountered in sediments, and especially in arenites, is their paramagnetic behaviour. Therefore, after adequate washing and sieving, magnetic separation permits rapid concentration of the granular pigment representative of the sampled outcrop. We are able now to study this concentrate.

A.1 Morphology of the green grains

Green grains display a variety of appearance which is essentially infinite. Several authors have proposed morphological classifications since Cayeux (1916) but, in order to understand the meaning of the green grains, only two kinds have to be distinguished. The basis for their distinction is the presence or absence of a recognisable "substrate of verdissement", defined as the host grain for the green mineralization process. Most research on sedimentary green pigments has not discussed why they occur in granular form. The reason is that the pigment is the result of the verdissement of various initial substrates themselves of granular form ; the most favourable have a size between 100 and 500 µm. Observations made on granular glaucony from ancient formations, as well as from present-day continental shelves, suggest that four kinds of substrate are common :

Internal moulds (or casts) of predominantly carbonate microfossil tests (e.g. foraminifera, ostracoda, small molluscs etc...). Internal moulds sometimes dominate in a sample but are frequently present in small amounts (Murray & Renard, 1891, pp. 378-391 ; Collet, 1908 ; Caspari, 1910 ; Wermund, 1961 ; Ehlmann, Hulings & Glover, 1963 ; Bjerkli & Ostmo-Saeter, 1973).

On present-day continental shelves, they are essentially most abundant furthest from the coast and diminish landwards.

Faecal pellets, which are initially more or less argillaceous or limy and contain variable amounts of organic matter. In glauconitized form, they frequently occur in great abundance in ancient and Recent glauconitic sediments (Takahashi & Yagi, 1929 ; Moore, 1939 ; Bell & Goodell, 1967 ;

Porrenga, 1967a ; Tooms, Summerhayes & McMaster, 1970 ; Giresse & Odin, 1973). Most pellets are more or less ellipsoidal
with a longest axis of 150-500 μm. Pryor (1975) shows that
most of the pellets are true faecal pellets produced in large
quantities by filter-feeding organisms in present-day shallow
seas. They also occur at greater depths locally (Moore, 1939),
but are essentially characteristic of the inner platform where
they are generated.

Biogenic carbonate debris, formed either by disarticula-
tion after disintegration of organic tissue or by biological
and mechanical break-up, is frequently found glauconitized
(Dangeard, 1928 ; Houbolt, 1957 ; Lamboy, 1974 ; Odin and
Lamboy, 1975). This substrate is usually found in abundance
at depths similar to those of faecal pellets where no bottom
currents have rolled mud pellets on present-day shelves and
where echinoderms are abundant as the benthic fauna.

Mineral grains and rock fragments, whether or not they
contain iron silica and aluminium as major ions, and whether
or not they are phyllitic, may become glauconitized (Cayeux,
1916 ; Wermund, 1961 ; Ojakangas & Keller, 1964 ; Odin, 1972 ;
Hein, Allwardt & Griggs, 1974). Glauconitized quartz, feldspar, biotite and muscovite, calcite, dolomite, phosphates,
volcanic glass shards, volcanic and plutonic rock fragments
as well as chert grains can be recognized.

On several areas of the present-day shelves (eg : California,
French Guyana), biotite flakes are the dominant substrates
of 2 different kinds of verdissement and seem to characterise
the proximity of sediment transport systems.

A rough appraisal of the observed substrates and chemistries
indicates that none is dominant nor, *a fortiori*, required
as a starting material for glauconitization or other verdissement process. Nevertheless, carbonate appears to represent
an especially favourable substrate, as noted by previous workers (Cayeux, 1916 ; Millot, 1964, p. 239 ; Lamboy, 1976).

The green grains within a bed may be derived from different
parent materials (substrate) each of which may itself be represented by several instances of the verdissement process ;
it is important to bear in mind that the bulk analysis of
a sample, even purified, is likely to be a mixture of initial substrates and authigenic minerals. What is remarkable,
looking at these green particles in present-day outcrops where the verdissement is under way, is that a complete evolutionary series may be identified from the initial substrates
to the wholly green grains. At this latest stage of evolution,
it is frequently very difficult to recognize the original

substrate. The grains without a recognisable substrate of
verdissement therefore come from other grains which, at an
earlier stage of evolution, would have been recognizable.
In ancient series, this evolved stage is more frequent than
at the present-day sea bottom.

These preliminary observations assist the determination of
the depositional environmental conditions. Dominantly detrital substrates (especially feldspars or micas) may be interpreted as the result of the close proximity either of a river
mouth or of an actively eroding coastline. Dominantly coproli-
tic substrates indicate the vicinity of a muddy sea bottom
deposit rich in worms or other mud eating organisms. This
is usually the case on the inner part of the shelf fed with
clay rich waters. Dominantly calcareous substrates reflect
the absence of detrital particles brought by rivers on the
inner part of the shelf. Finally, the zone where internal
moulds form the dominant substrates of verdissement marks
relatively deep sea conditions on the outer part of the shelf.

A.2 Mineralogy

The second point requiring investigation is the precise mineralogical composition of the green grains. Three main groups
of minerals can be identified.
        a) Chlorite : a peculiar form of green grains encounte-
red in the sands frequently shows an accordion-like shape.
Accordion-like grains may suffer three verdissements as distinguished later. The first is alteration to chlorite. This
is easily identifiable using the X ray diffraction (XRD) technique on powder mounts. Two peaks are observed at 14 Å and
7 Å and usually the remaining part of the initial substrate
is shown at 10 Å (Figure 1). These peaks are nearly always
very high and sharp and the peak at 14 Å does not disappear
when heating the material at 490° C for 4 hours, clearly indicating that the green booklets are micas altered to chlorite.

Very recently, our attention turned to a completely different
geological setting of chlorite. The substrates consist of
blackened carbonate debris ; a green colour is shown locally
in parts of these substrates, sometimes as infilling. XRD
shows peaks at 14 Å and 7 Å, together with peaks of magnesian calcite. The green pigment has not yet been isolated
for detailed study. Two outcrops are known from the present
day : Nouvelle Calédonie (study under way with C. Froget)
and Tunisia (Burollet et al. 1979 ; A. Jourdan, pers. comm.).
The importance of these cases appears to be limited. The paragenesis is tentatively designated as "marine granular chlorite".

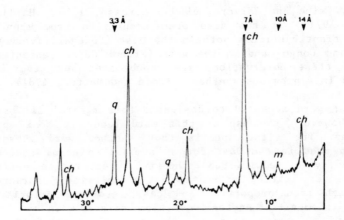

Figure 1 : X ray diffractogram of chloritized mica.
Peaks of chlorite (ch), biotite (m) and quartz (q)
are shown.

b) Marine "berthierine": The green colour of accordion-
like grains as well as internal moulds, faecal pellets, carbonate debris or grains of unknown initial substrate, is commonly due to the presence of a 7 Å layer lattice silicate
(Figure 2).

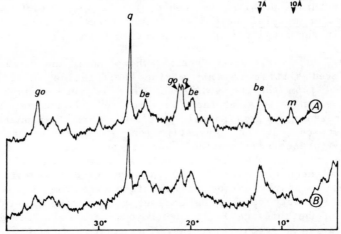

Figure 2 : X ray diffractograms of marine "berthierine" from French Guyana. A : green grains ; B :
green grains altered to gœthite (go) with traces
of the initial substrate : quartz (q) and biotite
(m).

This mineral has been identified from present-day marine environment (Von Gaertner and Schellmann, 1965 ; Porrenga, 1967b ;

Rohlich, Price and Calvert, 1969 ; Giresse, 1969 ; Hardjossoe-
sastro, 1971). It has also been identified from many parts
of the Atlantic borders both in the east (Senegal, Ivory Coast,
Gaboon and Congo) and in the west (French Guyana essentially).
Most of these observations are unpublished but several were
gathered in a previous synthesis (Odin and Matter, 1981).

The usual word accepted to designate this mineral is "berthierine". However, the use of this word leads to some specific
questions. The first one is that another word (chamosite)
has been used in the past for a "similar" mineral ; according
to Brindley et al. (1968) the term berthierine designates
a 7 Å layer lattice silicate, not a chlorite, characterized
by a high iron proportion and has historical priority over
that of chamosite. Another question arises when we look at
the chemical composition of the mineral berthierine. This
mineral (like the 7 Å "chamosite") was defined from iron ore
deposits known to be mostly formed in a lacustrine to lagoo-
nal environment ; it is characterized by a substitution of
the aluminium by ferrous iron, i.e. berthierine is a triocta-
hedral mineral. From the few chemical analyses of marine "berthierine" presently available (Odin and Matter, 1981, p. 625
recently complemented with confirmatory results) it is clear
that the iron is mostly ferric and, consequently, that the
marine green 7 Å mineral is essentially dioctahedral. This
architectural peculiarity would justify, in fact, the use
of another mineral name. That is the subject of future improve-
ments in the knowledge of this pigment (Odin, 1984).

However, as yet, green grains which could be unambiguously
considered similar to this marine "berthierine" have not been
recorded from ancient sediments. This is very surprising con-
sidering that this marine "berthierine" is rather common on
the present-day continental platform. Three explanations may
be proposed. 1. The investigation of green grains similar
to the present-day marine "berthierine" has not been done
or the grains, when encountered, were mistakenly interpreted
as of glauconitic nature ; 2. The present-day genesis of this
mineral is nearly unique in the history of the earth ; 3. After their early genesis on the sea bottom, the mineral formed
(marine "berthierine") is destroyed, or modified during a
very early diagenetic process. The third alternative is presently favoured but further studies are required.
    c) Glauconitic minerals. The word glaucony proposed ear-
lier (Odin and Létolle, 1978 ; Odin and Matter, 1981) will
be used here to designate (as in French "glauconie") the green
grains facies as a whole. Glaucony is the name of the pigment
whatever the precise mineralogy of the green glauconitic minerals of the particles. However, the marine authigenic green
minerals giving the colour to the glaucony (the green grains)

are of various sorts, from an end member which is a smectite (*glauconitic smectite*) to an end member which is a clay size mica (*glauconitic mica*). This glauconitic mica is the mineral "glauconite" of the mineralogists ; however, the same word (glauconite) is also used in English to designate the green grains as a whole (the glaucony). This source of confusion is very difficult to remove and sometimes precludes a correct understanding of the sediment. For this reason the term "glauconite" should be discontinued, substituting term <u>glaucony</u>

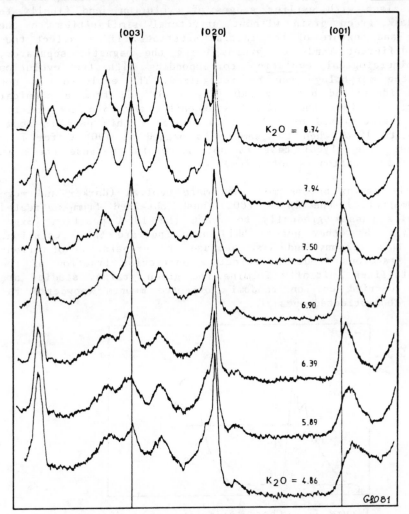

Figure 3 : X ray diffractograms of Cretaceous and Palaeogene glauconies as a function of their K_2O content (modified from Odin and Dodson, 1982). Samples with less than 4.8% K_2O are shown in figure 6.

(plural glauconies) when one does not intend to designate
a precise mineralogical component for this pigment. The term
glauconitic minerals should be used for designating their
mineralogical components.

The samples obtained from present-day outcrops show an evolution in morphology from an original substrate, to a pale
green still recognizable substrate, to a green cracked grain
in which the original substrate may only be recognized by
comparison with earlier stages of evolution, and finally to
a dark green grain without structural similarities to the
form and texture of the initial substrate. If we select these different kinds of grains using the magnetic separator,
a mineralogical evolution corresponding with the evolution
of the morphology can be recognized. The evolution is now
well identified both by XRD (Figure 3) and from a chemical
point of view. The initial glauconitic minerals are smectites characterized by a broad 14 Å peak : this peak is modified to 10 Å when the sample is heated at 490° C for four
hours. (The glycol test is not successful because the main
interlayer cation is potassium).

As the grains become more and more evolved (darker and more
paramagnetic) the 14 Å (001) peak obtained from untreated
samples passes gradually to reach the 10 A position (Figure 3). The other peaks (hkl) also indicate this evolution.
For this reason, and also because the preparation of oriented fractions tends to select a particular fraction of the
crystallites initially forming the green grains, studies are
best carried out on randomly oriented mounts (powders) rather than oriented ones.

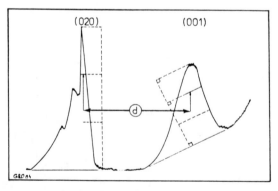

Figure 4 : Characterization of a diffractogram obtained from a glaucony. The distance (d) between
the middle of the peak (020) and the middle of the
peak (001) is a function of the potassium content
(modified from Odin, 1982, p. 394).

Thanks to the varying position of the (001) peak, compared with the very stable position of the (020) peak (at 4.53 A), it is possible to define the stage of evolution of the glauconitic minerals of the grains by the <u>distance</u> between these two peaks (Figure 4). Moreover, the position of the (001) peak, and the stage of evolution of the glauconitic minerals, are simply correlated with their potassium content. Therefore, the distance between the two peaks (001)-(020) is a simple function of the potassium content of the minerals, allowing the successful delineation of the potassium content (Odin, 1982, chapter 20 and Figure 5). An example of the obtained precision is given in Figure 6.

We have followed above the evolution of the authigenic green minerals of the grains but, fundamentally, the green grains comprise a second component : the initial substrate. This component (inherited) follows an evolution opposite to that of the glauconitic minerals. The latter show more and more characteristic peaks whereas the peaks of the minerals of the substrate progressively disappear.

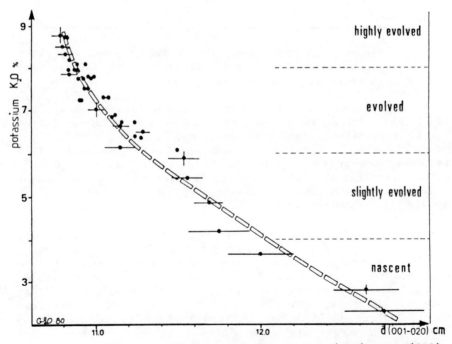

Figure 5 : Distance between peaks (001) and (020) as a function of the potassium content of glauconies. The analytical uncertainties on the potassium content and the distance between peaks is given for several samples (modified from Odin, 1982, p. 394).

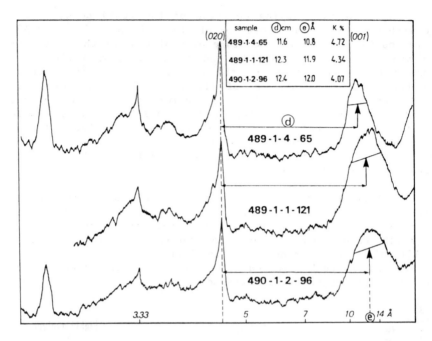

Figure 6 : Comparison between slightly-evolved glauconies. The X ray diffractograms give informations
coherent with the chemical data.

From the integration of the data obtained from the morphological and mineralogical observations, 4 stages of evolution
can be identified : *"nascent"* glaucony, *"little-evolved"* glaucony, *"evolved"* glaucony, *"highly-evolved"* glaucony. This
is summarized in the figures 5 and 7. Although this subdivision is artificial because the evolution is <u>continuous</u>, these
4 terms are convenient for designating the stage of evolution
of the grains observed in a particular sediments and distinguish their significance. "Nascent" glaucony is formed with
grains very pale green in colour and for which the substrate
is nearly undeformed ; the authigenic minerals are smectites.
In "little-evolved" glaucony the grains are green, the form
and texture of the initial substrate is still recognizable,
the diffractograms only show traces of the mineralogical nature of the substrate, the authigenic minerals show a broad
peak at about 12-13 A and the K_2O content is around 5%. In
"evolved" glaucony, the texture and the form of the substrate become difficult to recognize, fissures appear on the surface of the grains, diffractometry shows very little or no
traces of the substrate, the authigenic minerals display a
peak at about 10.5-11 A and the K_2O content is higher than
6%. Finally, "highly-evolved"glaucony is usually formed with
dark green grains, without any trace of the substrate ; the

authigenic minerals show a clear diffraction at about 10 A
and the K_2O content is higher than 8%.

Obviously, a sediment may contain grains at different stages
of evolution ; but usually (except when mixing occurs) one
stage is clearly dominant. In order to use correctly the information gathered at this stage of the study, it is necessary
to remember what is known about the genetic processes and
environments of the different sorts of grains identified above.

B) GENETIC PROCESSES

B.1 Chlorite genesis

The green mica flakes for which the mineral chlorite has been
identified by diffractometry are known to be the result of
an alteration : chloritization of biotite flakes occuring
in soils on the continent. The significance of these green
particles is not a problem provided that they have been mine-
ralogically investigated : they are inherited in the marine
deposits.

The marine granular chlorite has a completely different meaning. Our present knowledge is restricted to 2, possibly 3,
Recent marine outcrops. They seem to have in common a very
restricted distribution probably related to very shallow water, 2 to 40 m deep, either in shallow water Posidonia facies
(Tunisia) or in a reef setting close to the open ocean (Nouvelle Calédonie). A common factor appears to be a somewhat
high water temperature.

B.2 Marine "berthierine" genesis

Our present understanding of marine "berthierine" is uniquely
based on samples from the present-day continental platforms.
On these platforms, the geographical distribution is very
significant. The marine "berthierine" has essentially been
collected from tropical latitudes ; there, the presently studied main outcrops (off Senegal, Guinea, Ivory Coast, Nigeria,
Gaboon, Congo and French Guyana) show a variety of initial
substrates but the general environment always seems to be
related to river mouths. However, the verdissement occurs
in a more distal setting than that where most of the terrigenous detritus is deposited. This area is termed sub-deltaïc
(Odin, 1975 ; Odin and Matter, 1981). In this sub-deltaïc
environment, initial substrates were easily identified in
several areas : Congo - Gaboon (dominantly faecal pellets),
Senegal (dominantly internal moulds) and French Guyana (dominantly mica flakes frequently chloritized). In French Gu-

yana carbonate debris and faecal pellets are also affected
by the verdissement. It is very difficult to observe the evolution of the berthierinization process for two reasons. The
first one is that, in the two areas where detailed results
are available, the initial substrate (faecal pellets made
of kaolinite off the Congo or chloritized biotite off French
Guyana) already displays a peak at 7 A (i.e. in the same place as the authigenic 7 Å mineral). As a result, the progressive evolution may only be seen by looking at the morphology
of the grains, the shape of the 7 A peak, and the evolution
of the possible other peaks characteristic of the substrate.
Unambiguous criteria of evolution are therefore very difficult
to obtain. Secondly, unlike the evolution of substrates affected by glauconitization, the berthierinization process seems
relatively rapid. This was deduced from the general absence
of intermediate stages of evolution identifiable according
to the morphology and particularly the colour ; sedimentological observations also show that the marine "berthierine"
was obtained very soon after the deposition of the substrates when the Holocene transgression occured. The duration
of evolution is estimated at about 10^3 years.

However, micro environmental characteristics (nature of the
substrates, physical micro-environment) appear to be similar for both the berthierinization and the glauconitization
processes. The fundamental process of genesis of the marine
"berthierine" is therefore similar to that accepted for the
glauconitic minerals : i.e. it is a neoformation (crystal
growth) of an original marine mineral in a semi-confined micro-environment such as the pores of a substrate becoming
degraded on the sea-floor (Odin and Matter, 1981). There are
some small differences between the two verdissements. One
of these is, as indicated above, that berthierinization requires an environment more acid (i.e. under the influence
of continental waters) compared to glauconitization, which
occurs in the more basic open sea, far from the coast. Another difference is that the initial substrates of berthierinization are constantly smaller (about 100-200 um) than those
favourable for glauconitization (about 200-500 um). This means
that the 7 A green mineral grows in a more open (more dilu-
ted ?) environment compared with the glauconitic minerals.
This is in good agreement with observations on soils on the
continent. The clay mineral T.O.T. (3 layers : Tetrahedral +
Octahedral + Tetrahedral) structures (smectites) are formed
in a less open (more confined) and more basic (presence of
calcium) environment compared with the T.O. (2 layers : Tetrahedral + Octahedral) clay mineral structures (kaolinite)
which are formed in more acid soils with a larger circulation
of water. This will be better understood after a brief discus-

sion concerning our present knowledge of the process of glauconitization.

B.3 Glaucony genesis : mechanism and environment

Most of the following details have been discussed by Odin
and Matter (1981). As emphasized above, the most common habit of glaucony is granular because the most favourable substrate for the verdissement is itself granular. Initial granular substrates are always highly porous. It is in these
pores of 5-10 um size, which may extend across the entire
grain, that growth of the first green glauconitic crystals
is inferred. They are iron-rich glauconitic smectites with
potassium as the main interlayer cation. The first blades
grow perpendicular to the surface of the substrate, and develop into a box-work-like fabric by coalescence. Finally,
the entire pore space is filled, at which stage the grains
already appear green (nascent stage).

The initial substrate disappears, often by dissolution. While the last smectites are growing into the remaining pore
space of the substrate grain, the earlier smectites are being recrystallized. Chemical analyses of this stage reveal
a high iron content whereas the potassium content is just
beginning to increase.

The persistence of the substrate depends on its ease of alteration in the marine environment (Lamboy, 1976). Initial
carbonate substrates alter more quickly and are therefore
more favourable to glauconitization than are, for example,
quartz substrates which resist alteration. Therefore, glauconitization of quartz substrates is slow and frequently incomplete.

The disappearance of the substrate may be observed by SEM
and XRD. Nevertheless, pieces of the substrate may still be
present even if they cannot be detected by SEM and XRD. The
argon content of faecal pellets (essentially kaolinite) from
the Gulf of Guinea has been measured at different stages of
evolution (Odin and Dodson, 1982). There is no visible trace
(XRD) of the substrate remaining when pellets contain more
than 4-5% K_2O. However, in spite of this, there is nearly
10% of the initially present inherited argon in the pellets
even with K_2O as high as 6.6% (Odin and Dodson, 1982). This
proves that, even in quite evolved glauconitized substrates,
some of the initial minerals may have retained their argon.

When, according to the XRD pattern, the parent material has
become a minor admixture in the grain, important textural

changes are observed under the SEM : (1) progressive loss
of the primary texture ; (2) growth of larger and better shaped crystallites which form lamellae or lepispheres resem-
bling rosettes ; and finally (3) deformation of the grain
which takes on a bulbous, cracked habit.

The deformation of the initial substrate grain results from
the displacive growth of glauconitic crystals. In its last
stage of evolution, the grain is generally embedded beneath
an outer crust of little-evolved glauconitic minerals. At
the end of this process, a smooth crust is formed over the
initially irregular surface, increasing the roundness of the
grains (Lamboy & Odin, 1975). This sequence of events has
been observed on grains collected from the modern shelf floor
which have therefore been in permanent contact with the marine environment.

The reason why crystal growth occurs in grains on the sea
floor and usually not in the bulk sediment has been discussed
in Odin and Matter (1981). Glauconitization in common with
several other geochemical processes, occurs in a specific
environment at the boundary between, but different from, the
sea water above (too open to permit crystal growth) and the
underlying sediment below (too confined to permit the ion
exchanges necessary for crystal growth). The *semi-confined*
environment of the pores of the substrate acts as a passage-
way for the ions derived from the sea-water, from the initial
substrate of the grain, and from the underlying sediment.
The three sources combine in the substrate of verdissement
to supply the ions required for the glauconitic mineral crystal
growth and recrystallization. The influence of the quality
of the semi-environment is best illustrated by the rather
limited variety of the grain size of the substrate favourable
for the glauconitization process. Grains larger than about
1 mm or smaller than about 0.1 mm clearly show a much less
evolved stage of glauconitization compared to the more evolved stage displayed by substrates of intermediate dimensions.
This also applies to berthierinization but, as said above,
the semi-confinement favourable for the genesis of the marine
"berthierine" mineral is less confined with initial substrates systematically smaller (about 100-200 um) allowing easier
cationic exchanges.

Compared to berthierinization, the glauconitization process
appears to need a much longer time. We have a pedagogical
example on the present-day shelf of Congo. This shelf has
been submitted to a global transgression during the last 18.000
years. The substrates favourable for glauconitization were
therefore immersed more and more recently from the ocean towards the coast. XRD on samples collected at different depths

shows that deeper grains (300 to 200 m) are more evolved than
shallower ones. The shallower they are, the less evolved they
are (Giresse and Odin, 1973). Moreover, at the lowest level
of regression, about 110 metres below present-day sea level,
a red belt of oxidized pellets occurs. This clearly shows
that a subaerial oxidation occured on already glauconitized
(little-evolved glaucony) substrates. Consequently, we can
estimate the full process of glauconitization to take about
10^5 years. This, together with the fact that the whole process requires full connection with sea water allows us to
presume a **lack of deposition** of at least this duration for
any evolved glaucony. Finally, the process of glauconitization stops at any stage if burial occurs by preventing the
necessary ionic exchange between sea water and the evolving
green grains.

We have now gathered enough fundamental data to recognize
the different kinds of green grains in arenites and to understand how they formed. We will be best able to interpret these grains if we examine the global environment of their deposition according to some well known examples.

C) SIGNIFICANCE OF THE DIFFERENT GREEN GRAINS IN ARENITES

There are two different possible meanings of green grains
observed in arenites. The first one, inheritance, also applies
to other components of any sand. Inheritance may be of different natures : it could be extrabasinal, the grains being
transported and therefore not contemporaneous with deposition.
However it could be intrabasinal, the grains being remanié
from the bed rock at the sea floor. The second possibility,
authigenesis, is the most frequently encountered in ancient
series. It corresponds to an intrabasinal genesis, the grains
being non-transported (*in situ*) and formed contemporaneously
with the deposit. However, some grains may be intrabasinal
and non-transported, but slightly older than the sediment
in which they are found because of their relict nature : after
a long break in sedimentation, green grains developed at the
beginning of the hiatus may be mixed with much younger fossils
when deposition restarts.

Obviously, an intermediate possibility exists and must be
borne in mind -perigenic green grains (Lewis, 1964) which
are formed in one place (on a plateau for example) but deposited (accumulated) elsewhere, where bottom currents are less
effective. These grains are intrabasinal, but transported
and nearly contemporaneous. In any case, green grains, when
placed into one of the categories listed above, may give valuable information on both the depositional environment and
on the regional setting.

C.1 Inheritance

The most obvious case of continental inheritance described
above is the presence of chloritized micas. Because mica flakes may be altered in three possible ways, it is necessary
to investigate systematically the mineralogical nature of
these accordion-like green grains before making any interpretation. Chloritized micas clearly indicate a continental origin
followed by fluvial transportation to the sea and therefore,
the presence of a landmass in the vicinity of the sampled
formation.

Another case of inheritance is that of glaucony. This case
is especially difficult to identify when glauconitic formations are altered on the sea floor itself. Because the components of the glauconitic grains are obviously in equilibrium
with the sea water environment, they do not alter and therefore may well be remanié in younger marine arenite formations
without physical traces of the process. A series of examples
of remanié glauconies from ancient formations was reported
by Odin and Rex (1982). In any case, reworking is not easy
to recognize by sedimentological criteria ; consequently,
radiometric investigation is one of the few possibilities
able to prove that, under certains conditions, the glaucony
has been inherited from older intrabasinal or extrabasinal
levels. This kind of investigation is especially useful if
one considers also the example of a continental sand. The
immediate conclusion when looking at the presence of glaucony in a sand is to presume that the sand has been deposited
in the sea. However, if the radiometric data prove that the
green grains are remanié, the studied sand may well be of
continental origin. The proof may be obtained easily because,
even when submitted to moderate continental alteration, the
glaucony more or less retains its original apparent radiometric age (Odin and Rex, 1982).

Another example of suspected inheritance has also been investigated recently, from the present-day continental platform
off South Africa. A series of sediments previously examined
by Birch et al. (1976) were used to try to recognize whether
or not glauconies on the Atlantic margin of South Africa (30°-
40° S) are Recent or remanié from the underlying Cretaceous
to Tertiary indurated bed rocks. Table 1 shows a series of
presently unpublished results obtained from glauconies either
separated from unconsolidated sediment collected at the sea
floor or obtained from the underlying indurated bedrock. For
example, G 634 is a glaucony extracted from the unconsolidated
sediment 1113G, located itself above the bedrock 1113D of
early Cretaceous age and from which the glaucony G 635 was
obtained. Glaucony G 636 is separated from sediment 1331C,

Table 1 : Radiometric results on glauconies from the South African shelf. Samples provided by G. Birch (G.B.) ; Ar measurements done in Berne (Geological unit) by one ofus, K measurements by Madeleine Lenoble (Département Pétrographie, Paris). Note of interrogation for one measurement of sample G. 637 is due to an imperfect record. G.B. sample numbers (left column) indicate the nature of the glaucony collected either from an indurated bedrock (noted D) or from unconsolidated sediments (noted G or GC).

G.B. sample	Glaucony	K %	Ar nl.g^{-1}	% rad	T apparent	Presumed age
1113 G	G 634	6.01 ± 0.08	21.66 ± 0.10	87.8	90.5 ± 1.2	(above 1113D)
1113 G	G 634	6.01 ± 0.08	21.58 ± 0.10	90.7	90.1 ± 1.2	"
1113 D	G 635	4.77 ± 0.07	20.46 ± 0.09	90.5	107.1 ± 1.6	Barremian/Alb.
1331 G	G 636	6.74 ± 0.09	11.31 ± 0.09	85.3	42.7 ± 0.7	(above 1331 D)
1331 D	G 637	6.01 ± 0.09	10.45 ± 0.40(?)	78.0	44.2 ± 1.8(?)	(unknown)
1331 D	G 637	6.01 ± 0.09	9.70 ± 0.06	90.2	41.1 ± 0.67	(unknown)
1266 GC	G 638	6.02 ± 0.09	23.53 ± 0.11	93.8	97.9 ± 1.5	(above Aptian/ Santonian)
1764 D	G 639	5.75 ± 0.12	5.60 ± 0.02	74.6	24.9 ± 0.4	(mid Miocene)
1764 D	G. 639	5.75 ± 0.12	5.59 ± 0.03	80.5	24.9 ± 0.4	(mid Miocene)
1310 D	G. 640	7.20 ± 0.10	2.220± 0.015	70.9	7.91± 0.1	(Pliocene)
286 G	G 641	6.76 ± 0.10	0.949± 0.024	37.6	3.60± 0.11	above Mid Mioc
110 G	G 642	7.01 ± 0.10	1.43 ± 0.03	57.7	5.24± 0.13	(unknown)
112 G	G 643	7.04 ± 0.10	2.21 ± 0.03	61.1	8.07± 0.14	"
2710 G	G 644	6.78 ± 0.07	5.74 ± 0.04	73.7	21.7 ± 0.3	"
2246 G	G 645	7.18 ± 0.07	1.56 ± 0.03	50.2	5.59± 0.11	"
2683 G	G 646	6.64 ± 0.09	5.32 ± 0.03	71.4	20.5 ± 0.3	"
175 G	G 647	6.72 ± 0.10	1.50 ± 0.03	60.8	5.74± 0.15	"
2443 G	G 648	6.82 ± 0.10	1.69 ± 0.03	57.8	6.35± 0.15	"

whereas the glaucony G 637 is from the underlying bedrock
(1331D), at the same location, and for which no stratigraphic
information is available. G 638 is from another unconsolidated sediment (1266GC) located above Cretaceous bedrock, G 640
is from Pliocene bedrock ; no corresponding sediments were
available. G 641 to G 648 are all glauconies from unconsolidated sediments located above bedrock of unknown age, except
for G 641, located above Mid Miocene bedrocks. In these samples, there is a clear general relationship between the radiometric ages obtained from the glaucony extracted from the
bedrock and those from the unconsolidated sediments above
(eg couple G 634 and G 635 ; G 636 and G 637 ; G 638 and the
assumed age of the bedrock below). There is a problem with
G 641 because although its apparent age of 3.6 Ma is much
younger than Mid Miocene (about 10-15 Ma) another presumably
similar sample, analyzed by another laboratory, gave a Miocene
apparent age (G. Birch, personal communication, 1983). This
is probably an error of identification for this particular
sample. Another problem is that sample G 640 gave an apparent
age which is older than the stratigraphical age presently
accepted for the horizon (i.e. younger than 6 Ma). However,
forgetting these specific details and taking into account
the very important fact that all glauconies extracted from
the unconsolidated sediments are evolved, or highly-evolved
(K_2O from 7.2 to 8.6%) implying that they did not contain
any inherited argon from their initial substrates when buried,
it is clear that the grains on the present day continental
platform off South Africa are inherited.

The second main conclusion concerning this particular continental shelf is that, although the glauconies are highly-evolved,
not one gave a zero age ; the apparently younger glauconies
have apparent ages of 3.6 to 5.5 Ma. Therefore, in spite
of the fact that they are still in contact with the sea-water,
the green grains must be practically closed to the sea water
environment ; their evolution is finished, and they have begun
to accumulate the radiogenic isotopes (argon 40) from the
radioactive decay of the *in situ* potassium 40 isotopes. This
means that these glauconies are **relict** on a shelf on which
no active sedimentation has occurred since end Miocene to
Pliocene times. Consequently, this study shows that the glauconitization processes are not active presently in this area
although it is very active to the North, between Congo (5° S)
and Morocco (about 30° N). Further north, off NW Spain (43° N)
similar apparent ages of about 5.8 Ma were measured on evolved
glauconies (Odin and Dodson, 1982). Therefore these glauconies
were interpreted as relict.

A relict glaucony may be recognized by its generally highly-
evolved state : the grains are usually very dark and frequently

coated with a thin layer of material which imparts a bright
shine to the grains. This kind of glaucony (authigenic but
relict) indicates a very long period of non-deposition between one and several million years in duration.

C.2 Authigenesis

The genesis of authigenic marine chlorite is not very well
known principally because it is not frequently encountered
in present-day seas. Consequently we herein confine our remarks to authigenic marine "berthierine" and glaucony.

As emphasized above, genesis of marine "berthierine" on the
present-day continental platform seems strictly linked with
the influence of river mouths. The area covered by the active genetic process is therefore somewhat restricted. Outcrop
depths are generally less than about 50 metres. However, these
grains may be observed in deeper water for different reasons.
For example, off the Amazon River mouth, some marine "berthierine" has been found at depths greater than 100 metres because
of transportation by bottom currents perpendicular to the
coast related to the Amazon canyon (study in progress with
M. Chagnaud and M. Pujos). A second possibility, illustrated
off French Guyana, is that during a transgression, the marine
"berthierine" initially formed at shallow depths under the
influence of the considerable mass of the Amazon River waters
is now found at greater depth due to the Holocene sea level
rise. This marine "berthierine", now widespread on the platform,
is relict. A good criterion of their relict character is the
fact that, due to its formation at relatively shallow depth,
there is a good chance that the marine "berthierine" alters
to gœthite as soon as the organic matter has been destroyed.
When this occurs, the environment which, initially, in spite
of the proximity of oxygenated waters was reducing because
of the presence of organic matter, becomes more agressive
causing the destruction of the previously formed T.O. phyllites.
From the pH of the waters, we may infer that marine "berthierine" is formed in relatively acid conditions.

By recognizing the main initial substrate of glauconitization as described above, we may infer eustatic changes which
occured prior to verdissement, as follows. The nature of the
initial substrate gives an idea of the depth at which the
substrate was formed - dominant shelly debris indicate shallow depth and dominant planktonic tests indicate deeper water.
Also the glauconitization process is known to occur at a depth
sufficiently far from the coast to prevent oxidizing influences and excess sediment deposition, i.e. at more than about
50-80 metres depth. Giresse et al. (1980) suggest that the
amplitude of the sea level rise can be inferred by conside-

ring the difference between the depth at which the substrate formed and the depth of verdissement (at least 100 m). Glauconitized substrates of shelly debris, formed under tidal influences indicate an obvious sea level rise (more than 50-80 metres) whereas concentrations of green internal moulds of planktonic faunas indicates glauconitization has not necessarily occurred through a sea level rise. In general, the glauconitization process is favoured by a transgressive phase. Verdissement follows the transgression essentially because favourable substrates, previously in the tidal zone, suddenly become available at depths suitable for glauconitization. Also, transgressions frequently cause periods of sediment starvation in the basin by partially inundating the source areas. However, the presence of glauconitized particles is not necessarily linked to a transgression, and interpretation must depend on the nature of the grains which have undergone verdissement. Regressive phases may also be recognized from the study of glaucony. For example, regression may introduce gœthite to the green grains as demonstrated by the "red belt" found today in many places off Africa at a depth of about 110 m. This is caused by the increasingly oxidizing conditions which follow a regression thus altering the previously formed glaucony. In the case of Africa, sea levels have fallen by more than 50-80 metres.

The precise mineralogical nature of the authigenic components of the glaucony also provides evidence for palaeoenvironmental reconstructions. We know that nascent glaucony forms in about 10^3 years whereas evolved to highly-evolved glaucony takes about 10^5 years : we also know that the glauconitization process requires open exchange with sea-water. It follows that the stage of evolution recognized in the green grains is a "measure" of the duration of the lack of deposition marked by the glauconitic horizon. A highly-evolved or a fortiori a relict glaucony marks a hiatus of about 10^6 years provided the facies is in situ (see figure 7).

Glaucony has also been used as an indicator of depth. However, glaucony is not a precise and reliable depth indicator. It is present today on shelves between 50 and 500 m. Abundant authigenic glauconitic grains have been recovered at greater depths (up to 800 m), from submarine highs which may have been tectonically subsiding, and from active continental margins. At shallower depths turbulent, and therefore well-oxygenated, waters are unfavourable to glauconitization. It appears that glaucony is best able to develop in water depths of 50-500 m in the open marine environment.

This limited zone of formation of glaucony on passive margins can be tentatively explained by the balance between detrital

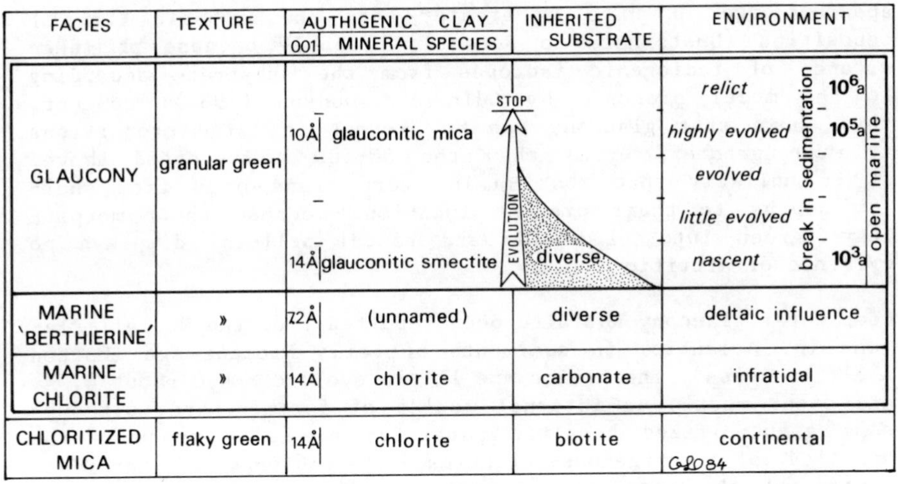

FACIES	TEXTURE	AUTHIGENIC CLAY 001 MINERAL SPECIES		INHERITED SUBSTRATE	ENVIRONMENT	
GLAUCONY	granular green	10Å	glauconitic mica	relict	relict	10⁶ₐ
				highly evolved	highly evolved	10⁵ₐ
				evolved	evolved	
		14Å	glauconitic smectite	little evolved	little evolved	
			diverse	nascent	nascent	10³ₐ
MARINE 'BERTHIERINE'	»	22Å	(unnamed)	diverse	deltaic influence	
MARINE CHLORITE	»	14Å	chlorite	carbonate	infratidal	
CHLORITIZED MICA	flaky green	14Å	chlorite	biotite	continental	

Figure 7 : Green grains ; variety, composition and geological meaning (modified from Odin and Matter, 1981). Note that the facies glaucony (the presently best known) is varied itself and may have several slightly different meanings.

influx and winnowing by marine currents. In the zone close to the shore, detrital influx exceeds erosion. Despite the presence of suitable substrates (e.g. faecal pellets) glaucony cannot form because of relatively high net accumulation. Beyond about 50 m water depth, detrital influx is less and, in addition, winnowing causes a continual redistribution of sediment towards the shelf edge. The progradational features recognizable on seismic profiles across the outer shelf confirm the operation of this process. It is in this zone, near the shelf edge, that the substrates are exposed for sufficiently long times to be glauconitized. At greater depths the lower energy conditions and the accumulation of winnowed sediment inhibits glauconitization of the substrate.

The rounded aspect of the grains does not indicate a highly turbulent environment, merely the length of evolution. However, this does not exclude the possibility of bottom current causing moderate transport of the grains, thus prolonging contact with sea water facilitating further ionic exchange. In this sense, glaucony may be an indicator of winnowing.

The existence of deep-water glaucony off Western North America has recently been discussed by Odin and Stephan (1981). The glaucony is present at several places off North and Central America at depths in excess of 1000 m. Radiometric dating on several samples, all nascent to little-evolved, gave

apparent ages 3 to 5 Ma greater than the probable time of
deposition (Quaternary to Recent), probably because of inheritance of radiogenic isotopes from the substrate according
to the model proposed by Odin and Dodson (1982). However,
this shows that glaucony **can be present** in large proportions
at much greater depths than the 50-500 m suggested above.
It is unlikely that most grains were transported from shallow depths to their present location, because their morphology (green internal moulds and faecal pellets) displays no
evidence of attrition.

Deep water glaucony has also been reported from the Rockall Plateau (N. Atlantic) in sediments of early Miocene age (Morton
& al. in press). The grains are little evolved : K_2O about 4.5%,
and occur mostly as internal moulds of foraminifera, although
some glauconitized basaltic particles are also present. The
scarcity of terrigenous clastics, the absence of turbidite
units and the deep-water nature of the associated foraminiferal assemblages (indicating water depths between 1600-2500 m)
all point strongly towards the authigenesis of these particles
in deep water. The glaucony concentrations occur immediately
above a major hiatus believed to be the result of the activity of strong bottom water currents related to the overflow
of cold Norwegian Sea waters over the Iceland-Faeroe Ridge.

Consequently, the use of the presence of glaucony as a depth
indicator is very limited when used alone. Furthermore, glauconitic green grains could be transported easily from the
shelf into deep waters by gravity-flow processes, adding to
the difficulty of using this facies as a depth indicator.

We are now sufficiently aware of the characteristics of green
grains found in arenites to be able to distinguish different
categories. We have discussed the processes and environments
of genesis and are therefore able to indicate their significance. A common point of all these various green particles
and their alteration products is their high iron content,
and we therefore discuss their significance in terms of the
geochemical path of iron in the sea.

C.3 **Iron rich minerals in the geochemical path of iron in the sea**

This model has already been presented on several occasions
(Odin, 1975 ; Odin and Matter, 1981) and is reinterpreted
here. Five zones of iron activity are distinguished (Figure 8).
    Zone 1 is the area of passive iron, characterized by
rapid deposition of particulate and colloidal iron by gravity
settling, flocculation, precipitation or by filter-feeding
organisms (Glangeaud, 1941 ; Pryor, 1975). An important frac-

tion of the continent-derived iron is thus largely immobili-
zed near river outflows and, because of high accumulation
rates, has no chance to react with sea-water. However, in
the presence of organic matter in the sediment, iron is reduced and becomes soluble as Fe^{2+}. This explains the much
greater concentrations of iron in the interstitial waters
of marine sediments than in the adjacent sea-water (Goldhaber
& Kaplan, 1974). This iron is either precipitated in the sediment, e.g. as pyrite, or migrates back into the sea-water,
to feed other zones (small arrows on figure 8).

    Zone **2** appears to be restricted to warm climates and
is well illustrated on the Atlantic Ocean borders. On the
shore and the nearshore seafloor of the tropics, chemical
or biochemical iron precipitates have been observed in less
than 10 m water depth. In addition, quartz, oolites and pellets are commonly coated by iron oxides, hydroxides or carbonates (Giresse, 1969 ; Bouysse, Kudrass & Le Lann, 1977).
On a global scale at the present-day these precipitates and
oxidized particles are of minor importance, but this might
not have been the case throughout the geological past (Borchert, 1965). During a regressive phase, previously formed
marine "berthierine" as well as glaucony may be oxidized in
this zone.

    Zone **3** is characterized by the formation of marine "ber-
thierine" provided that organic matter is sufficient to prevent oxidation. The formation of marine "berthierine" occurs
within the influence of river mouths and seems to be restricted today to depths of less than about 40-50 metres in tropical seas. The marine granular chlorite quoted above probably lie in a subzone of this area (3' on figure 8).

    Zone **4** consists of two sub-zones and covers a much wi-
der area than zones 1-3. Its primary geochemical feature is
the formation of glauconitic minerals. One sub-zone encompas-
ses the outer continental shelf and upper slope which are
still reached by continental iron. The deep water glaucony
at the foot of the continental slope off western North America is probably related to this sub-zone. The other sub-
zone of glaucony formation comprises some oceanic highs and
midocean ridges and rises which receive essentially juvenile
iron. The Miocene glaucony of the Rockall area probably belongs to this sub-zone with high Fe concentrations carried
in solution from the basaltic Iceland-Faeroe Ridge.

    Finally **Zone 5** covers vast expanses of the deep sea floor
where iron is found incorporated into ferromanganese nodules
and encrustations and also as iron smectites frequently green
in colour. Most of this iron is probably derived from juvenile sources.

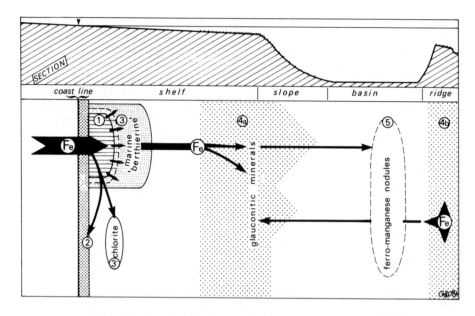

Figure 8 : Location of green sheet silicates in
the geochemical path of iron in the sea. 1 - accumulation of detrital iron ; 2 - chemical precipitation
of hydroxides and oxidation of detrital iron minerals ; 3 - authigenic growth of marine "berthierine"
(3' chlorite) ; 4 - authigenic growth of glauconitic minerals ; growth of ferromanganese nodules
(modified from Odin, 1975).

Obviously, iron is present in different forms in each of the
five zones. When dealing with ancient formations this information provides a rough indication as to the palaeogeographic
framework within which an iron-bearing marine sediment was
deposited.

This arrangement of zones is usually disturbed by eustatic
sea level changes. Longshore currents inhibit glauconitization
on their upstream side as illustrated off the Congo and Amazon
Rivers. Another modification occurs when the depositional
cone (zone 1) is only found at the base of the slope, through
the existence of an active canyon facing the river mouth ;
in this case, supply of iron for authigenesis is only in the
form of ions or as long-terms suspended matter. This appears
to be the case off Monterey Bay (Normark, this volume), where
glaucony is forming on the shelf.

DISCUSSION AND CONCLUSIONS

This paper intended to show 1 - the present state of the art
concerning the green grains frequently observed in arenites ;
2 - the methods involved in their study and 3 - what kind
of information may be obtained from them. Apart from a few
samples of exceptional characteristics, meaningful conclusions may only be drawn from data obtained by sampling a vertical sequence over a wide area, thus giving the information
needed for the reconstruction of the basin at a given time
and its evolution during a given period. Green grains are
very powerful sources of information if one "reads" them carefully, as proposed in figure 7. They may also be considered
as an integrant part of the characteristics of the evolutionary sequence in the sense of Goldschmidt (see figure 9).

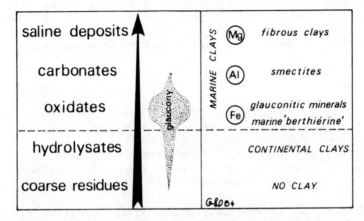

Figure 9 : Location of glaucony (dotted area) and
marine berthierine in the Goldschmidt's sedimentary sequence. The location of the diverse clay minerals is emphasized in the right hand column (modi-
ied from Odin and Létolle, 1978).

Although standard analytical procedure cannot be followed
in all cases, because each horizon is particular, it is always necessary to carry out analysis by magnetic separation,
fractionation and X-ray diffractometry to determine the nature of the green grains.

From the data obtained with these techniques and other data
from the other components of the sediment, hypotheses on their
genetic environment may be drawn. Radiometric measurements
help to choose between several possible hypotheses especially
in the case of true glaucony. The term glaucony (or its
equivalent) must only be used after determination of t h e
characteristic diffractometric properties of the glauconitic mi-

nerals. In the majority of cases, the term will be merited,
but if not, the information is equally valuable. When the
facies has been recognized, it is of great interest to determine the stage of evolution of the grains because the interpretation of a glauconitic horizon may depend on this criterion. In summary not all green grains are necessarily glauconies. If glaucony is formally identified, its significance
may be multiple. The more precise the investigation, the more
detailed will be the conclusions obtained from this facies.

ACKNOWLEDGEMENTS

I am grateful to Professor E. Jäger (director), J.C. Hunziker
and the entire team of the Geochronology Unit of the Institute
of Geology at Berne, for making available their K-Ar dating
facilities. The collaboration of the Department of Oceanography
at the University of Bordeaux, especially the numerous samples
provided by Dr.M. Pujos, is also gratefully acknowledged.

REFERENCES

BELL, D.L. & GOODELL, H.G., 1967, A comparative study of glauconite and the associated clay fraction in modern marine
sediments : Sedimentology, 9, p. 169-202.
BIRCH, G.F., WILLIS, J.P. & RICKARD, R.S., 1976, An electron
microprobe study of glauconites from the continental margin off the west coast of S. Africa : Mar. Geol., 22, p.
271-384.
BJERKLI, K. & ÖSTMO-SAETER, J.S., 1973, Formation of glauconie in foraminiferal shells on the continental shelf off
Norway : Mar. Geol., 14, p. 169-178.
BORCHERT, H., 1965, Formation of marine sedimentary iron ores:
in Chemical Oceanography, 2 (Ed. by J.P. Riley and G. Skir-
row), p. 159-201, Academic Press, London.
BOUYSSE, P., KUDRASS, H.B. & LE LANN, F., 1977, Reconnaissance sédimentologique du plateau continental de la Guya-
ne française : Bull. B.R.G.M., 2, p. 141-179.
BRINDLEY, G.W., BAILEY, S.W., FAUST, G.T., FORMAN, S.A. &
RICH, C.I., 1968, Report of the Nomenclature Committee (66-
67) of the Clay Minerals Society : Clays Clay Miner., 16,
p. 322-324.
BUROLLET, P.F., CASSOUDEBAT, M. et DUVAL, F., 1979, La Mer
Pélagienne, Constituants lithologiques, microfaciès : Géo-
logie Méditerranéenne, 6, 1, p. 83-110.
CASPARI, W.A., 1910, Contributions to the chemistry of submarine
glauconite : Proc. Roy. Soc. Edimb., 30, p. 364-373.
CAYEUX, L., 1916, Introduction à l'étude pétrographique des
roches sédimentaires : p. 241-252. Imprimerie Nationale,
Paris, 524 p.

COLLET, L.W., 1908, Les dépôts marins : p. 132-194. Doin, Paris, 325 p.

DANGEARD, L., 1928, Observations de géologie sous-marine et d'océanographie relatives à la Manche : Ann. Inst. océanogr. VI, 1, chap. XI, p. 199-211.

EHLMANN, A.J., HULINGS, N.C. & GLOVER, E.D., 1963, Stages of glauconite formation in modern foraminiferal sediments : J. sedim. Petrol., 33, p. 87-96.

GIRESSE, P., 1969, Etude des différents grains ferrugineux authigènes des sédiments sous-marins au large du delta de l'Ogooué (Gabon) : Sciences Terre, XIV, p. 27-62.

GIRESSE, P., LAMBOY, M. & ODIN G.S., 1980, Evolution géométrique des supports de glauconitisation, reconstitution de leur paléo-environnements : Ocean. Acta, 3, p. 251-260.

GIRESSE, P. & ODIN G.S., 1973, Nature minéralogique et origine des glauconies du plateau continental du Gabon et du Congo : Sedimentology, 20, p. 457-488.

GLANGEAUD, L., 1941, Sur la formation et la répartition des faciès vaseux dans les estuaires : C. r. hebd. Séanc. Acad. Sci. Paris, p. 1022-1024.

GOLDHABER, M.B. & KAPLAN, I.R., 1974, The sulfur cycle : in The Sea (Ed. by E.G. Goldberg), 5, p. 569-655. Wiley, New York.

HARDJOSOESASTRO, R., 1971, Note on chamosite in sediments of the Surinam shelf : Geologie Mijnb., 50, p. 29-33.

HEIN, J.R., ALLWARDT, A.O. & GRIGGS, G.B., 1974, The occurence of glauconite in Monterey Bay, California. Diversity origins and sedimentary environmental significance : J. sedim. Petrol., 44, p. 562-571.

HOUBOLDT, J.J.H.C., 1957, Surface sediments of the Persian Gulf near the Qatar Peninsula : Ph. D. thesis. University of Utrecht, Mouton, Den Haag, 113 p.

LAMBOY, M., 1974, La glauconie du plateau continental au Nord-Ouest de l'Espagne dérive d'anciens débris coquilliers : C. r. hebd. Séanc. Acad. Sci., Paris, 280, p. 157-160.

LAMBOY, M., 1976, Géologie marine du plateau continental au N.O. de l'Espagne : Thèse Doctorat d'Etat, University of Rouen, 283 p.

LAMBOY, M. & ODIN, G.S., 1975, Nouveaux aspects concernant les glauconies du plateau continental Nord-Ouest espagnol : Rev. Géogr. phys. Géol. dyn., XVII, p. 99-120.

LEWIS, D.W., 1964, "Perigenic" ; a new term : J. Sed. Petrol., 34, p. 875-876.

MILLOT, G., 1964, Géologie des Argiles : Masson, Paris, 499 p.

MOORE, M.B., 1939, Faecal pellets in relation to marine deposits : in Recent marine Sediments. A Symposium (Ed. by P.D. Trask), p. 516-524, Am. Ass. Petrol. Geol., Tulsa.

MORTON, A.C., MERRIMAN, R.J. and MITCHELL, J.G. in press, in Roberts D.G. et al., ed., Init. Rep. D.S.D.P. 81, Washington.

MURRAY, J. & RENARD, A.F., 1891, Report on Deep-Sea Deposits Based on the Specimens Collected During the Voyage of H.M.S. "Challenger" in the years 1872-1876 : H.M.S.O., London, 525 p.

ODIN, G.S.,1972, Observations nouvelles sur la glauconie en accordéon : processus de genèse par néoformation : Sedimentology, 19, p. 285-294.

ODIN, G.S., 1975, Migrations du fer des eaux continentales jusqu'aux eaux océaniques profondes : C. r. hebd. Séanc. Acad. Sci. Paris, 281, p. 1665-1668.

ODIN, G.S. (ed.), 1982, Numerical Dating in Stratigraphy, John Wiley Publ., Chichester, 2 vol., 1094 p.

ODIN, G.S., 1984, "Marine berthierine", further data and questions on its mineralogy and destiny : Abstracts, 21st Ann. Meet. Clay-Min. Soc., Oct. 84, Baton Rouge.

ODIN, G.S. and DODSON, M.H., 1982, Zero isotopic age of glauconies : in Odin, ed., 1982, p. 277-306.

ODIN, G.S. and LAMBOY, M., 1975, Sur la glauconitisation d'un support carbonaté d'origine organique : débris d'Echinodermes du plateau continental N.O. espagnol : Bull. Soc. géol. Fr., 17, p. 108-115.

ODIN, G.S. & LETOLLE, R., 1978, Les glauconies et aspects voisins ou confondus : signification sédimentologique : Bull. Soc. géol. Fr., XX, p. 553-558.

ODIN, G.S. & MATTER, A., 1981, De glauconiarum origine : Sedimentology, 28, p. 611-641.

ODIN, G.S. and REX, D.C., 1982, Potassium argon dating of washed, leached, weathered and reworked glauconies : in Odin ed., 1982, p. 363-386.

ODIN, G.S. & STEPHAN, J.F., 1981, The occurence of deep water glaucony from Eastern Pacific : in Watkins et al. ed., Init. Rep.D.S.D.P. 66, Washington, p. 419-428.

OJAKANGAS, R.W. & KELLER, W.D., 1964, Glauconitization of rhyolite sand grains : J. sedim. Petrol., 34, p. 84-90.

PORRENGA, D.H., 1967a, Clay mineralogy and geochemistry of recent marine sediments in tropical areas : Publ. Fysisch-Geographisch Lab. Univ. Amsterdam, 9, 145 p. Dort-Stolk, Amsterdam.

PORRENGA, D.H., 1967b, Glauconite and chamosite as depth indicators in the marine environment : Mar. Geol., 5, p. 495-501.

PRYOR, W.A., 1975, Biogenic sedimentation and alteration of argillaceous sediments in shallow marine environments : Bull. geol. Soc. Am. 86, p. 1244-1254.

ROHRLICH, V., PRICE, N.B. & CALVERT, S.E., 1969, Chamosite in the recent sediments of Loch Etive (Scotland) : J. sedim. Petrol., 39, p. 624-631.

TAKAHASHI, J.I. & YAGI, T., 1929, The peculiar mud grains in the recent littoral and estuarine deposits with special reference of the origin of glauconite : Econ. Geol., 24, p. 838-852.

TOOMS, J.S., SUMMERHAYES, C.P. & McMASTER, R.L., 1970, Marine geological studies on the north west African margin : Rabat-Dakar : in´ The Geology of the East Atlantic Continental Margin (Ed. by F.M. Delany). Inst. Geol. Sci. Rep.n° 70/16, p. 9-25.

VON GAERTNER, H.R. & SCHELLMANN, W., 1965, Rezente Sedimente im Küstenbereich der Halbinsel Kaloum, Guinea : Tschermaks miner. petrogr. Mitt., 10, p. 349-367.

WERMUND, E.G., 1961, Glauconite in early Tertiary : sediments of Gulf Coastal Province : Bull. Am. Ass. Petrol. Geol., 45, p. 1667-1696.

READING PROVENANCE FROM MODERN MARINE SANDS

Renzo VALLONI

Istituto di Petrografia, Università di Parma,
Via Gramsci 9, 43100 Parma, Italy

ABSTRACT

Data concerning the composition of sands from modern
continental margins has enabled an actualistic scheme of
provenances to be formulated. In order to avoid secondary
controls of detrital modes some of the variables have been
standardized. This means that optical-microscope analyses can
be interpreted in terms of provenance only when (1) they are
conducted according to the Gazzi-Dickinson method and in such
detail as to include secondary parameters, (2) they deal with
terrigenous, texturally immature and medium- to fine-grained
sands, and (3) they are obtained from deep-sea deposits which
have not been subjected to diagenetic modifications. Source-area
types have been identified on the scale of large plate features
and named First Order Provenances (FOP); the same size-scale
should be considered when using such provenances for paleogeographic reconstructions.

1. INTRODUCTION

The formulation of the plate-tectonic theory has given a
new impulse to investigations in the field of tectonics and
sedimentation (Dott, 1978). The compositional variations of
flysch successions have found a new interpretation in the works
of Mitchell and Reading (1969), Dewey and Bird (1970), Dickinson
(1974), Crook (1974) and Bhatia (1983); some broad - scale
relationships have in fact been identified between source and
depositional areas.

The compositional study of modern sediments is very much on

309

G. G. Zuffa (ed.), Provenance of Arenites, 309–332.
© *1985 by D. Reidel Publishing Company.*

the increase and a large number of analyses which are both
homogeneous and detailed are already to be found in the
literature. Moreover, modern settings are especially suited
for conducting provenance-oriented studies because there is no
major diagenetic interference and because the relationships
between sources and deposits can be physically verified.

The grouping of sediment-producing areas according to
plate-interaction conditions into a limited number of categories
characterized by distinct lithologic and structural features
provides an actualistic model of provenances. A similar approach
has been used by Inman and Nordstrom (1971) and Reading (1972)
to investigate world coasts and continental margins,respectively.

Several factors capable of modifying detrital modes may
intervene during sediment dispersal. Particularly significant
problems (Curray, 1975) are sea-level fluctuations, due to which
the present shelf areas are mostly covered with relict sands and
the deep-sea turbidite systems become especially active during
the low-stand phases, and cascade deposition, which may cause
mixing of extrabasinal and intrabasinal materials in the absence
of bypassing via submarine canyons (McMillen et al., 1982).

When comparing data from different compositional studies,
it is advisable to know whether or not the main variables (e.g.
depositional environment) are constant, even if the influence
of some processes is not known exactly. This is the case, for
instance, with basins where temporary storage sites are to be
observed (Ricci Lucchi, this volume; Cleary et al., 1984); in
the present study it was conventionally decided to disregard
detrital modes from impure allochemical rocks (Folk, 1968).

2. PREVIOUS WORK ON SAND COMPOSITION

The inclusion of some data from modern basins in the study
of arenites on a large scale dates back to the works of Crook
(1974) and Schwab (1975), followed by those of Potter (1978)
and Dickinson and Suczek (1979). Studies exclusively based on
the survey of modern marine sands have been conducted by
Ingersoll and Suczek (1979), Dickinson and Valloni (1980),
Valloni and Maynard (1981), Maynard et al. (1982), and Valloni
and Mezzadri (1984).

Although all investigations are based on the assumption
that some typical geotectonic settings reflect distinct
compositional suites, they may differ widely in purpose. Detrital
modes have been correlated to variables such as tectonics and
climate (e.g. Potter, 1978), basin type (e.g. Maynard et al.,
1982), and provenance (e.g. Dickinson and Suczek, 1979). The
provenance-oriented syntheses based primarily on actualistic
data are listed in Table 1 .

Table 1. Modern sands and provenance: A summary of published schemes

PROVENANCE TYPE	DIAGNOSTIC GRAIN CONTENT		AUTHOR and compositional parameters
	High	Characteristic	
Passive (Atlantic) C.M. Interior	Q	—	K.A.W. Crook
	Ls	Lm-rich	1974
C.M. (Andean) Magmatic Arc	—	Q-intermediate	Quartzose grains
	Lm or LV	—	content
Island Arc and Obducted Lithosph.	—	Q-poor	and other
	Lv or P/F	—	Key-grain types
Suture Belt	Lsm	—	
	Ls + Lm	—	
Rifted Continental Margin	Qp	—	Ingersoll & Suczek
	Ls + Lm	—	1979
Mixed Rifted C.M.and Magmatic Arc	Lvm + Lsm	—	
	Lv + Ls	—	Qp Lvm Lsm
Magmatic Arc	Lvm	Lsm-rich	and
	Lv	Lm-rich	Lm Lv Ls
			proportions
Mixed Magm.Arc and Subduct.Compl.	Lsm	Qp-rich	
	Ls + Lm	—	
Cratonic Block	Q	—	
	—	Low C/Q and P/F	
Craton plus Rift Belt	Q	—	
	—	Qp-poor	
	—	Intermediate P/F	Dickinson & Valloni
Craton plus Orogenic Belt	—	F-intermediate	1980
	Qp + Ls	—	
	—	Intermediate P/F	Q F L and
			Qp Lv Ls
Transform Arc Orogen	F	—	proportions
	Ls	—	
	C/Q	—	C/Q, P/F and V/L
			ratios
Continental Margin Arc	F + L	—	
	C/Q and P/F	—	
Oceanic Island Arc	L	—	
	P/F and V/L	—	
Intraplate Archipelago	L	—	
	Lv	—	
	P/F	—	
Cratonic Block	Q + F	—	
	Lm + Ls	—	
	—	Low P/F and V/L	Valloni & Mezzadri
Sedimentary Orogenic Belt	Q + L	—	1984
	Lm + Ls	—	
	—	Intermediate P/F	Q F L and
Volcanic Orogenic Belt	—	Q-intermediate	Lm Lv Ls
	Lv	—	proportions
	—	Intermediate P/F	C/Q, P/F and V/L
			ratios
Volcanic Arc	F + L	—	
	Lv	—	

2. 1.   Marine   Sands   and   Provenance

     Table 1 provides an overall view of the established
provenance types as well as the type and amount of their
diagnostic grains. Every single provenance has been described
as major, or fundamental, or essential in order to emphasize
its relationship to features generated by major-plate interactions. These features are both morphologic (e.g. topographic
maturity; high vs. low elevation), lithologic (e.g. supracrustal
vs. basement-rock exposure), and tectonic (e.g. transform vs.
consumption plate-boundary).
     The number of provenances defined in each scheme mainly
depends upon how extensive a sampling is and how detailed the
analyses are, and, obviously, on the size-scale used. The
provenance schemes of Table 1 provide either distinctive
compositional fields (the two first) or average values (the
third) for each provenance. In the last study, four provenances
are determined which, however, are merely the result of the
samples' natural tendency to separate into four clusters (suites)
on the diagrams. In general, researchers use a limited number
of primary (e.g. Q F L) and secondary (e.g. P/F, V/L; Lm Lv Ls)
petrologic parameters. The work of Zuffa (this volume) discusses
grain varieties and parameters in detail.

3.   GRAIN   TYPES   AND   PETROLOGIC   PARAMETERS

     Since the operational grain definitions and the computational
parameters are crucial both for comparing data and for determining
provenances, the procedure followed in this work needs to be
summarized.
     Modal analyses of arenites can be carried out by means of
two different methodologies according to the way in which the
rock fragments are considered. Zuffa (this volume) explains the
characteristics and advantages of each of the two procedures.
     Most analyses conducted on modern deep-sea sands make use
of the Gazzi-Dickinson methodology (1966, 1970), whereby the
large crystals within polymineralic particles are counted as
single minerals. The major grain categories, together with the
initially-proposed parameters and some minor integrations, are
illustrated in Figure 1. The peculiarities of the metodology
followed here are: 1. extrabasinal (terrigenous) carbonate grains
are considered to be an integral part of the lithic family; 2.
the types of grain to be included in the ambiguous category of
polycrystalline quartz are further clarified; 3. the total lithic-
grain proportions are expressed by Lq Lv Ls.
     In Figure 1 only essential grains are considered.
Conventionally various types of particles are disregarded in
computing detrital modes; micas, heavy minerals and all types of
intrabasinal grains are considered to be accessory for

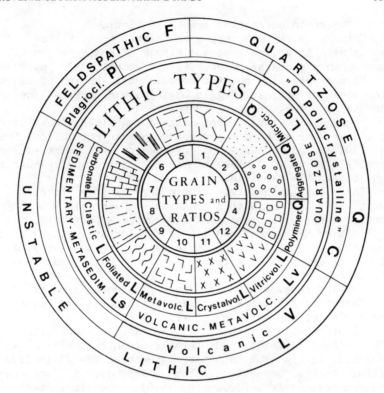

Figure 1. Essential framework constituents with their grain
parameters. See text for the description of grain
categories (1-12). In the outermost circle are the
primary parameters (Q, F, L). Parameters C and V
(see Dickinson, 1970, for original definition), which
represent the same grain types as Lq and Lv, respectively, are considered obsolete by some researchers;
however, C and V (as well as P) have traditionally
been utilized to compare the basic grain families
within a primary parameter simply using a ratio (e.g.
polycrystalline quartzose on total quartzose grains =
C/Q). The lithic types of the next circle can be
described by using grain proportions (e.g. Lq,Lv,Ls)
and various grain ratios (e.g. clastic L on total
sedimentary-metasedimentary L = ClL / Ls).

provenance-oriented parameters. The criteria used in listing the
grain categories also aim (1) to stress the opportunity to calculate the incidence of every single lithic type present (e.g.
CaL / Ls) and (2) to allow the computation of primary parameters
in more than one way (e.g. Q F L and Q F R).

        The grain categories are the following.
1. Unitary quartz and composite quartz (cf. Potter, 1978):
monomineralic grains composed either of a single crystal or of
more than one coarse crystal (Blatt et al., 1972, p. 271);
2. Microcrystalline quartz: monomineralic grains of sedimentary
(chemical or organic) origin having a microcrystalline or cryptocrystalline texture, i.e. chert;
3. Aggregate quartz: coarse-grained monomineralic multicrystal
units with or without a pronounced planar fabric (cf. Graham et
al.,1976), and fine-grained pure-quartz or quartz-dominated
grains with various fabrics (e.g. microgranular-quartzose
unintelligible particles);
4. Polymineralic quartz: single crystal or multicrystal quartz
units (> 0.062 mm) within granular - phaneritic - porphyritic
sedimentary, metamorphic and igneous rock fragments;
5. Unitary K-feldspar and lithic K-feldspar: single-crystal
monomineralic grains and single-crystal or multi-crystal units
(> 0.062 mm) within igneous , metamorphic and sedimentary rock
fragments;
6. Unitary plagioclase and lithic plagioclase: as above;
7. Carbonate lithic: extrabasinal carbonate and metacarbonate
particles of any texture (Zuffa, 1980);
8. Clastic-texture lithic;   9. Foliated-texture lithic;   10.
Metavolcanic lithic;   11. Crystalvolcanic lithic;   12. Vitricvolcanic lithic: The last five categories (8-12) include those
lithic grains, or parts of them, which have a fine-grained or
vitric texture (cf. Dickinson, 1970). The last two separate
microcrystalline from holohyaline textures. For practical reasons,
serpentinite rock fragments (Zimmerle, 1968) are assigned to the
metavolcanics and glass shards are included in the vitricvolcanic
lithic grains.

4. SAMPLES

        All analytical data utilized here have already been
published. A total of 694 analyses were considered: 117 of
these were taken from Valloni and Mezzadri (1984), the other 577
were selected from 28 different studies in the literature.
Figure 2 shows the geographic distribution of the samples; their
location is indicated with different symbols depending on whether
they correspond to single-sample sites or multi-sample sites or
to several samples taken in different areas of the same basin.
Cores are mostly from DSDP drills and Lamont-Doherty pistons and
the great majority of them are Quaternary in age with some
extending back to the Pliocene and Miocene.
        In order to standardize the samples, only materials taken
in marine environments at a depth of at least some hundreds of
meters were used. For each single case the authors have studied
sands of medium, medium-fine and fine grain-size. Within this

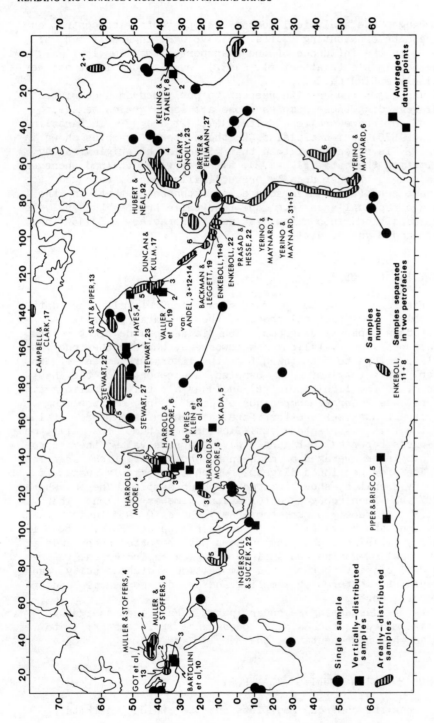

Figure 2. Location and number of samples; cf. data plots of figure 3.

size range, the analytical method used prevents size-control on
the results of counting (Zuffa, this volume; Ingersoll et al.,
1984), and the influence of the transport mechanisms is likely
to be negligible (Blatt et al., 1972), even though it cannot be
precisely estimated.

The major change with respect to published data consisted
in disregarding those samples whose accessory grains exceeded
50% of the framework (e.g. sites 127, 128 and 130, in Bartolini
et al., 1975), which explains why the number of samples shown in
Figure 2 is often lower than that given in the original study.
In plotting the data (Figure 3), averages were obtained whenever
the samples, whether areally or vertically spaced, belonged to
the same petrofacies. In some cases it happened that in the
same geographical area (e.g. the Peru-Chile trench),the analyses
could be divided into a few distinct petrofacies (31+15 in
Yerino and Maynard, 1984).

5. DETRITAL MODES

Figure 3 shows the primary parameters of all the data
available. For practical reasons, the 82 data plots were
reported on separate diagrams. The diagram on the right refers
to data obtained by Valloni and coworkers, whereas the one on
the left refers to data taken from the literature. Ten of the
latter, for a total of 162 samples, are reported on the inset
diagram, as the lithic pole (R) includes all polymineralic grains
and prevents accurate comparison of data. The overwhelming
majority of analyses are therefore homogeneous and where obtained
with the texture-oriented Gazzi-Dickinson system. Not all the
data-plots reported in Figure 3 have the same value; they represent single or averaged or mean-and-standard deviation analyses.

The aforementioned diagrams, which are commonly used to
classify arenites, show two zones with no samples. The one where
$F \gtrsim 80\%$, has been recognized for a long time as a zone where
samples very rarely occur; the other, where $Q > 95\%$ (quartz sands
of Valloni and Mezzadri, 1984; quartzarenite of Folk, 1968), is
rather surprising since many authors have reported such sands in
modern shallow-marine and ancient deep-sea environments. Future
work will clarify whether this compositional discontinuity can
be explained in terms of sampling and/or environmental and/or
diagenetic control.

Before discussing secondary parameters and assigning
detrital modes to each single provenance it is essential to
consider how representative the samples are.

6. MODIFICATION DURING TRANSPORT

Among the numerous modifications to which detrital

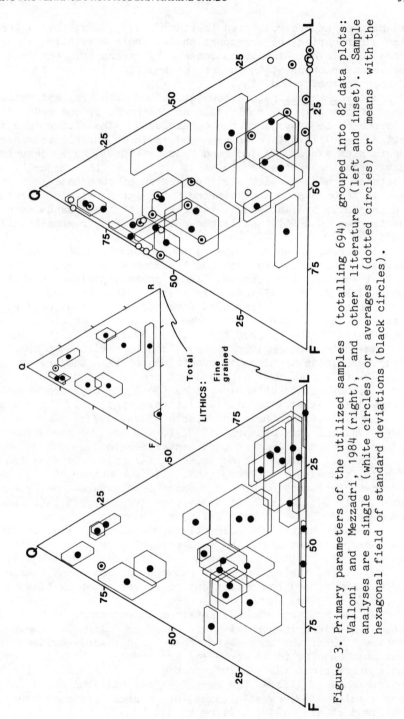

Figure 3. Primary parameters of the utilized samples (totalling 694) grouped into 82 data plots: Valloni and Mezzadri, 1984 (right), and other literature (left and inset). Sample analyses are single (white circles) or averages (dotted circles) or means with the hexagonal field of standard deviations (black circles).

assemblages may be subjected (Suttner, 1974; Basu, this volume),
some occur during the transport phase. Their cumulative effect
tends to eliminate labile components by means of chemical or
mechanical processes; the degree of modification is a function
of both intensity and time.

The mechanical maturation (abrasion) can be evaluated in
terms of the textural modifications undergone by the sediment if
a second-cycle provenance can be excluded. The latter (see also
Potter, 1978) is not to be regarded as indispensable to generate
quartz-rich sands. Valloni and Mezzadri (1984) have described
as matured these first-cycle sands in which the majority of
quartzose grains display a marked degree of roundness; in this
case it becomes much more difficult to interpret provenance.
The presently available data indicate that these sands occupy a
quartz-rich (Q > 65%) and lithic-poor (F/L > 1) compositional
field.

The areas in which deep-sea sands subjected to mechanical
maturation are reported, include the Amazon Cone and Atlantic
Central America, the Sohm and Biscay abyssal plains, the Nile
Cone and North West Africa; this distribution of the samples
lends itself to a few comments. Advanced maturation primarily
occurs in passive margins of large continents but the samples
from Central America prove that it could merely be the result
of long-protracted residence in a beach environment before
final deposition. Climatic conditions of high temperature and
rainfall (Potter, 1978) also suggest some relationship to these
samples; however notable exceptions (Sohm, Biscay) are to be
observed in basins with similar settings (passive). The eolian-
sand turbidite (Sarnthein and Diester-Haas, 1977; Sarnthein,
1978; Valloni and Mezzadri, 1984, DSDP site 139) are widespread
in some portions of modern continental margins.

Chemical processes also cause differential elimination
effects in grain types (Basu, this volume; Suttner et al., 1981),
and their influence on detrital modes (Cleary and Conolly, 1971;
Franzinelli and Potter, 1983) may prove very significant (Harrel
and Blatt, 1978). The detrital modes obtained from deep-sea
sands cannot, however, be directly compared with climatic zones,
since deep-sea sedimentation is a pulsating process related to
sea-level stand (Damuth and Kumar, 1975; Boggs, 1984; Mutti, this
volume); thus, the relationship between climate and basin
sedimentary record could turn out to be very complex (Damuth and
Fairbridge, 1970).

As far as deep-sea sedimentation is concerned, however, sand
composition is only partially affected by climatic control.
Irrespective of the basin type, there is now ample evidence that
sediments heading to the deep sea are usually subjected to a
sedimentary cycle that includes rapid funnelling across littoral
environments (Hollister and Heezen, 1964; Milliman et al., 1975;
Wang et al., 1982) even in cases of scarce sediment supply
(Milliman et al., 1972; Kelling et al., 1975).

7. INFLUENCE OF BASIN SETTING

Although standardized in terms of water depth the sands
discussed in this paper were deposited in different basins which
have been classified according to the type of plate or crustal
juncture present. These basins can therefore be considered as
vast depositional compartments which could be further subdivided;
e.g., an arc-trench system into forearc, slope and trench basins.
The distribution of samples in the various depositional compartments needs some clarification.

One of the limits imposed by the sampling is that only
basins with oceanic affinity are considered. Figure 4 which
reports the seven key-types underlines the fact that the first
six are distally joined to an oceanic-intraplate basin which
represents an end-member. The physiographic features allowing
the materials shed from highlands to reach an oceanic-intraplate
basin may be represented by a large,deep cone or by the development of a spreading center in a marginal sea; in active (strike-
slip and consumption) continental margins, morphotectonic
barriers may drive detritus onto the oceanic plate along a
fracture zone or allow it to cross a filled trench and invade
the outer swell.

As this work focuses on the sedimentary fill of basins, it

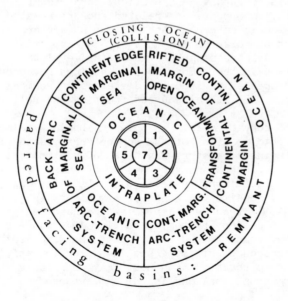

Figure 4. Key depositional settings with oceanic affinity;
          types 1-7 account for 93% of the data plots. The
          remnant-ocean settings (outermost circle) are a
          composite category.

has been decided to subdivide the depositional settings   as   in
Figure 4; the distribution pattern of the available data   is   as
follows: 1. Rifted, 33 % ; 2. Transform, 11 % ;   3. Continental
arc-trench, 13 % ;   4. Oceanic arc-trench, 12 % ;   5. Back-arc
of marginal sea, 10 % ; 6. Continent edge of marginal sea, 7 % ;
7. Oceanic intraplate, 7 % ; other ensialic deep-water settings
account for the remaining 7 % . Many   important   depositional
compartments do not appear in the above list, namely those filled
with continental and shallow-water deposits or having the shape
of narrow troughs (ensialic). The disregarded basins frequently
correspond to an early or late phase in the evolution of a
continent edge and can be related to continental rifting
(Protoceanic, Aulacogen), active subduction (Retroarc Foreland,
Intra-arc) or collision orogens (Foredeep, Peripheral Foreland).
The fact that some basin types are not represented in the
samples poses no problems to the outcome of the present research
which is primarily directed towards characterizing source-areas
(provenance).

The criterion adopted in distinguishing the categories of
basins shown in Figure 4 is not only provenance-oriented but is
also based upon the assumption that basins have a transversal
feed, so that the source of detritus may be identified with the
landmass paralleling the basin (Sibley and Pentony, 1978). This
makes it possible to subdivide a critical compartment, such as
the marginal sea, into continent-related and arc-related basins.
The frequently-mentioned remnant-ocean basin appears in a
separate area of the diagram since it is a composite setting
occurring at an advanced stage of the basins' evolution. Deciphering provenance in the remnant-ocean settings may prove to be
highly complex since a longitudinal dispersal is commonly added
to dispersal from the lateral sources fronting the relict ocean
(Velbel, 1980; Moore et al., 1982).

Examples of complex relations between basin sedimentary
fill and geographic position of the source area created by long-
distance longitudinal transport can be seen in almost all basin
types. Rifted continental margins may be affected by relatively
frequent megaevents (Elmore et al., 1979; Cita et al., 1984);
marginal seas may receive the sands of big rivers (e.g. Griffin
and Goldberg, 1969); trenches (Yerino and Maynard, 1984; Ross,
1971) and remnant oceans (Bartolini et al., 1972; Ingersoll and
Suczek, 1979) may transfer sediments along tectonic strike for
hundreds or thousands of kilometers. The creation of such
complexities due to sediment transport may seriously hinder
paleogeographic reconstructions in the ancient record.

8. PROVENANCE TYPES

Sediment producing areas are classified according to their
relative positions with respect to plate-junctures and to the

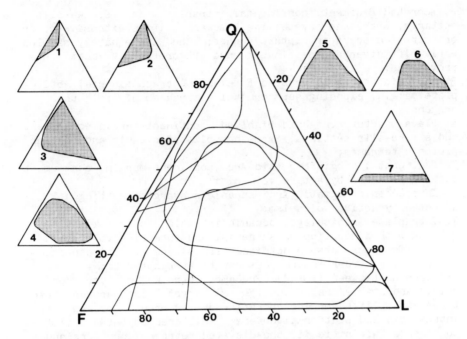

Figure 5. QFL fields of First Order Provenances. 1. Cratonic
        basement; 2. Accreted basement; 3. Foldthrusted fore-
        land; 4. Plate-juncture highland; 5. Continental arc;
        6. Oceanic arc; 7. Archipelago. Overlapping of prove-
        nances is remarkable.

type of crust involved. As a consequence, the individual source
areas refer to distinctive geotectonic regimes. Reading (1972)
and Dickinson and Valloni (1980) have proved that the source -
basin relations are positively interpreted when the distances
involved are in the order of hundreds (sometimes thousands) of
kilometers. On this scale several distinct morphotectonic units
can be identified which remain more or less unchanged. When
dealing with provenance, the time factor should also be taken
into consideration. In general, the source area types described
here can be identified when the distances involve at least a few
hundred of kilometers for a time span of at least a dozen million
years. As these source-area types are related to large plate
interactions, they are defined as First Order Provenances (FOP).

8.1. First Order Provenances

    The First Order Provenances are as follows.
1. Cratonic Basement: continental intraplate settings, where
cratonic crystalline rocks are exposed;

2. Accreted Basement: continental intraplate and strike-slip
settings, where the crystalline basement exibits extensive covers
or accretion scars or rupture wounds; the inadequately sampled
early-rift divergent settings could theoretically meet this
provenance;

3. Orogenic Belt Foldthrusted Foreland: intracontinental active
plate margin, exposing supracrustal terranes of fold-thrust
orogens;

4. Plate-Juncture Orogenic Highland: continent-margin convergent
and strike-slip settings, exposing uplifted oceanic and volcanoplutonic terranes;

5. Continental Margin Magmatic Arc: continent-margin convergent
setting affected by active volcanism;

6. Intraoceanic Magmatic Arc: oceanic convergent setting
affected by active volcanism;

7. Intraplate Archipelago: oceanic intraplate or strike-slip
setting with discontinuous subaerial and submarine volcanoes.

Differences between detrital assemblages of the aforementioned First Order Provenances can be attributed to:
differences between (FOP 1) Precambrian rocks of unbroken
shield and stable craton and (FOP 2) crystalline basement fractured and uplifted or incorporating deeply eroded orogens, arc
intrusions, and platformal covers; differences between (FOP 3)
continental affinities in the elevated retroarc or foreland
belts, with coarse siliciclastic terranes accompanied by portions
of their own carbonate-siliceous substrate or crystalline
basement slabs, and (FOP 4) oceanic affinities in suture zones
of collision orogens, uplifted subduction complexes, and trans-
pressional variants of arc orogens; differences in the
elevation and magnitude of the plutonic roots and metamorphic
envelopes of (FOP 5) continent-margin and (FOP 6) intraoceanic
arcs.

To delimit the compositional range pertaining to the various
provenances, two triangular diagrama accompanied by three grain
ratios were used (Figures 5 and 6). On the diagram showing
primary parameters (QFL) the data-plot contours for each provenance showed conspicuous overlapping. Only some small areas of
the diagram of Figure 5 can help decipher provenance; these
areas are located near the QL border and close to the F pole
(> 60%). The detrital modes expressed by means of the primary
parameters are not sufficiently detailed to enable the compositional peculiarities of First Order Provenances to be determined.

The overlapping of compositional fields on the QFL diagram
cannot be attributed to a scattering of data due to unusual
composition. The QFL diagram was cleared of a fair number of
samples whose secondary parameters did not pass a uniformity
test with respect to the most frequent values for the specific
provenance involved. Such atypical compositions may be related
to smaller-sized and shorter-lived sources, namely, Second Order
Provenances (SOP) which, however, would appear negligible on a
regional scale.

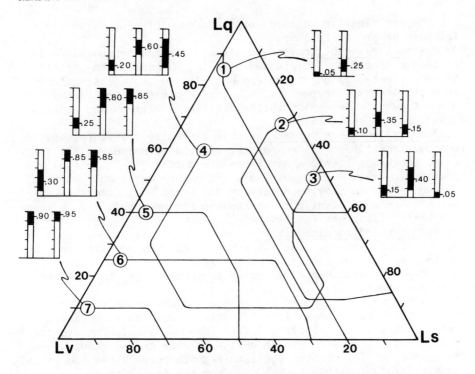

Figure 6. Diagnostic grain proportions and grain ratios (see
fig. 1) of First Order Provenances (1-7, see fig. 5).
C/Q = left, P/F = middle, and V/L = right bars; bar
fill represents two standard deviations of the indicated mean value. Lq = quartzose lithics, Lv =
volcanic-metavolcanic lithics, Ls = sedimentary-
metasedimenatry lithics.

8.2. Deciphering Provenances

Figure 6 shows the field of each First Order Provenance
with respect to the main lithic categories. In this case several
areas are identified which enable provenances to be attributed
to samples. Provenances then become virtually unequivocal when
use is also made of the grain ratios (C/Q, P/F, and V/L) reported outside the diagram. This suggests that the first-level
subdivisions of primary parameters (Lq Lv Ls ; C/Q, P/F, V/L)
are barely sufficient to decipher an FOP . When a large set
of samples is available, some uncertainties can be removed by
taking into account the trend of lithic-grain proportions
(Galloway, 1974). Until further data are obtained, the remaining overlap zones may reasonably be considered as fields of
composite provenance (Mack, 1984).

        The distinctive features of the composition of First Order
Provenances are as follows.
1. <u>Cratonic basement provenance</u> is represented by quartzofeldspathic compositions (QF suite) with low C/Q and with lithic
types dominated by quartzose grains. It can be distinguished
from provenance of type 2 also by using the C/Q and P/F ratios
where C and P represent polycrystalline quartzose and plagioclase grains, respectively.
2. <u>Accreted basement provenance</u> has generally quartzofeldspathic
compositions (QF suite) characterized by a relatively high V/L
even though the lithic-grain types are mostly quartzose (Lq) and
sedimentary-metasedimentary (Ls). It can be distinguished from
type-3 provenance by means of the V/L ratio and, sometimes, by
the abundance of clastic and carbonate grains within the sedimentary-metasedimentary family.
3. <u>Foldthrusted foreland provenance</u> is basically constituted by
quartzolithic compositions (QL suite) with very low V/L and a
sedimentary (Ls)-dominated lithic association.   It can be
directly distinguished from type-4 provenance in the lithic
diagram (Lq Lv Ls).
4. <u>Plate-juncture highland provenance</u> is characterized by
feldspatholithic compositions (FL suite) with characteristic
intermediate values of P/F and V/L. The C/Q ratio,which averages
0.2 , tends to be either high or low. These grain ratios seem
to be more efficient than the lithic-grain proportions (Lq Lv Ls)
in distinguishing this provenance from the following.
5. <u>Continental-arc provenance</u> has primarily a lithovolcanic
composition (LV suite) in which the three . grain ratios tend
toward very high values. Within the lithic family, the prevail-
ing volcanic-metavolcanic (Lv) grains may be indifferently
accompanied by a quartzose (Lq) or a sedimentary-metasedimentary
(Ls) lithic assemblage. A gradual transition into type-6
provenance is apparent. For the purpose of discrimination, it
is therefore necessary to consider the trend of the P/F values
and to ascertain whether the volcanic (Lv) grains tend to be
accompanied by both quartzose (Lq) and sedimentary (Ls) grains
or whether they are accompanied by sedimentary grains only.
6. <u>Oceanic-arc provenance</u> is essentially constituted by lithovolcanic (LV) compositions in which the P/F ratio can reach
very high values. This is also true for the C/Q ratio, which,
though averaging 0.3 , tends either to very high or very low
values. A clear-cut tendency to form Lv-Ls assemblages often
develops in the lithic family.
7. <u>Archipelago provenance</u> is characterized by peculiar Q-poor
and F-rich compositions (Okada, 1973), and very common vitricvolcanic grains. For the remaining the lithics are tholeiite and
alkaline basalts.

ACKNOWLEDGMENTS

The author is grateful to Maria A. Calzolari, Paola Conti
and Luigi Pingani for their precious help and to Denise Scott,
Winfried Winkler and Gian G. Zuffa for reviewing the manu-
script.
This study was supported by grants from the Italian National
Research Council (CNR) and the Italian Ministry of Education.

REFERENCES

Bachman, S. B., and Leggett, J. k., 1982, Petrology of Middle
    America Trench and slope sands, Guerrero Margin, Mexico:
    Init. Rep. Deep Sea Drill. Proj., Washington, D.C., U.S.
    Government Printing Office, v. 66, p. 429-436.

Bartolini, C., Gehin, C., and Stanley, D. J., 1972, Morphology
    and Recent Sediments of the Western Alboran Basin in the
    Mediterranean Sea: Mar. Geol., v. 13, p. 159-224.

Bartolini, C., Malesani, P. G., Manetti, P., and Wezel, F. C.,
    1975, Sedimentology and Petrology of Quaternary Sediments
    from the Hellenic Trench, Mediterranean Ridge and the Nile
    Cone from D.S.D.P., Leg 13, cores: Sedimentology, v. 22,
    p. 205-236.

Bhatia, M. R., 1983, Plate tectonics and geochemical composition
    of sandstone: Jour. Geology, v. 91, p. 611-627.

Blatt, H., Middleton, G., and Murray, R., 1972, Origin of
    Sedimentary Rocks: Prentice-Hall, Englewood, New York, 634 p.

Boggs, S., Jr., 1984, Quaternary sedimentation in the Japan arc-
    trench system: Geol. Soc. America Bull., v. 95, p. 669-685.

Breyer, J. A., and Ehlmann, A. J., 1981, Mineralogy of arc-
    derived sediment: siliciclastic sediment on the insular
    shelf of Puerto Rico: Sedimentology, v. 28, p. 61-74.

Campbell, J. S., and Clark, D. L., 1977, Pleistocene Turbidites
    of the Canada Abyssal Plain of the Arctic Ocean: Jour. Sed.
    Petrology, v. 47, p. 657-670.

Cita, M. B., Beghi, C., Camerlenghi, A., Kastens, K. A., McCoy,
    F. W., Nosetto, A., Parisi, E., Scolari, F., and Tomadin,
    L., 1984, Turbidites and megaturbidites from the Herodotus
    abyssal plain (Eastern Mediterranean) unrelated to seismic
    events: Mar. Geol., v. 55, p. 79-101.

Cleary, W. J., and Conolly, J. R., 1971, Distribution and
   genesis of quartz in a piedmont-coastal plain environment:
   Geol. Soc. America Bull., v. 82, p. 2755-2766.

Cleary, W. J., and Conolly, J. R., 1974, Petrology and origin
   of deep-sea sands: Hatteras Abyssal Plain: Mar. Geol.,
   v. 17, p. 163-279.

Cleary, W. J., Curran, H. A., and Thayer, P. A., 1984, Barbados
   Ridge: Inner trench slope sedimentation: Jour. Sed.
   Petrology, v. 54, p. 527-540.

Crook, K. A. W., 1974, Lithogenesis and Geotectonics: The
   Significance of Compositional Variations in Flysch Arenites
   (Graywackes), in Dott, R. H., and Shaver, R. H., eds.,
   Modern and Ancient Geosynclinal Sedimentation: Soc. Econ.
   Paleontologists Mineralogists Spec. Publ. 19, p. 304-310.

Curray, J. R., 1975, Marine Sediments, Geosynclines and Orogeny,
   in, Fischer, A. G., and Judson, S., eds., Petroleum and
   Global Tectonics: Princeton University Press, Princeton,
   p. 157-222.

Damuth, J. E., and Fairbridge, R. W., 1970, Equatorial Atlantic
   Deep-Sea Arkosic Sands and Ice-Age Aridity in Tropical South
   America: Geol. Soc. America Bull., v. 81, p. 189-206.

Damuth, J. E., and Kumar, N., 1975, Amazon Cone: Morphology,
   Sediments, Age, and Growth Pattern: Geol. Soc. America
   Bull., v. 86, p. 863-878.

Dewey, J. F., and Bird, J. M., 1970, Mountain Belts and the
   New Global Tectonics: Jour. Geophys. Research, v. 75,
   p. 2625-2647.

Dickinson, W. R., 1970, Interpreting detrital modes of graywacke
   and arkose: Jour. Sed. Petrology, v. 40, p. 695-707.

Dickinson, W. R., 1974, Plate tectonics and sedimentation, in
   Dickinson, W. R., ed., Tectonics and Sedimentation: Soc.
   Econ. Paleontologists Mineralogists Spec. Publ. 22, p. 1-27.

Dickinson, W. R., and Suczek, C. A., 1979, Plate tectonics and
   sandstone compositions: Am. Assoc. Petroleum Geologists
   Bull., v. 63, p. 2164-2182.

Dickinson, W. R., and Valloni, R., 1980, Plate settings and
   provenance of sands in modern ocean basins: Geology, v. 8,
   p. 82-86.

Dott, R.H., Jr., 1978, Tectonics and Sedimentation a Century
    Later: Earth Sci. Rev., v. 14, p. 1-34.

Duncan, J. R., and Kulm, L. D., 1970, Mineralogy, provenance,
    and dispersal history of Late Quaternary deep-sea   sands
    in Cascadia basin and Blanco fracture zone off Oregon:
    Jour. Sed. Petrology, v. 40, p. 874-887.

Elmore, R. D., Pilkey, O. H., Cleary, W. J., and Curran, H. A.,
    1979, Black Shell turbidite, Hatteras Abyssal Plain,
    western Atlantic Ocean: Geol. Soc. America Bull., v. 90,
    p. 1165-1176.

Enkeboll, R. H., 1982, Petrology and provenance of sands and
    gravels from the Middle America Trench and trench slope,
    southwestern Mexico and Guatemala: Init. Rep. Deep Sea
    Drill. Proj., Washington, D. C., U. S. Government Printing
    Office, v. 66, p. 521-530.

Folk, R. L., 1968, Petrology of Sedimentary Rocks: Hemphill's,
    Austin, 170 p.

Franzinelli, E., and Potter, P. E., 1983, Petrology, Chemistry,
    and Texture of Modern River Sands, Amazon River System:
    Jour. Geology, v. 91, p. 23-39.

Galloway, W. E., 1974, Deposition and Diagenetic Alteration of
    Sandstone in Northeast Pacific Arc-Related Basins: Implica-
    tions for Graywacke Genesis: Geol. Soc. America Bull.,
    v. 85, p. 379-390.

Gazzi, P., 1966, Le arenarie del flysch sopracretaceo dell'Ap-
    pennino modenese; correlazioni con il flysch di Monghidoro:
    Mineralog. et Petrog. Acta, v. 12, p. 69-97.

Got, H., Monaco, A., Vittori, J., Brambati, A., Catani, G.,
    Masoli, M., Pugliese, N., Zucchi-Stolfa, M., Belfiore, A.,
    Gallo, F., Mezzadri, G., Vernia, L., Vinci, A., and Bonaduce,
    G., 1981, Sedimentation on the ionian active margin
    (Hellenic Arc) - Provenance of sediments and mechanisms of
    deposition: Sed. Geology, v. 28, p. 243-272.

Graham, S. A., Ingersoll, R. V., and Dickinson, W. R., 1976,
    Common provenance for lithic grains in Carboniferous sand-
    stones from Ouachita Mountains and Black Warrior basin:
    Jour. Sed. Petrology, v. 46, p. 620-632.

Griffin, J. J., and Goldberg, E. D., 1969, Recent Sediments of
    Caribbean   Sea,   in McBirney, A. R., ed., Tectonic Relations
    of Northern Central America and the Western Caribbean - the
    Bonacca Expedition: Am. Assoc. Petroleum Geologists Memoir
    11,   p. 258-268.

Harrell, J., and Blatt, 1978, Polycrystallinity: effect on
    the durability of detrital quartz: Jour. Sed. Petrology,
    v. 48, p. 25-30.

Harrold, P. J. and Moore, J. C., 1975, Composition of deep-sea
    sands from marginal basins of the northwestern Pacific:
    Init. Rep. Deep Sea Drill. Proj., Washington, D. C., U. S.
    Government Printing Office, v. 31, p. 507-514.

Hayes, J. B., 1973, Petrology of indurated sandstones, Leg 18,
    Deep Sea Drilling Project: Init. Rep. Deep Sea Drill.
    Proj., Washington, D. C., U. S. Government Printing Office,
    v. 18, p. 915-924.

Hollister, C. D., and Heezen, B. C., 1964, Modern Graywacke-
    Type Sands: Science, v. 146, p. 1573-1574.

Hubert, J. F., and Neal, W. J., 1967, Mineral composition and
    dispersal patterns of deep-sea sands in the western North
    Atlantic petrologic province: Geol. Soc. America Bull.,
    v. 78, p. 749-771.

Ingersoll, R. V., Bullard, T. F., Ford, R. L., Grimm, J. P.,
    Pickle, J. D., and Sares, S. W., 1984, The effect of grain
    size on detrital modes: A test of the Gazzi-Dickinson point-
    counting method: Jour. Sed. Petrology, v. 54, p. 103-116.

Ingersoll, R. V., and Suczek, C. A., 1979, Petrology and
    provenance of Neogene sands from Nicobar and Bengal fans,
    DSDP sites 211 and 218: Jour. Sed. Petrology, v. 49, p.
    1217-1228.

Inman, D. L., and Nordstrom, C. E., 1971, On the Tectonic and
    Morphologic Classification of Coasts: Jour. Geology, v. 79,
    p. 1-21.

Kelling, G., Sheng, H., and Stanley, D. J., 1975, Mineralogic
    Composition of Sand-Sized Sediment on the Outer Margin off
    the Mid-Atlantic States: Assessment of the Influence of the
    Ancestral Hudson and Other Fluvial Systems: Geol. Soc.
    America Bull., v. 86, p. 853-862.

Kelling, G., and Stanley, D. J., 1972, Sedimentation in the
    Vicinity of the Strait of Gibraltar, in Stanley, D. J., ed.,
    The Mediterranean Sea: A Natural Sedimentation Laboratory:
    Dowden, Hutchinson & Ross, Inc., Stroudsburg, p. 489-519.

Klein, G., de Vries, McConville, R. L., Harris, J. M., and
    Steffensen, C. K., 1980, Petrology and diagenesis of sand-
    stones, Deep Sea Drilling Project Site 445, Daito Ridge:
    Init. Rep. Deep Sea Drill. Proj., Washington, D. C., U. S.
    Government Printing Office, v. 58, p. 609-616.

Mack, G. H., 1984, Exceptions to the relationship between plate
    tectonics and sandstone composition: Jour. Sed. Petrology,
    v. 54, p. 212-220.

Maynard, J. B., Valloni, R., and YU, H. S., 1982, Composition
    of modern deep-sea sands from arc-related basins, in Leggett,
    J. K., ed., Trench-Forearc Geology: Sedimentation and tec-
    tonics on Modern and Ancient Active Plate Margins: Geological
    Society of London Spec. Publ. No 10, Blackwell Scientific
    Publications, Oxford, p. 551-561.

McMillen, K. J., Enkeboll, R. H., Moore, J. C., Shipley, T. H.,
    and Ladd, J. W., 1982, Sedimentation in different tectonic
    environments of the Middle America Trench, Southern Mexico
    and Guatemala, in Leggett, J. K., ed., Trench-Forearc
    Geology: Sedimentation and Tectonics on modern and ancient
    Active Plate Margins: Geological Society of London Spec.
    Publ. No 10, Blackwell Scientific Publications, Oxford,
    p. 107-119.

Milliman, J. D., Pilkey, O. H., and Ross, D. A., 1972, Sediments
    of the Continental Margin off the Eastern United States:
    Geol. Soc. America Bull., v. 83, p. 1315-1334.

Milliman, J. D., Summerhayes, C. P., and Barretto, H. T., 1975,
    Quaternary Sedimentation on the Amazon Continental Margin:
    A Model: Geol. Soc. America Bull., v. 86, p. 610-614.

Mitchell, A. H. G., and Reading, H. G., 1969, Continental
    margins, geosynclines and ocean floor spreading: Jour.
    Geology, v. 77, p. 629-646.

Moore, G. F., Curray, J. R., and Emmel, F. J., 1982, Sedimentation
    in the Sunda Trench and forearc region, in Leggett, J. K.,
    ed., Trench-Forearc Geology: Sedimentation and Tectonics on
    Modern and Ancient Active Plate Margins: Geological Society
    of London Spec. Publ. No 10, Blackwell Scientific Publica-
    tions, Oxford, p. 245-258.

Müller, G., and Stoffers, P., 1972, Deep sea sands of the Black
    Sea: Composition and origin: 24 th Int. Geol. Congr. Proc.
    Section 8, p. 90-99.

Müller, G., and Stoffers, P., 1974, Mineralogy and Petrology of
    Black Sea Basin Sediments, in Degens, E. T., and Ross, D. A.,
    eds., The Black Sea: Geology, Geochemistry and Biology:
    Am.Assoc. Petroleum Geologists Memoir 20, p. 200-248.

Okada, H., 1973, Abyssal feldspathic sediments in the north-
    western Pacific: Init. Rep. Deep Sea Drill. Proj.,
    Washington, D. C., U. S. Government Printing Office, v. 20,
    p. 359-362.

Piper, D. J. W., and Brisco, C. D., 1975, Deep-water continental
    margin sedimentation, DSDP Leg 28, Antarctica: Init. Rep.
    Deep Sea Drill. Proj., Washington, D. C., U. S. Government
    Printing Office, v. 28, p. 727-755.

Potter, P. E., 1978, Petrology and Chemistry of Modern Big
    River Sands: Jour. Geology, v. 86, p. 423-449.

Prasad, S., and Hesse, R., 1982, Provenance of detrital sedi-
    ments from the Middle America Trench transect off Guatemala,
    Deep Sea Drilling Project Leg 67: Init. Rep. Deep Sea
    Drill. Proj., Washington, D. C., U. S. Government Printing
    Office, v. 67, p. 507-514.

Reading, H. G., 1972, Global tectonics and the genesis of flysch
    successions: 24 th Int. Geol. Congr. Proc. Section 6, p.
    59-65.

Ross, D. A., 1971, Sediments of the Northern Middle America
    Trench: Geol. Soc. America Bull., v. 82, p. 303-322.

Sarnthein, M., 1978, Neogene sand layers off northwest Africa:
    Composition and source environment: Init. Rep. Deep Sea
    Drill. Proj., Washington, D. C., U. S. Government Printing
    Office, Supplement to v. 38-41, p. 939-959.

Sarnthein, M., and Diester-Haas, L., 1977, Eolian-sand
    turbidites: Jour. Sed. Petrology, v. 47, p. 868-890.

Schwab, F. L., 1975, Framework mineralogy and chemical composi-
    tion of continental margin-type sandstone: Geology, v. 3,
    p. 487-490.

Sibley, D. F., and Pentony, K. J., 1978, Provenance variation
    of turbidite sediments, Sea of Japan: Jour. Sed. Petrology,
    v. 48, p. 1241-1248.

Slatt, R. M., and Piper, D. J. W., 1974, Sand-silt petrology and sediment dispersal in the Gulf of Alaska: Jour. Sed. Petrology, v. 44, p. 1061-1071.

Stewart, R. J., 1976, Turbidites of the Aleutian abyssal plain: Mineralogy, provenance, and constraints for Cenozoic motion of the Pacific plate: Geol. Soc. America Bull., v. 87, p. 793-808.

Stewart, R. J., 1977, Neogene Turbidite Sedimentation in Komandorskiy Basin, Western Bering Sea: Am. Assoc. Petroleum Geologists Bull., v. 61, p. 192-206.

Stewart, R. J., 1978, Neogene Volcaniclastic Sediments from Atka Basin, Aleutian Ridge: Am. Assoc. Petroleum Geologists Bull., v. 62, p. 87-97.

Suttner, L. J., 1974, Sedimentary Petrographic Provinces: An Evaluation, in Ross, C. A., ed., Paleogeographic Provinces and Provinciality: Soc. Econ. Paleontologists Mineralogists Spec. Publ. 21, p. 75-84.

Suttner, L. J., Basu, A., and Mack, G. H., 1981, Climate and the origin of quartz arenites: Jour. Sed. Petrology, v. 51, p. 1235-1246.

Vallier, T. L., Harold, P. J., and Girdley, W. A., 1973, Provenances and Dispersal Patterns of Turbidite Sand in Escanaba Trough, Northeastern Pacific Ocean: Mar. Geol., v. 15, p. 67-87.

Valloni, R., and Maynard, J. B., 1981, Detrital modes of Recent deep-sea sands and their relation to tectonic setting: a first approximation: Sedimentology, v. 28, p. 75-83.

Valloni, R., and Mezzadri, G., 1984, Compositional suites of terrigenous deep-sea sands of the present continental margins: Sedimentology, v. 31, p. 353-364.

Van Andel, Th. H., 1964, Recent marine sediments of Gulf of California, in Van Andel, Th. H., and Shor, G. G., eds., Marine Geology of the Gulf of California: A symposium: Am. Assoc. Petroleum Geologists Memoir 3, p. 216-310.

Velbel, M. A., 1980, Petrography of subduction zone sandstone - A discussion: Jour. Sed. Petrology, v. 50, p. 303-304.

Wang, Y., Piper, D. J. W., and Vilks, G., 1982, Surface textures of turbidite sand grains, Laurentian Fan and Sohm Abyssal Plain: Sedimentology, v. 29, p. 727-736.

Yerino, L. N., and Maynard, J. B., 1984, Petrography of modern
    marine sands from the Peru-Chile Trench and adjacent areas:
    Sedimentology, v. 31, p. 83-89.

Zimmerle, W., 1968, Serpentine graywackes from the North Coast
    basin, Colombia, and their geotectonic significance: N. Jb.
    Miner. Abh., v. 109, p. 156-182.

Zuffa, G. G., 1980, Hybrid arenites: their composition and
    classification: Jour. Sed. Petrology, v. 50, p. 2129-2137.

INTERPRETING PROVENANCE RELATIONS FROM DETRITAL MODES OF SANDSTONES

William R. Dickinson

Laboratory of Geotectonics, Department of Geosciences,
University of Arizona, Tucson, Arizona 85721, USA

ABSTRACT

   Detrital modes of sandstone suites primarily reflect the different tectonic settings of provenance terranes, although various other sedimentological factors also influence sandstone compositions. Comparisons of sandstone compositions are aided by grouping diverse grain types into a few operational categories having broad genetic significance. Compositional fields associated with different provenances can then be displayed on standard triangular diagrams.

   The major provenance types related to continental sources are stable cratons, basement uplifts, magmatic arcs, and recycled orogens. Each provenance type contributes distinctive detritus preferentially to associated sedimentary basins that occupy a limited number of characteristic tectonic settings in each case. Sands of composite provenance can be described as mixtures of quartzose sand from stable cratons, quartzofeldspathic sand from basement uplifts or arc plutons, feldspatholithic sand from arc volcanics, and quartzolithic sands of several types from different kinds of recycled orogens that yield varying proportions of quartzose and lithic grains. Proportions of contributions from different provenance types can be estimated from mean compositions for ideal derivative sands represented by points or restricted areas on triangular plots.

   Evolutionary trends in sandstone composition within individual basins or sedimentary provinces commonly reflect changes in tectonic setting through time, or erosional modification of provenance terranes. Forearc sandstone suites typically evolve

G. G. Zuffa (ed.), Provenance of Arenites, 333–361.
© *1985 by D. Reidel Publishing Company.*

from feldspatholitic petrofacies of volcaniclastic nature, through lithofeldspathic petrofacies of volcanoplutonic origin, to quartzofeldspathic petrofacies of plutonic derivation. Foreland sandstone suites commonly evolve from rift-related quartzofeldspathic petrofacies, through quartzose petrofacies of passive continental margins, to quartzolithic petrofacies derived from recycled orogens.

INTRODUCTION

The use of quantitative detrital modes, calculated from point counts of thin sections, to infer sandstone provenance is well established (Dickinson and Suczek, 1979). The tectonic setting of the provenance apparently exerts primary control on sandstone compositions (Dickinson et al, 1983), although relief, climate, transport mechanism, depositional environment, and diagenetic change all can be important secondary factors. Studies of sand compositions on modern sea floors have yielded analogous results (Dickinson and Valloni, 1980; Valloni and Maynard, 1981). As might be expected, fields of compositional variation for oceanic sands in general reflect a greater degree of mingling of detritus from different types of provenance than is the case for continental sands.

The generation of reproducible detrital modes requires the establishment of clearcut operational definitions for different grain types (Dickinson, 1970). There is no limit to the number of categories of grains that can be counted, but effective regional or global comparisons of sandstone compositions are aided by grouping grain types into a few general categories. It is universal practice to describe monocrystalline particles as mineral grains, and polycrystalline particles as lithic or rock fragments. Mineral grains are subdivided according to mineral species, whereas lithic fragments are classified, as are rocks, on the joint criteria of overall chemical composition, the mineral facies of modal constituents, and their internal texture and fabric.

There are two methodological approaches to the treatment of rock fragments during point counting (Zuffa, 1980; Ingersoll et al, 1984). On the one hand, all parts of all polycrystalline particles can be counted as lithic fragments. This seemingly rational approach leads to severe dependence of calculated sandstone compositions on clastic grain size. Coarse-grained source rocks, composed of crystals larger than the matrix limit (0.0625 mm), can then occur partly as lithic fragments in coarse sandstones, but generally disintegrate to form only individual mineral grains in fine sandstones. The severity of this procedural problem can be gauged from the fact that the sand range spans between one and two orders of magnitude in grain size.

Dependence of calculated sandstone compositions on clastic grain size can be reduced markedly by adopting a different point counting procedure that restricts lithic fragments to microcrystalline aphanitic materials containing no crystals larger than the matrix limit (= sand framework limit). Single crystals larger than this size (0.0625 mm) are then counted as mineral grains, regardless of whether they actually occur as separate clastic particles or as constituent crystals within polycrystalline particles. For example, phenocrysts within volcanic rock fragments are counted as mineral grains if their mean visible diameters exceed 0.0625 mm. Only when the microscope crosshair centers above part of the groundmass of a volcanic rock fragment, or a smaller microphenocryst, is the point counted as a lithic fragment. Similarly, constituent crystals within granitic rock fragments and coarser grained quartzite fragments are commonly large enough to be tallied as mineral grains for calculation of detrital modes.

There are two key points to be made about this procedure. First, raw counts should be performed in a way that retains full information about the occurrence of specific types of mineral grains within different kinds of lithic fragments. The information obtained can thus be recast into any desired format. Second, the method is merely a means for normalization of data with respect to grain size, and cannot correct for inherent or genetic variations in grain proportions with changing grain size. For example, where the quartz/feldspar ratio is actually different in sands of different grain size, this strictly procedural method does not improve compositional comparisons between sandstones of contrasting mean grain size. For this reason, it is good practice where possible to restrict provenance studies to sandstones of comparable grain size. Favorable counting statistics and ease of identification are best achieved jointly in thin sections of medium-grained to coarse-grained sandstone (mean grain size near 0.5 mm).

GRAIN TYPES

Table 1 presents the classification of grain types used for discussions in this paper. Detrital modes are recalculated to 100 per cent as the sum of Qm, Qp, P, K, Lv, and Ls. Intrabasinal grains (Zuffa, 1980) are ignored. In this summary, the occurrence of heavy minerals is also ignored, not because they convey no information about provenance, but simply because their different response to hydrodynamic and geochemical influences makes their volumetric distribution so variable.

TABLE 1. CLASSIFICATION AND SYMBOLS OF GRAIN TYPES

A. Quartzose Grains (Qt = Qm + Qp)
        Qt = total quartzose grains
        Qm = monocrystalline quartz (>0.625 mm)
        Qp = polycrystalline quartz (or chalcedony)

B. Feldspar Grains (F = P + K)
        F = total feldspar grains
        P = plagioclase grains
        K = Kspar grains

C. Unstable Lithic Fragments (L = Lv + Ls)
        L = total unstable lithic fragments
        Lv = volcanic/metavolcanic lithic fragments
        Ls = sedimentary/metasedimentary lithic fragments

D. Total Lithic Fragments (Lt = L + Qp)
        Lc = extrabasinal detrital limeclasts
             (not included in L or Lt)

Extrabasinal carbonate grains or detrital limeclasts (Lc) are not recalculated with other lithic fragments because of their vastly different geochemical response during weathering and diagenesis, as well as the ease of confusion with intrabasinal carbonate grains (intraclasts, bioclasts, ooliths, peloids). Calclithite is thus regarded here as a variety of calcarenite, rather than litharenite. Where the proportion of detrital limeclasts is discussed, percentage of total framework is reported, as is the case for micas (M). This practice is adopted provisionally, with the knowledge that inclusion of Lc within the Ls population may help to clarify provenance relationships in some studies (Mack, 1984).

Distinctions between Qm and Qp vary depending upon the grain size conventions used by different operators during point counting. When lithic points are restricted to microcrystalline aggregates of aphanitic materials, most Qp grains are chert or metachert, although foliated or microgranular metaquartzite fragments are abundant in some rocks. On the other hand, if points for coarse crystals in rock fragments are not reapportioned as mineral grains, Qp as reported may also include composite quartz grains and orthoquartzite fragments. Severe compaction may develop polygonal textures within single quartz crystals, which then resemble composite or aggregate grains but are not lithic fragments.

Reliable identification of P and K, and their distinction from quartz, requires routine staining for both feldspars.

Perthite is counted as K, and albite as P, but diagenetic albite may replace either original feldspar in some cases. Albite takes neither feldspar stain. Systematic replacement of Kspar by diagenetic albite may frustrate attempts to recover the detrital ratio of Kspar to plagioclase in some instances (Dickinson et al, 1982; Walker, 1984). However, albitization of plagioclase is a much more common process at moderate stages of diagenesis (Boles, 1982).

Unstable lithic fragments are identified by joint observation of texture and mineralogy. Volcanogenic (Lv) types display microlitic (lathlike), microgranular-felsitic, or vitric textures, and feldspar-rich or mafic mineralogy. Metamorphic variants of Lv may be foliated, but otherwise it is difficult or impossible to distinguish between transported fragments of low-grade metavolcanic rock and volcanic rock fragments that have been diagenetically altered in place within a sandstone. Felsitic volcanic rock fragments can only be distinguished from chert with confidence if they take one or both of the feldspar stains, contain microphenocrysts, or have groundmasses coarse enough to show the contrast in refringence between tiny quartz and feldspar crystallites. In general, chalcedonic volcanic rock fragments that are thoroughly silicified cannot be distinguished with confidence from sedimentary varieties of chert grains (Qp).

Sedimentary and metasedimentary lithic fragments (Ls) include chiefly microclastic siltstone, massive cryptocrystalline argillaceous grains, phyllosilicate-rich shaly or slaty grains showing mass extinction effects, and foliated or microgranular aggregates of quartz and mica from phyllite and/or hornfels. Firm distinctions between these various subtypes are commonly complicated by the presence of texturally and compositionally transitional grains. Gradations from chert (Qp) to argillite (Ls) are also common, and are best resolved by relegating siliceous argillite to Ls. The various Ls subtypes are more quartz-rich and/or phyllosilicate-rich than the various Lv subtypes.

In some sandstones, the presence of shaly intraclasts derived from syndepositional erosion of interbedded fine-grained sediments presents a special problem of interpretation. Where the true nature of such argillaceous grains can be established, they should be excluded from the detrital framework for most types of provenance studies. Their inclusion as part of the Ls population introduces unwanted bias into results. In some instances, apparently detrital limeclasts (Lc) may be of similar intrabasinal origin.

Some workers systematically separate metamorphic rock fragments (Lm) from volcanic and sedimentary rock fragments for

diagnostic purposes (Ingersoll and Suczek, 1979). Consistent
values for Lm are difficult to achieve, especially in studies
involving multiple operators, because criteria are subtle or
ambiguous in many instances. No attempt is made to resolve Lm
values in the compilations reported here. Where such attempts
are made, it is important to report the criteria used to define
the threshold of metamorphism for counting purposes.

In determining detrital modes, it is vital to assess the
effects of diagenesis and outcrop weathering on the framework.
Every effort should be made to obtain fresh, unweathered samples. Where secondary porosity of any origin has developed,
allowances for its effect on framework composition must be made
in all interpretations of point-count data. Intrastratal solution presents the most severe potential problems for interpretation in cases where recompaction of the framework has removed
all direct evidence for dissolution of selected grains. The
correct classification of altered grains also involves questions
of widely varying difficulty. The parent grains from which many
alterites were derived can be inferred without serious doubt,
but the antecedents of some may be indeterminate.

PROVENANCE TYPES

Compositional fields characteristic of different provenances are well shown on one or more of four triangular diagrams
(Dickinson and Suczek, 1979; Dickinson et al, 1979, 1983a):
QtFL, with emphasis on maturity; QmFLt, with emphasis on source
rock; QpLvLs, with emphasis on lithic fragments; and QmPK, with
emphasis on mineral grains. Table 2 lists major provenance
types in terms of overall tectonic setting within or adjacent to
continental blocks, and gives key aspects of derivative sand
compositions.

Figure 1 shows nominal fields proposed within QtFL/QmFLt
diagrams for discrimination of sands derived from various types
of provenances in continental blocks, magmatic arcs, and recycled orogens. In this context, continental blocks are tectonically consolidated regions composed essentially of amalgamations
of ancient orogenic belts that have been eroded to their deep-
seated roots and lack any relict genetic relief. Magmatic arcs
are belts of positive relief composed dominantly of penecontemporaneous associations of cogenetic volcanic and plutonic
igneous rocks, together with associated metamorphic wallrocks,
produced by continuing subduction along arc-trench systems.
Recycled orogens include the deformed and uplifted supracrustal
strata, dominantly sedimentary but also volcanic in part, exposed in varied fold-thrust belts of orogenic regions. Figure 2
shows the reported distribution of mean detrital modes for sand-

TABLE 2. MAJOR PROVENANCE TYPES AND KEY COMPOSITIONAL
ASPECTS OF DERIVATIVE SANDS

Provenance Type	Tectonic Setting	Derivative Sand Composition
stable craton	continental interior or passive platform	quartzose sands (Qt-rich) with high Qm/Qp and K/P ratios
basement uplift	rift shoulder or transform rupture	quartzofeldspathic (Qm-F) sands low in Lt with Qm/F and K/P ratios similar to bedrock
magmatic arc	island arc or continental arc	feldspatholithic (F-L) volcaniclastic sands with high P/K and Lv/Ls ratios grading to quartzofeldspathic (Qm-F) batholith-derived sands
recycled orogen	subduction complex or fold-thrust belt	quartzolithic (Qt-Lt) sands low in F and Lv with variable Qm/Qp and Qp/Ls ratios

stone suites actually derived from the various types of provenance terranes.

Stable Cratons

The main sources for craton-derived quartzose sands are low-lying granitic and gneissic exposures, supplemented by recycling of associated flat-lying platform sediments. The sands either accumulate as platformal successions deposited within continental interiors, or are transported chiefly to passive continental margins and the cratonal flanks of foreland basins.

Ultimate sources for quartz grain populations can be inferred to some degree from the frequency and nature of undulose extinction and polycrystallinity (Basu et al, 1975). However, both properties can be affected markedly by grain deformation during diagenesis (Graham et al, 1976). Textural details of polycrystalline quartzose grains offer more reliable guides to specific source rock types in favorable instances (Young, 1976). High Qm/Qp ratios in the most mature quartzose sands indicate that monocrystalline quartz has greater potential for survival in the sedimentary cycle than polycrystalline lithic fragments. Experimental evidence that microcrystalline chert is more resistant to abrasion that monocrystalline quartz indicates

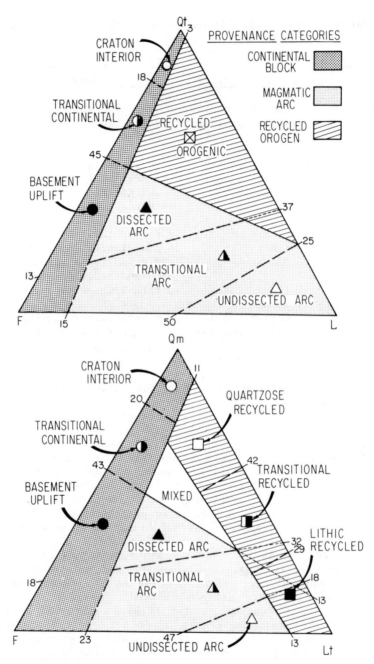

Figure 1. Provisional compositional fields indicative of sand derivation from different types of provenances (from Dickinson et al, 1983a).

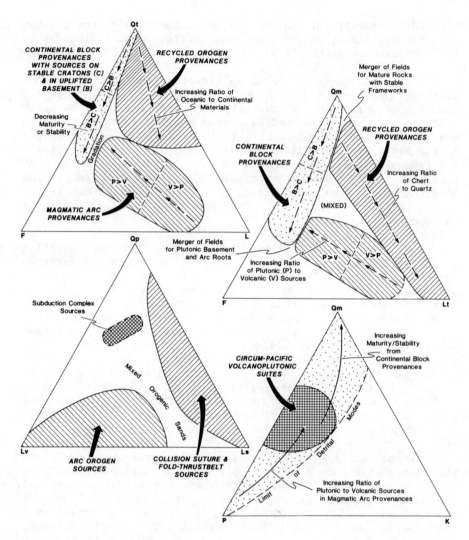

Figure 2. Actual reported distribution of mean detrital modes
for sandstone suites derived from different types of
provenances plotted on standard triangular diagrams;
after Dickinson and Suczek (1979) as modified by
Dickinson (1982; et al, 1982; and unpublished compilations).

that this capacity for survival stems fundamentally from greater chemical stability, rather than from greater mechanical durability (Harrell and Blatt, 1978).

Many have speculated that multicyclic reworking on cratons might be required to develop mature quartzose sands. However, recent work has shown conclusively that quartzose sand is being produced as first-cycle sediment from deeply weathered granitic and gneissic bedrock exposed in tropical lowlands of the modern Amazon basin (Franzinelli and Potter, 1983). Tributaries draining Precambrian basement (n=21), as well as Precambrian basement plus Paleozoic platform cover (n=13), carry sand containing 90-95% quartz, much more Kspar (1-5%) than plagioclase (which occurs in trace amounts only), and less than 5% lithic fragments on average.

The importance of intense weathering for concentrating quartz in relation to feldspar and/or lithic fragments deserves special emphasis. In most cases, fluvial transport alone is apparently ineffective as an influence on the proportion of quartz in sand. More than 600 km of transport on the Platte River in the Great Plains proves insufficient to change either the Q/F or K/P ratio of the stream sands to any significant degree (Breyer and Bart, 1978). On the Amazon, the Q/Lt ratio does not change appreciably for 2500 km from the rugged headwaters to the junction with the first major tributary whose drainage basin lies wholly within the deeply weathered Amazon lowlands (Franzinelli and Potter, 1983). For 1500 km below that point to the mouth, however, the quartz content rises markedly, presumably from steady dilution with quartzose tributary sands.

The necessity of low relief to allow prolonged weathering also deserves emphasis as a requirement for the development of first-cycle quartzose sand. Even where the climatic potential for intense equatorial weathering exists, tropical highlands develop quartz-rich regoliths only on restricted interfluves with gentle local relief; both fluvial and littoral Holocene sands derived from drainage basins in tropical highlands with high relief are quartz-poor (Ruxton, 1970). In evaluating the combined effects of climate and relief on the production of quartzose sands, however, it is important to include an evaluation of the possibility of appreciable weathering during temporary storage on low-lying floodplains along the continental dispersal path.

Basement Uplifts

Fault-bounded basement uplifts along incipient rift belts and transform ruptures within continental blocks shed arkosic sands mainly into adjacent linear troughs or local pull-apart

TABLE 3. AVERAGE Qm/(Qm+F) AND K/(K+P) RATIOS FOR
SELECTED FIRST-CYCLE ARKOSIC SAND AND SANDSTONE SUITES
IN SOUTHERN CALIFORNIA
See Table 1 for symblols of grain types.

Description	N	Qm/(Qm+F)[*]	K/(K+P)[*]	Reference
Eocene, Santa Ynez Mountains	25	0.39	0.35	Helmold, 1980
Upper Cretaceous, Simi Hills	20	0.39	0.38	Carey, 1981
Pliocene, Ridge Basin	60	0.43	0.41	Link, 1982
Holocene, Salton Basin	9	0.45	0.40	Van de Kamp, 1973
Cretaceous, Santa Monica Mountains	11	0.48	0.40	Carey, 1981

[*]Mean values within 0.40-0.45 for Qm/(Qm+F) and 0.35-0.40 for K/(K+P).

basins. Similar detritus can be derived from basement uplifts
within broken foreland provinces, and from eroded plutons in
deeply dissected magmatic arcs (see discussions below). A spectrum of lithic-poor quartzofeldspathic sands forms a roughly
linear array on QtFL and QmFLT diagrams (Figs. 1,2) linking
these arkosic sands with the craton-derived quartzose sands that
plot near the Qt and Qm poles. This spectrum of sands reflects
derivation from varius tectonic elements of continental blocks
where basement rocks are exposed. Where erosion has been insufficient to remove cover rocks from basement, uplifts may shed
sands having affinity with detritus derived from magmatic arcs
or recycled orogens (Mack, 1984).

The composition of the arkosic end member of this spectrum
of sand types is approximated well by the mean of first-cycle
arkose in southern California (Table 3). Although granitic rock
fragments may be abundant in coarser size grades, the proportion
of true lithic fragments (see discussion above) is generally
less than 10% and averages about 5%. Quartz forms about 40-45%
of the quartzofeldspathic grain population, and Kspar averages
35-40% of the total feldspar present. These are values appropriate to areas of rugged relief in arid or semiarid climatic
zones. The spectrum of sands derived from continental blocks
can be viewed as forming a mixing array or evolutionary trend
connecting this ideal arkosic composition to the cratonal
quartzose sand composition (Fig. 3).

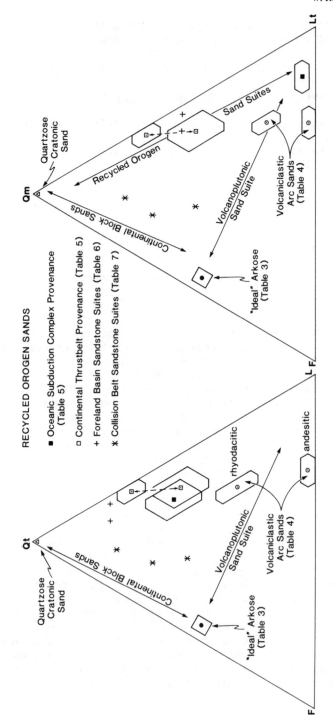

Figure 3. Compositional plots of key sandstone suites (see Tables 3-7); fields surrounding some points depict calculated standard deviations (see Tables 3-7); trend line for volcano-plutonic sand suites from Dickinson, 1982; see text for discussion.

Humid weathering raises the proportion of quartz relative to feldspar. In the southern Appalachians, first-cycle arkosic sands may have quartz:feldspar ratios as high as 2:1 (Basu, 1976). Plagioclase is less stable than Kspar in all climatic regimes, but the disparity is much more marked for humid areas. In stream sands from semiarid regions, the fraction of total feldspar that is plagioclase may be nearly as high as in the parent granitic rock; however, in stream sands from humid regions, the fractional value may fall to only a third of that for the parent rock (James et al, 1981). Consequently, Kspar:plagioclase ratios as high as 3:1 have been reported for some arkosic sands (Basu, 1976). The tendency for the Kspar:plagioclase ratio to rise as the quartz content of typical sands increases is shown by the QmPK diagram of Figure 2.

Magmatic Arcs

The most characteristic sands derived from active magmatic arcs constructed parallel to subduction zones are volcaniclastic materials erupted and eroded from stratovolcano chains and associated ignimbrite plateaus. Deep dissection of arc assemblages can expose their batholithic roots, and may give rise to quartzofeldspathic sands indistinguishable from the arkosic debris produced by other basement uplifts. Mixing of volcanic and plutonic contributions results in a spectrum of volcanoplutonic sands (Dickinson, 1982), which range from feldspatholithic to lithofeldspathic in composition, but display consistently high Lv/Lt ratios coupled with low to moderate quartz contents (Figs. 1,2). Arc-derived debris is typically deposited in forearc or interarc basins, but may also reach foreland basins locally. Sandstones incorporated into subduction complexes are commonly arc-derived variants of the volcanoplutonic suite deposited in trenches or on trench slopes (Dickinson, 1982). However, subduction complexes may also include quartzose turbidites derived from distant continental blocks, and rafted into subduction zones as sediment cover on an oceanic plate (Velbel, 1980).

The mean composition of the volcaniclastic end member of the array of volcanoplutonic sand types can be estimated from data for appropriate first-cycle sand suites from the circum-Pacific region (Table 4). Generally andesitic sands (columns A-D) are most abundant, and are used here to calculate an average modal composition. More silicic sands (columns E-F) include both rhyolitic and dacitic varieties. Volcanic rock fragments in the more common andesitic (to basaltic and/or dacitic) sands are dominantly microlitic varieties, whereas those in the rhyodacitic sands have felsitic or pyroclastic textures.

TABLE 4. AVERAGE MODAL COMPOSITIONS OF SELECTED
VOLCANICLASTIC SAND AND SANDSTONE SUITES
DERIVED FROM MAGMATIC ARC

A-D, normal mixed andesitic stratovolcano sources
E-F, special silicic ignimbrite sources
Frameworks also include 1-5% heavy minerals (mainly pyriboles)
    and traces of mica.
Ranges (±) are standard deviations for reported data.
Table 1 gives definitions of symbols for grain types.

	A	B	C	D	Ave	E	F
N	27	55	6	11	(A-D)	7	19
Qm	2	3	tr	3	2	19±5	18±3
Qp	4	1	tr	?	1	1	18±3
P	32±8	24±5	30±2	25±5	28	8±2	19±5
K	1	tr	tr	tr	tr	11±3	2
Lv	60±8	71±6	67±4	69±5	67	59±8	42±7
Ls	1	1	2	3	2	2	1

A - Neogene marine sands, Atka Basin, Aleutian Ridge (Stewart,
    1978)
B - Jurassic-Cretaceous   Yahgan   Formation,   Tierra   del   Fuego
    (Winn, 1978)
C - Lower Jurassic Mowich Group and Middle Jurassic Snowshoe
    Formation, central Oregon (Dickinson et al, 1979)
D - Lower Triassic North Range Group, Hokonui Hills, Southland,
    New Zealand (Boles, 1974)
E - Holocene fluvial sands derived from Tertiary ignimbrite
    field in Sierra Madre Occidental of northwest Mexico (Webb
    and Potter, 1971)
F - Cretaceous Sandebugten sandstones of South Georgia Island
    derived from Jurassic Tobifera Volcanics of Patagonia (Winn,
    1978)

    As expected, the rhyodacitic volcaniclastic sands are more
quartzose than the andesitic sands; the rhyodacitic field plots
near a mixing line between the andesitic composition and the
quartz pole (Fig. 3).   The array of circum-Pacific volcanoplutonic sands lies along a mixing path between the volcaniclastic
sands and the ideal arkosic composition (Fig. 2).   Surprisingly,
the position of the dominant volcanoplutonic compositional trend
(Fig. 3) seemingly suggests mixture of arkosic materials mainly
with silicic rather than andesitic volcaniclastic debris.   This
inference is probably spurious, for it fails to take into

account the likelihood of minor admixtures of recycled quartzose
contributions from metamorphic wallrocks of the magmatic arcs.

Recycled Orogens

Orogenic recycling occurs in several tectonic settings
where stratified rocks are deformed, uplifted, and eroded: (a)
subduction complexes, where oceanic and trench slope deposits
are exposed as thrust panels, recumbent isoclines, and melange
belts along tectonic ridges between trenches and forearc basins;
(b) backarc thrustbelts, where folded sedimentary and metasedimentary strata of continental derivation override the flanks of
retroarc foreland basins; and (c) suture belts, where structurally juxtaposed sequences of both oceanic and continental types
can provide sources of sediment for transverse dispersal systems
that feed adjacent peripheral foreland basins, and for longitudinal dispersal systems that feed nearby remnant ocean basins.
Orogenic highlands also give birth to major river systems that
can transport recycled orogenic sediment across the surfaces of
adjacent continental blocks and into distant basins having a
variety of tectonic settings (Potter, 1978).

By its nature, sediment from recycled orogens includes
various proportions of materials whose compositions reflect
ultimate derivation from cratonic, arkosic, or volcaniclastic
sources, modified in part by metamorphic processes. In addition
are materials generated by sedimentary processes, acting alone
or in combination with diagenesis and metamorphism.   These
latter are lithic fragments of chert and metachert, or pelitic
debris such as shale, argillite, slate, and phyllite. Given the
potential diversity of recycled orogenic sediment, it is a
severe challenge to devise a scheme for its identification and
classification that has empirical validity for interpretation of
the sedimentary record. As yet (see above), there is also no
consensus on the way in which detrital limeclasts (Lc) should be
treated during recalculation of detrital modes (Mack, 1984).

Sands derived from fold-thrust systems of indurated sedimentary and low-grade metamorphic rocks have consistently low
contents of feldspar and volcanic rock fragments (Dickinson and
Suczek, 1979). Consequently, they form a quartzolithic array of
compositions that plot near the Qt-L, Qm-Lt, and Qp-Ls legs of
standard triangular diagrams (Figs. 1,2).   Van Andel (1958)
early called attention to the abundance of this type of sand in
the orogenic region of western Venezuela. In the modern Andean
foreland (Franzinelli and Potter, 1983), the least mature sands
derived directly from the adjacent highlands contain roughly
equal amounts of quartz grains and rock fragments (with only
5-10% feldspar grains).

TABLE 5. AVERAGE MODAL COMPOSITIONS OF SELECTED
SANDSTONE SUITES DERIVED FROM DEFORMED
AND UPLIFTED SUPRACRUSTAL SOURCES

A, Upper Triassic Vester Formation in central Oregon derived
    from   sand-poor   subduction   complex   of   chert-argillite-
    greenstone oceanic facies (Dickinson et al, 1979)
B and C, middle (B) and lower (C) Siwalik Group (Mio-Pliocene)
    in peninsular India derived from Himalayan fold-thrust
    system of sedimentary and metasedimentary strata (Parkash et
    al, 1980)
Ranges (±) are standard deviations for reported data

	N	Qm	Qp	Qt	F[*]	Lv	Ls	Lt
A	11	5	45±10	50±11	13±6	27±7	10±2	82±7
B	15	43±11	5	48±9	10±5	tr	42±7	47±8
C	14	59±7	7	66±5	3	tr	31±6	38±9

[*]Plagioclase (P) in A is 12±6; P and K not reported separately
for B and C.

Relative proportions of resistant quartzose grains and
unstable lithic grains are highly variable in recycled orogenic
sands.   Modern sands derived from such sources have Qt/L ratios
that range from about 3:1 in humid regions to about 1:3 in semi-
arid regions (Suttner et al, 1981a; Mack, 1981); polycrystalline
Qp and Ls in the same sands tend to be subequal in abundance.
Within the Amazon drainage, where both rugged highlands and
tropical lowlands are present, the full range of Qm/Lt ratios in
fluvial sands is virtually from zero to 100 (Franzinelli and
Potter, 1981).   In all these Holocene assemblages, the content
of total feldspar averages about 5% or less.

Clastic sequences in foreland basins commonly display
interstratified or intertonguing petrofacies with contrasting
Qt/L and/or Qm/Lt ratios (e.g., Putnam, 1982).   In some cases,
the salient differences can be attributed to the mingling of
contributions from disparate sources.   For example, Carboniferous foreland clastics of the Trenchard Group in England are
composed of a northern quartzarenite facies (Qt87, F8, L5),
derived from the gentle cratonal flank of the basin, and a
southern litharenite facies (Qt46, F5, L49) derived from the
orogenic flank (Jones, 1972).   In the Carboniferous Pottsville
Group of West Virginia, however, an upward stratigraphic transition from litharenite (Qt73, F2, L25) to quartzarenite (Qt95,
F0, L5) apparently was produced by weathering and recycling of
detritus from the same source on a broad coastal plain at the

TABLE 6. AVERAGE MODAL COMPOSITIONS OF SANDSTONE
SUITES FROM SELECTED FORELAND BASINS

A - Mississippian Antler deltaic (and turbidite) clastics
(Chainman Shale and Diamond Peak Formation) derived from
sand-rich chert-argillite subduction complex in peripheral
foreland basin of central Nevada (Dickinson et al, 1983b).
B - Upper Cretaceous deltaic (and contourite) clastics (Cody
Shale and Parkman Sandstone) derived from sedimentary-
metasedimetnary thrustbelt in retroarc foreland basin of
central Wyoming (Hubert et al, 1972).
C - Cretaceous-Paleocene deltaic (and associated) clastics
(Difunta Group) derived mainly from arc volcanics in
retroarc foreland basins of northeast Mexico (McBride et al,
1975).

	N	Qm	Qp	Qt	F	Lv	Ls	Lt
A	18	47	26	73	2	tr	25	51
B	35	48	24	72	7	tr	21	45
C	81	32	4	36	28	32	4	40

margin of the foreland basin (Houseknecht, 1980). In some
sequences, recycled sands derived from uplifted quartzose strata
cannot be distinguished compositionally from craton-derived
sands, especially where humid weathering and intense reworking
enhance the quartz content of the sands (Mack et al, 1981; Mack,
1984).

Chert-rich sand frameworks are characteristic in some
recycled orogenic suites (Dickinson and Suczek, 1979). The
supracrustal chert sources are typically either radiolarites
within deformed oceanic assemblages, or replacement nodules in
platform carbonate successions. The Qm/Qp ratios in derivative
sands are largely a function of the extent to which turbidite or
platform sandstones are intercalated within the chert-bearing
sections. Table 5 shows the extremes to which the Qm/Qp ratio
may vary (Fig. 3) by comparing (A) one sandstone suite derived
from a chert-rich subduction complex composed of deep marine
facies, deposited on an oceanic substratum and lacking any
intercalated sandstones, with (B,C) two horizons within a sandstone suite derived from uplifted sandy clastics and meta-
clastics detached structurally from the edge of a continental
block. By contrast, Table 6 shows the compositional similarity
(Fig. 3) of two otherwise dissimilar foreland sandstone suites
derived from (A) an allochthon of overthrust oceanic facies in a
subduction complex emplaced adjacent to a peripheral foreland
basin that developed during arc-continent collision, and (B)

TABLE 7. AVERAGE MODAL COMPOSITIONS OF SELECTED SAND AND
SANDSTONE SUITES DERIVED FROM COLLISON OROGENS

A - Cenozoic sand, Indus Cone, Arabian Sea (Suczek and
    Ingersoll, 1984)
B - Neogene sand, Bengal-Nicobar Fan, Indian Ocean (Ingersoll
    and Suczek, 1979)
C - Pennsylvanian Haymond Formation, Marathon Basin, Texas
    (McBride, 1966)
A-B derived from Himalayan-Tibetan orogen, and C derived from
    Ouachita system
Ranges (±) are standard deviations for reported data

	N	Qm	Qp	Qt	P	K	F	Lv	Ls	Lt
A	15	44±6	1	45	21±3	11±3	32	2	21±6	24
B	22	56±6	1	57	19±4	9±3	28	1	13±5	15
C	33	68±7	3	71	14±6	3	17	tr	12±5	15

thrust sheets of miogeoclinal strata including carbonates within
an intracontinental foldbelt that formed adjacent to a retroarc
foreland basin. The chert is presumably of different origins in
the two cases, but both source sequences contain a sufficient
proportion of interbedded quartzose sandstone to impart essentially the same Qm/Qp ratio to derivative recycled sands.

Table 6 also gives the mode of an arc-derived sandstone
suite (C), rich in feldspars and volcanic rock fragments,
deposited within a retroarc foreland basin. Although generally
not abundant within foreland regions, volcaniclastic detritus
can be important in the evolution of retroarc foreland basins
located close to arcs where backarc deformation is either minor
or fails to create a drainage divide between eruptive centers
and the foreland basin (Misko and Hendry, 1979). Basement uplifts in broken foreland provinces may also contribute arkosic
debris locally to foreland basins. Both influences are present
jointly in some foreland regions (Suttner et al, 1981b;
Schwartz, 1982).

Table 7 indicates the nature of representative sands and
sandstones derived from intercontinental collision orogens;
similar sandstones may be derived from arc collision belts
(Hiscott, 1978). Their recycled affinity is indicated by the
consistently high Ls/Lv ratios. However, their coordinate feldspar content indicates that uplifted basement rocks or arc
batholiths are exposed within the orogenic highlands in addition
to deformed supracrustal strata. Longitudinal dispersal of
collision-belt sands into remnant ocean basins is characteristic

of many orogenic systems (Graham et al, 1975), but transverse dispersal into foreland basins also occurs.

Composite Provenances

Triangular compositional diagrams provide a convenient means to plot graphically in a quantitative format. Any point within the plot can obviously be interpreted as a certain mixture of the three entities represented by the poles of the diagram. In principle, any such point can also be interpreted as a particular mixture of various other arbitarily chosen entities whose compositions can be represented by other points within the diagram.

As noted previously, the array of quartzose and arkosic sands derived from various parts of continental blocks can be regarded as various mixtures or evolutionary stages along a linear trend connecting the ideal arkose composition to the quartz pole (Fig. 3). Similarly, the array of volcanoplutonic sands derived from magmatic arcs can be viewed as mixtures of ideal volcaniclastic debris and the ideal arkose composition. The exact proportions required for the appropriate mixtures can be scaled from the triangular plot as an inverse function of relative distances between the relevant points. More generally, any point within the interior of the diagram can be interpreted in terms of mixing any three compositional end members distributed near the perimeter of the diagram. Of course, any one of such controlling end members may itself be a mixture of two or three other compositions.

If the compositions of end members that are truly diagnostic of provenance type can be accurately defined, the actual parentage of specific sandstone suites can thus be inferred. For example, the three sandstone suites derived from collision orogens (Table 7, Fig. 3) can be described alternately as mixtures of (1) various pairs of sands derived from recycled orogens and continental blocks, or (2) three end members representing (a) quartzose cratonic sand (at or near the quartz pole), (b) ideal arkose from uplifted plutonic basement, and (c) some ideal recycled sand of quartzolithic composition. The particular degree of "mixing" defined compositionally by the graphical display of the plot need not in all cases be attributed to actual mechanical mixing. In the case of the sands derived from collision belts, for example, the variation in quartz content may be attributed either to different proportions of recycled quartzose sand derived from the orogenic belts in question, or to contrasting climatic influences within those same orogenic belts, rather than literally to an admixture of first-cycle cratonic detritus.

A significant task for the future is the use of sandstone detrital modes to help establish the nature of composite orogenic provenances composed of various genetic associations of source rock types. For example, a recent study of Paleogene fluvial sandstones within the complex Cascades arc orogen of the Pacific Northwest indicates that various parts of the sequence were derived from recycled orogenic sources, an eroded magmatic arc, and uplifted continental basement (Johnson, 1984). Sedimentary mixing of such detritus, followed by wider dispersal, would give rise to sands of composite provenance. In the Apennines, synorogenic turbidite sandstones derived mainly from the complex Alpine collision orogen have compositions indicative of derivation from similarly compound sources within a composite provenance (Valloni and Zuffa, 1984).

PETROFACIES EVOLUTION

Spatial patterns of correlative petrofacies within a sedimentary basin reflect simultaneous contributions of detritus from different sources. Successive time-dependent petrofacies reflect either the evolution of individual provenance terranes or changes in disperal patterns through time. Significant contrasts in petrofacies imply major differences in diagenetic processes within basin fill. For example, porosity-depth relationships are strikingly different for typical quartzose, feldspathic (arkosic), and lithic sandstones (Fig. 4). Stratigraphic variations in petrofacies within forearc and foreland regions suggest that an integrated view of basin evolution requires attention to petrofacies evolution, both as a record of tectonic events and as a means of understanding diagenetic conditions.

The main provenance for typical forearc basins is the adjacent magmatic arc. Consequently, forearc petrofacies generally reflect the igneous and morphologic evolution of the arc. Where eruptive rejuvenation of the surficial volcanic edifice counteracts uplift and erosion, volcaniclastic sandstones may be deposited across the forearc over long periods of time (Cawood, 1983). Volcaniclastic sandstone suites commonly become more quartzose upsection, because the normal evolution of many magmatic arcs leads to the eruption of more felsic materials as time passes (Korsch, 1984). Moreover, there is commonly a tendency for uplift and erosion to expose more and more of the plutonic roots of an arc as time passes. Consequently, forearc sandstones of the volcanoplutonic suite (Figs. 1-3) generally become more arkosic and less lithic with time (Dickinson and Rich, 1972). Recycled orogenic sources in exposed subduction complexes may influence petrofacies locally (Tennyson and Cole, 1978), and major changes in drainage patterns within arcs may cause abrupt shifts in petrofacies within adjacent forearc

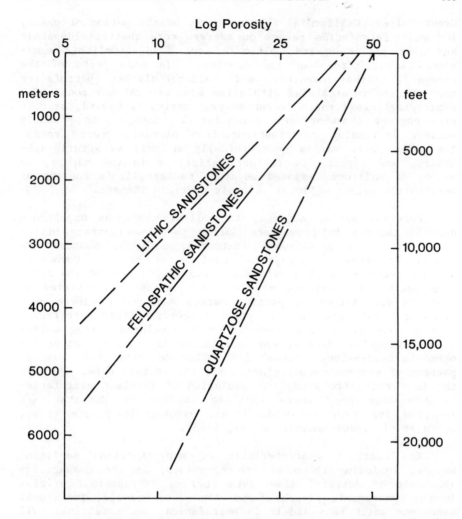

Figure 4. Graph showing typical relations of porosity to depth of burial for sandstone suites of varying composition; data from Galloway (1974), Zieglar and Spotts (1978), and Sclater and Christie (1980).

basins (Heller and Ryberg, 1983).   Transform tectonics during oblique subduction may lead to especially complex patterns of forearc petrofacies in both time and space (Pacht, 1984).

Ingersoll (1983) has recently presented a thorough analysis of the evolution of petrofacies within the large and complex

Great Valley (California) forearc basin over a period of nearly 100 ma. Petrofacies become on average more quartzofeldspathic and less lithic upward, although some feldspatholithic sandstones occur throughout the section. In many parts of the sequence, lithofeldspathic and feldspatholithic petrofacies variants are intercalated within the same general horizons. The Kspar:plagioclase ratio increases systematically upward, as does mica content somewhat more irregularly. Both trends probably reflect increasing contributions from plutonic source rocks. The Lv/Ls ratio varies geographically as well as stratigraphically, and appears to reflect variations in the nature and amount of wallrocks exposed adjacent to batholithic sources as well as the extent of local volcanic cover in the arc.

Foreland basins adjacent to collision orogens experience drastic changes in provenance during their sedimentary history. Prior to collisional tectonics, sediment sources are dominantly within the adjacent continental block. Commonly, arkosic petrofacies of a rift phase lie at the base of the foreland succession, and are overlain by quartzose petrofacies of platform successions or passive margin sequences. These pre-orogenic strata are then succeeded by quartzolithic petrofacies derived from recycled orogenic sources developed during collision orogenesis. Arc-derived petrofacies may form a part of the orogenic succession. Schwab (1981) has described this general pattern of successive petrofacies for the western Alps. Perhaps the best region to study the evolution of foreland petrofacies is the Apennines, where multiple sources of detritus are inferred for both pre-orogenic and synorogenic phases (i.e., Zuffa et al, 1980; Gandolfi et al, 1983).

An important characteristic of many foreland sandstone suites, including those of the Apennines, is the comparative abundance of detrital limeclasts (Lc) of extrabasinal origin. In the synorogenic flysch of the Alps, for example, individual sandstone units vary widely in petrofacies, but essentially all contain some proportion of limeclasts reworked from deformed and uplifted sedimentary strata exposed as a recycled orogenic provenance within the developing collision orogen (Hubert, 1967). Limeclasts are also common in the Mesozoic retroarc foreland of the Rocky Mountains (work in progress).

ACKNOWLEDGMENTS

Discussions with many colleagues over the years have influenced the thoughts expressed here, but I especially thank W. D. Darton, J. E. Decker, P. L. Heller, K. P. Helmold, R. V. Ingersoll, T. F. Lawton, G. H. Mack, P. E. Potter, P. T. Ryberg, C. A. Suczek, R. Valloni, and G. G. Zuffa for recent discussions and/or preprints. This paper was prepared with support from the Laboratory of Geotectonics in the Department of Geosciences at the University of Arizona.

REFERENCES

Basu, A., 1976, Petrology of Holocene fluvial sand derived from plutonic source rocks: implications to paleoclimatic interpretation: Jour. Sed. Petrology, v. 46, p. 694–709.

Basu, A., S. W. Young, L. J. Suttner, W. C. James, and G. H. Mack, 1975, Re-evaluation of the use of undulatory extinction and polycrystallinity in detrital quartz for provenance interpretation: Jour. Sed. Petrology, v. 45, p. 873–882.

Boles, J. R., 1974, Structure, stratigraphy, and petrology of mainly Triassic rocks, Hokonui Hills, Southland, New Zealand: N. Z. Jour. Geology and Geophysics, v. 17, p. 337–374.

Boles, J. R., 1982, Active albitization of plagioclase, Gulf Coast Tertiary: Am. Jour. Sci., v. 282, p. 165–180.

Bryer, J. A. and H. A. Bart, 1978, The composition of fluvial sands in a temperate semiarid region: Jour. Sed. Petrology, v. 48, p. 1311–1320.

Carey, S. M., 1981, Sandstone petrography of the Upper Cretaceous Chatsworth Formation, Simi Hills, California, in Link, M. H., R. L. Squires, and I. P. Colburn, eds., Simi Hills Cretaceous turbidites, southern California: Pacific Sec., Soc. Econ. Paleontologists and Mineralogists, Los Angeles, California, p. 89–97.

Cawood, P. A., 1983, Modal composition and detrital pyroxene geochemistry of lithic sandstones from the New England fold belt (east Australia), a Paleozoic forearc terrane: Geol. Soc. America Bull., v. 94, p. 1199–1214.

Dickinson, W. R., 1970, Interpreting detrital modes of graywacke and arkose: Jour. Sed. Petrology, v. 40, p. 695–707.

Dickinson, W. R., 1982, Compositions of sandstones in Circum-Pacific subduction complexes and fore-arc basins: Am. Assoc. Petroleum Geologists Bull., v. 66, p. 121–137.

Dickinson, W. R., L. S. Beard, G. R. Brakenridge, J. L. Erjavec, R. C. Ferguson, K. F. Inman, R. A. Knepp, F. A. Lindberg, and P. T. Ryberg, 1983a, Provenance of North American Phanerozoic sandstones in relation to tectonic setting: Geol. Soc. America Bull., v. 94, p. 222-235.

Dickinson, W. R., D. W. Harbaugh, A. H. Saller, P. L. Heller, and W. S. Snyder, 1983b, Detrital modes of upper Paleozoic sandstones derived from Antler orogen in Nevada: implications for nature of Antler orogeny: Am. Jour. Sci., v. 283, p. 481-509.

Dickinson, W. R., K. P. Helmold, and J. A. Stein, 1979, Mesozoic lithic sandstones in central Oregon: Jour. Sed. Petrology, v. 49, p. 501-516.

Dickinson, W. R., R. V. Ingersoll, D. S. Cowan, K. P. Helmold, and C. A. Suczek, 1982, Provenance of Franciscan graywackes in coastal California: Geol. Soc. America Bull., v. 93, p. 95-107.

Dickinson, W. R. and E. I. Rich, 1972, Petrologic intervals and petrofacies in the Great Valley Sequence, Sacramento Valley, California: Geol. Soc. America Bull., v. 83, p. 3007-3024.

Dickinson, W. R. and C. A. Suczek, 1979, Plate tectonics and sandstone compositions: Am. Assoc. Petroleum Geologists Bull., v. 63, p. 2164-2182.

Dickinson, W. R. and R. Valloni, 1980, Plate settings and provenance of sands in modern ocean basins: Geology, v. 8, p. 82-86.

Franzinelli, E. and P. E. Potter, 1983, Petrology, chemistry, and texture of modern river sands, Amazon River system: Jour. Geology, v. 91, p. 23-39.

Galloway, W. E., 1974, Deposition and diagenetic alteration of sandstone in northeast Pacific arc-related basins: implications for graywacke genesis: Geol. Soc. America Bull., v. 85, p. 379-390.

Gandolfi, G., L. Paganelli, and G. G. Zuffa, 1983, Petrology and dispersal pattern in the Marnoso-Arenacea Formation (Miocene, northern Apennines): Jour. Sed. Petrology, v. 53, p. 493-507.

Graham, S. A., W. R. Dickinson, and R. V. Ingersoll, 1975, Himalayan-Bengal model for flysch dispersal in the Appalachian-Ouachita system: Geol. Soc. America Bull., v. 86, p. 273-286.

Graham, S. A., R. V. Ingersoll, and W. R. Dickinson, 1976, Common provenance for lithic grains in Carboniferous sandstones from Ouachita Mountains and Black Warrior Basin: Jour. Sed. Petrology, v. 46, p. 620-632.

Harrell, J. and H. Blatt, 1978, Polycrystallinity: effect on the durability of detrital quartz: Jour. Sed. Petrology, v. 48, p. 25-30.

Heller, P. L. and P. T. Ryberg, 1983, Sedimentary record of subduction to forearc transition in the rotated Eocene basin of western Oregon: Geology, v. 11, p. 380-383.

Helmold, K. P., 1980, Diagenesis of Tertiary arkoses, Santa Ynez Mountains, California [PhD thesis]: Stanford University, Stanford, California, 225 p.

Hiscott, R. N., 1978, Provenance of Ordovician deep-water sandstones, Tourelle Formation, Quebec, and implications for initiation of Taconic Orogeny: Canadian Jour. Earth Sci., v. 15, p. 1579-1597.

Houseknecht, D. W., 1980, Comparative anatomy of a Pottsville lithic arenite and quartz arenite of the Pocahontas Basin, southern West Virginia: petrogenetic, depositional, and stratigraphic implications: Jour. Sed. Petrology, v. 50, p. 3-20.

Hubert, J. F., 1967, Sedimentology of Prealpine Flysch sequences, Switzerland: Jour. Sed. Petrology, v. 37, p. 885-907.

Hubert, J. F., J. G. Butera, and R. F. Rice, 1972, Sedimentology of Upper Cretaceous Cody-Parkman delta, southwestern Powder River Basin, Wyoming: Geol. Soc. America Bull., v. 83, p. 1649-1670.

Ingersoll, R. V., 1983, Petrofacies and provenance of late Mesozoic forearc basin, northern and central California: Am. Assoc. Petroleum Geologists Bull., v. 67, p. 1125-1142.

Ingersoll, R. V. and C. A. Suczek, 1979, Petrology and provenance of Neogene sand from Nicobar and Bengal fans, DSDP sites 211 and 218: Jour. Sed. Petrology, v. 49, p. 1217-1228.

Ingersoll, R. V., T. F. Bullard, R. L. Ford, J. P. Grimm, J. D. Pickle, and S. W. Sares, 1984, The effect of grain size on detrital modes: a test of the Gazzi-Dickinson point-counting method: Jour. Sed. Petrology, v. 54, p. 103-116.

James, W. C., G. H. Mack, and L. J. Suttner, 1981, Relative alteration of microcline and sodic plagioclase in semi-arid and humid climates: Jour. Sed. Petrology, v. 51, p. 151-164.

Johnson, S. Y., 1984, Stratigraphy, age, and paleogeography of the Eocene Chuckanut Formation, northwest Washington: Can. Jour. Earth Sci., v. 21, p. 92-106.

Jones, P. C., 1972, Quartzarenite and litharenite facies in the fluvial foreland deposits of the Trenchard Group (Westphalian), Forest of Dean, England: Sed. Geology, v. 8, p. 177-198.

Korsch, R. J., 1984, Sandstone compositions from the New England Orogen, eastern Australia: implications for tectonic setting: Jour. Sed. Petrology, v. 54, p. 192-211.

Link, M. H., 1982, Petrography and geochemistry of sedimentary rocks, Ridge Basin, southern California, in Crowell, J. C. and M. H. Link, eds., Geologic history of Ridge Basin, southern California: Pacific Sec., Soc. Econ. Paleontologists and Mineralogists, Los Angeles, California, p. 159-180.

Mack, G. H., 1981, Composition of modern stream sand in a humid climate derived from a low-grade metamorphic and sedimentary foreland fold-thrust belt of north Georgia: Jour. Sed. Petrology, v. 51, p. 1247-1258.

Mack, G. H.., 1984, Exceptions to the relationship between plate tectonics and sandstone composition: Jour. Sed. Petrology, v. 54, p. 212-220.

Mack, G. H., W. C. James, and W. A. Thomas, 1981, Orogenic provenance of Mississippian sandstones associated with southern Appalachian-Ouachita Orogen: Am. Assoc. Petroleum Geologists Bull., v. 65, p. 1444-1456.

McBride, E. F., 1966, Sedimentary petrology and history of the Haymond Formation (Pennsylvanian), Marathon Basin, Texas: Tex. Bur. Econ. Geol. Rpt. Inv. No. 57, 101 p.

McBride, E. F., A. E. Weidie, and J. A. Wolleben, 1975, Deltaic and associated deposits of Difunta Group (Late Cretaceous to Paleocene), Parras and La Popa basins, northeastern Mexico, in Broussard, M. L., ed., Deltas, models for exploration: Houston Geol. Soc., Houston, p. 485-522.

Misko, R. M. and Hendry, H. E., 1979, The petrology of sands in the uppermost Cretaceous and Palaeocene of southern Saskatchewan: a study of composition influenced by grain size, source area, and tectonics: Can. Jour. Earth Sci., v. 16, p. 38-49.

Pacht, J. A., 1984, Petrologic evolution and paleogeography of the Late Cretaceous Nanaimo Basin, Washington and British Columbia: implications for Cretaceous tectonics: Geol. Soc. America Bull., v. 95, p. 766-778.

Parkash, B., R. P. Sharma, and A. K. Roy, 1980, The Siwalik Group (molasse) -- sediments shed by collision of continental plates: Sed. Geology, v. 25, p. 127-159.

Potter, P. E., 1978, Petrology and chemistry of modern big river sands: Jour. Geology, v. 86, p. 423-449.

Putnam, P. E., 1982, Fluvial channel sandstones within upper Mannville (Albian) of Lloydminster area, Canada -- geometry, petrography, and paleogeographic implications: Am. Assoc. Petroleum Geologists Bull., v. 66, p. 436-459.

Ruxton, B. P., 1970, Labile quartz-poor sediments from young mountain ranges in northeast Papua: Jour. Sed. Petrology, v. 40, p. 1262-1270.

Schwab, F. L., 1981, Evolution of the western continental margin, French-Italian Alps: sandstone mineralogy as an index of plate tectonic setting: Jour. Geology, v. 89, p. 349-368.

Schwartz, R. K., 1982, Broken Early Cretaceous foreland basin in southwestern Montana: sedimentation related to tectonism, in Powers, R. P., ed., Geologic studies of the Cordilleran thrust belt: Rocky Mtn. Assoc. Geologists, Denver, Colorado, p. 159-183.

Sclater, J. G. and P. A. F. Christie, 1980, Continental stretching, an explanation of the post-mid-Cretaceous subsidence of the central North Sea Basin: Jour. Geophys. Research, v. 85, p. 3711-3739.

Stewart, R. J., 1978, Neogene volcaniclastic sediments from Atka Basin, Aleutian Ridge: Am. Assoc. Petroleum Geologists Bull., v. 62, p. 87-97.

Suczek, C. A. and R. V. Ingersoll, 1984, Petrology and provenance of Cenozoic sand from the Indus Cone and the Arabian Basin (DSD sites 221, 222, and 224): Jour. Sed. Petrology, in press.

Suttner, L. J., A. Basu, and G. H. Mack, 1981a, Climate and the origin of quartz arenites: Jour. Sed. Petrology, v. 51, p. 1235-1246.

Suttner, L. J., R. K. Schwartz, and W. C. James, 1981b, Late Mesozoic to early Cenozoic foreland sedimentation in southwest Montana, in Tucker, T. E., ed., Guidebook to southwest Montana: Montana Geol. Soc., Billings, p. 93-103.

Tennyson, M. E. and M. R. Cole, 1978, Tectonic significance of upper Mesozoic Methow-Pasayten sequence, northeastern Cascade Range, Washington and British Columbia, in Howell, D. G., and K. A. McDougall, eds., Mesozoic paleogeography of the western United States: Pacific Sec., Soc. Econ. Paleontologists and Mineralogists Pacific Coast Paleogeography Symp. 2, p. 499-508.

Valloni, R. and J. B. Maynard, 1981, Detrital modes of recent deep-sea sands and their relation to tectonic setting: a first approximation: Sedimentology, v. 28, p. 75-83.

Valloni, R. and G. G. Zuffa, 1984, Provenance changes for arenaceous formations of the northern Apennines, Italy: Geol. Soc. America Bull., v. 95, in press.

Van Andel, Tj. H., 1958, Origin and classification of Cretaceous, Paleocene, and Eocene sandstones of western Venezuela: Am. Assoc. Petroleum Geologists Bull., v. 42, p. 734-763.

Van de Kamp, P. C., 1973, Holocene continental sedimentation in the Salton Basin, California: a reconnaissance: Geol. Soc. America Bull., v. 84, p. 827-848.

Velbel, M. A., 1980, Petrography of subduction zone sandstones: Jour. Sed. Petrology, v. 50, p. 303-304.

Walker, T. R., 1984, Diagenetic albitization of potassium feldspar in arkosic sandstones: Jour. Sed. Petrology, v. 54, p. 3-16.

Webb, W. M. and P. E. Potter, 1971, Petrology and geochemistry of modern detritus derived from a rhyolitic terrane, western Chihuahua, Mexico: Bol. Soc. Geol. Mexicana, v. 32, p. 45-61.

Winn, R. D., Jr., 1978, Upper Mesozoic flysch of Tierra del Fuego and South Georgia Island: a sedimentological approach to lithosphere plate restoration: Geol. Soc. America Bull., v. 89, p. 533–547.

Young, S. W., 1976, Petrographic textures of detrital polycrystalline quartz as an aid to interpreting crystalline source rocks: Jour. Sed. Petrology, v. 46, p. 595–603.

Zieglar, D. L. and J. H. Spotts, 1978, Reservoir and source-bed history of Great Valley, California: Am. Assoc. Petroleum Geologists Bull., v. 62, p. 813–826.

Zuffa, G. G., 1980, Hybrid arenites: their composition and classification: Jour. Sed. Petrology, v. 50, p. 21–29.

Zuffa, G. G., W. Gaudio, and S. Rovito, 1980, Detrital mode evolution of the rifted continental-margin Longobucco Sequence (Jurassic), Calabrian Arc, Italy: Jour. Sed. Petrology, v. 50, p. 51–61.

SUBJECT INDEX

abandonment 53,
abrasion 23, 151, 253, 256, 318,
    marking 220,
abyssal
    hill 53,
    turbidite 34,
accessibility 41,
accessory grain 312,
accreted basement 322,
activator
    of luminescence 200,
    term 196,
active lobe 53,
age
    of the deposit 167,
    of the detritus 167,
Ager Valley, Pyrenees 168, 184, 186,
air-fall pyroclastic 255,
Albidona Formation, Italy 176,
albitization 96, 103, 109, 156,
    of k-feldspar 109, 157,
    of plagioclase 95, 106, 156,
alcali feldspar 141,
allochem 166,
allocycle mechanism 86,
alluvial fan 36,
Almond Formation 154, 155, 156,
Alpine Molasse 34,
al-Si ordering 146,
Amazon River 297,
anadiagenesis 262, 264, 266,
ancient turbidite system 68,
andalusite 100,
angle of repose 12,
amphibole 100, 102,
amphibole composition 270,
Antler Formation, Nevada 349,
apatite 100, 104,
Apennines 34, 36,
Apenninic-Adriatic Promontory 39,

363

INDEX OF NAMES